T0178228

Construction Management and Design of Industrial Concrete and Steel Structures

Construction Management and Design of Industrial Concrete and Steel Structures

Mohamed A. El-Reedy, Ph.D.

Consultant Engineer
Cairo, Egypt

CRC Press

Taylor & Francis Group

Boca Raton London New York

CRC Press is an imprint of the
Taylor & Francis Group, an **informa** business

CRC Press
Taylor & Francis Group
6000 Broken Sound Parkway NW, Suite 300
Boca Raton, FL 33487-2742

First issued in paperback 2019

ISBN-13: 978-1-4398-1599-1 (hbk)
ISBN-13: 978-0-367-38356-5 (pbk)

Library of Congress Cataloging-in-Publication Data

El-Reedy, Mohamed A. (Mohamed Abdallah)
    Construction management and design of industrial concrete and steel structures /
Mohamed El-Reedy.
        p. cm.
    "A CRC title."
    Includes bibliographical references and index.
    ISBN 978-1-4398-1599-1 (alk. paper)
    1. Building, Iron and steel. 2. Concrete construction. 3. Industrial buildings. 4.
Building--Superintendence. I. Title.

TH1611.E42 2011
693'.5--dc22

                                                                                    2010025888

Visit the Taylor & Francis Web site at
http://www.taylorandfrancis.com

and the CRC Press Web site at
http://www.crcpress.com

This book is dedicated to the spirits of my mother and father,

my wife, and my children Maey, Hisham, and Mayar.

Contents

Preface

The development of any country depends on the rate of industrial growth. Currently, there is a race in industrial projects worldwide. The development of the industry depends on the development of the energy reserve by investment in projects of oil and gas exploration, onshore and offshore, which require new facilities or rehabilitation of existing facilities. At the same time, there are projects that are running in parallel to deliver electricity from electrical power stations or through nuclear power plants.

In this book, the term *industrial structures* means all the reinforced concrete and steel structures from a small factory to a nuclear plant. This book will be an overview of industrial project management, design, construction, and eventually providing a maintenance plan. Industrial projects, in most cases, are huge and can cost a billion dollars for one project, so the client, engineering firm, and contractor are in the same boat until they achieve project success through a strong management system and technical competence. Therefore, this book discusses all items that interface among these main three partners.

In these types of projects, all the engineering disciplines are working together, but, unfortunately, the structural or civil engineers are usually the last ones to obtain the exact data from the other disciplines and the first ones to start on site. Therefore it is a challenge for the structural engineers to work fast and efficiently in this type of project.

This book focuses on the structural engineering of all of these projects. The aim of this book is to provide up-to-date methodology and industry technical practice and guidelines to design, construct, and maintain the reinforced concrete and steel structures in these industrial projects. The essential processes of protection, repair, and strengthening of the industrial structures necessitated by deterioration or a change in the mode of operation are illustrated in this book. It is intended to be a guidebook to junior and senior engineers who work in design, construction, repair, and maintenance of reinforced concrete and steel structures and to assist them through all of the stages of industrial projects.

The other challenge that faces structural engineers is that most of the undergraduate courses they studied in college focused mainly on real estate projects and housing. However, the characteristics of industrial projects are different. This book provides a guide for the project and construction manager to lead the project and to successfully achieve the owner's requirements. On the other hand, from a technical point of view, this book describes the first principle of the codes and standards that are usually used in industrial projects and the most applicable methods used in the design of the steel and reinforced concrete structures that serve the static equipment, tanks, towers, and vibrating equipment. This book describes current research and development in the design, construction, repair, and maintenance philosophy.

An overview of offshore structure design and construction is very important and provides the tools to check the design and to control the project in all of its phases. Recently, there is a trend toward maintaining the reliability of the structure from both safety and economic points of view by developing the structural integrity management system, which will also be a part of this book.

The last chapter describes the soil investigation tests that are essential to the industrial projects and provides the main key to selecting the most reasonable type of test and also the main features for the pile foundation design.

This book provides a practical guide to designing the reinforced concrete and steel structures and foundations in industrial projects with the principle of repairing the concrete structures and the methodology to deliver a maintenance plan for the concrete and steel structures serving onshore and offshore facilities.

Mohamed Abdallah El-Reedy
Cairo, Egypt
elreedyma@gmail.com

Author

Mohamed A. El-Reedy pursued a career in structural engineering. His main area of research is the reliability of concrete and steel structures. He has been a consultant to different engineering companies and oil and gas industries in Egypt as well as international companies such as the International Egyptian Oil Company (IEOC) and British Petroleum (BP). Moreover, he provides different concrete and steel structure design packages for residential buildings, warehouses, telecommunication towers, and electrical projects with WorleyParsons Egypt. He has participated in Liquefied Natural Gas (LNG) and Natural Gas Liquid (NGL) projects with international engineering firms. Currently, Dr. El-Reedy is responsible for reliability, inspection, and maintenance strategy for onshore concrete structures and offshore steel structure platforms. He has performed these tasks for hundreds of structures in the Gulf of Suez and in the Red Sea.

Dr. El-Reedy has consulted with and trained executives for many organizations, including the Arabian American Oil Company (ARAMCO), BP, Apache, Abu Dhabi Marine Operating Company (ADMA), the Abu Dhabi National Oil Company, King Saudi's Interior Ministry, Qatar Telecom, the Egyptian General Petroleum Corporation, Saudi Arabia Basic Industries Corporation (SAPIC), the Kuwait Petroleum Corporation, and Qatar Petrochemical Company (QAPCO). He has taught technical courses on repair and maintenance for reinforced concrete structures and advanced materials in the concrete industry worldwide, especially in the Middle East, Malaysia, and Singapore.

Dr. El-Reedy has written numerous publications and presented many papers at local and international conferences sponsored by the American Society of Civil Engineers, the American Society of Mechanical Engineers, the American Concrete Institute, the American Society for Testing and Materials, and the American Petroleum Institute. He has published many research papers in international technical journals and has authored four books about total quality management, quality management and quality assurance, economic management for engineering projects, and repair and protection of reinforced concrete structures. He received his bachelor's degree from Cairo University in 1990, his master's degree in 1995, and his PhD from Cairo University in 2000.

1

Introduction

Civilization and development in most countries depend on the sources of energy that feed all the industrial projects. At this time, the main sources of energy have been oil and gas. Therefore, oil and gas projects are critical from both safety and economic points of view. The petroleum industry is one of the richest in the world, so there is much research and development in this area to enhance the design, construction, and management of these projects.

Industrial projects have different characteristics from both management and technical points of view; for example, time is more important than cost. This principle differs from other projects. In addition, the industrial projects depend on different types of machines, cranes, vessels, tanks, and other specific equipment for each type of industry. These projects require concrete and steel structures for their equipment, which requires a special design procedure and philosophy, as these structures are at times under the effects of dynamic loading. Most industrial projects are located onshore, but many oil and gas projects have facilities and structures offshore and near-shore for activities such as exploration and loading of ships.

Management is critical to solving the interface between the different engineering disciplines that will work together in the engineering office and on-site. The electrical, mechanical, instrument, and civil engineers are focused on their concerns only, so the main challenge to management in any phase is to resolve conflict and create and maintain harmony among the team members to successfully complete the project in terms of time, cost, and quality.

Management of the projects is the main key to success. Imagine that you have very skilled team members but their objectives are not clear, there is conflict between members and a lack of cooperation. You cannot expect the project to be successful. The main tools and skills for construction management are discussed from a practical point of view in Chapter 2, as well as how to build teamwork and increase and monitor team performance in a professional manner.

In any university that graduates civil and structural engineers, most courses focus on the design of regular buildings for housing and real estate projects and their codes and standards. Industrial projects—such as oil, gas, and electrical power—have their own codes, standards, and concepts. The main differences are related to the loads that affect the structure in industrial projects. Chapter 3 defines the loads affecting the industrial project including the common codes, standards, and technical practices that are traditionally used in these types of projects. This chapter illustrates the required data from electrical and mechanical engineers and how the mode of operation influences the design load parameters. Moreover, in the case of a new plant in a new location that is outside of any major cities—as is usually true of oil and gas plant facilities—there are some data required from third parties such as MetOcean data (in the case of offshore or near-coast facilities) or information necessary to define the hazard area in case of earthquake. This chapter discusses in detail the loads affecting concrete and steel and their nature and how the designer can define the scope of the work professionally to a third party and thereby obtain useful data.

The main equipment in any industrial project is rotating equipment such as compressors, pumps, and power turbines. This type of equipment requires special precautions in the methodologies of design and construction, which will be discussed in Chapter 4.

It is traditional in these types of projects to use tanks. Chapter 5 provides the necessary guidelines and features in designing the reinforced concrete tanks that are usually used in this industry. It is common in the case of oil and gas that the steel tanks are designed by the static equipment designer. The key element are the mechanical valves with the instrumentation system that monitors and controls the levels. In minor cases, the tanks will be designed by structural engineers.

In Chapter 5, the main element of design of these steel tanks is discussed, emphasizing the essential precaution required during construction. The design of the reinforced concrete ring beam under the steel tank is discussed using a real example. In the industrial plant, there are usually retaining walls, and in most cases, these walls are located around the tanks as a safety requirement in case of a tank leak. Hence, the design of retaining walls is also presented in this chapter.

The static equipment such as the separators, steel towers, knock-out drums, and heaters are designed by static equipment specialist engineers, and they also provide the required data to design the foundations under this type of equipment. The design of these foundations will be illustrated in Chapter 6 by defining the data and the philosophy of operation of each piece of equipment.

Steel structures are usually used in industrial projects because they can be erected quickly and because of their capital cost value over time. The structures' requirements for maintenance and protection are easily met in industry as there is usually a professional crew available to conduct maintenance and ensure structural integrity over time.

In the case of the steel structures for pipe racks, some precautions are required when choosing the structure system and estimating how the loads from pipes and electrical cable trays will affect the structures. All these factors are discussed in Chapter 7. All steel structures and static equipment will be fixed to the concrete foundations by anchor bolts. The design of these anchor bolts will also be illustrated in detail in this chapter.

In general, industrial projects experienced fast growth after World War II, and after the mid-1950s, there was also fast growth in oil production worldwide. As a result, there are now many mature facilities worldwide, and therefore, it is usually required to assess the steel and concrete structures to define any problems and determine if the structures can accommodate the existing load.

In some cases, there are changes in the mode of operation or a need to install a new piece of heavy equipment, so it is necessary to evaluate and assess the existing structure to determine if it can carry the heavier load. The method for evaluating the existing structures in industrial plants will be discussed in Chapter 8.

Chapter 9 presents the method for protecting the foundations under the equipment and the main reinforced concrete structures from corrosion, as most of the facilities are near the shoreline. Factories located inside cities are subject to the effects of carbonation, so the advantages and disadvantages of each type of protection against each cause of corrosion are discussed from both technical and economic viewpoints.

The processes of repairing the reinforced concrete structures and strengthening their members to resist higher loads are presented in Chapter 10. The methods of repair are chosen based on their fast application. Aesthetics is not a main concern, as we are not working in a shopping mall or hotel building. Repair and strengthening will involve using steel sections or carbon fiber to reduce the risk as much as possible.

The integrity management system to maintain mature structures is the most recent management policy that depends on risk-based inspection and maintenance. Risk-based and underwater inspections in the case of offshore structures are discussed in Chapter 11.

Chapter 12 discusses the offshore structures used in oil and gas projects in shallow and deep water. The loads, features of design, and method for reviewing the design of a fixed offshore structure will be illustrated. The construction phase has special features as did the design phase. Therefore, the steps of construction and the loads affecting the structure during transportation, lifting, and installation are presented from a practical point of view.

Chapter 13 presents the geotechnical investigation tools and methods used to obtain the required data necessary to design the foundation for static and rotating equipment as well as the foundation for reinforced concrete and steel structures. The geotechnical investigation and the design and construction of the piles are usually performed by a third party. This chapter provides a method for preparing a precise scope of work for the third party, presenting the main concept of soil investigation and pile design, so that the required accurate data may be obtained.

2

Construction Management
for Industrial Projects

2.1 Introduction

The subject of project management has become one of the most common themes in the recent past because of the increase in the number of mega-projects worldwide. The development of modern technology in all areas of knowledge requires new methods of project management to cope with the rapid advances.

The concept of the project is very different from how the daily routine operation works; therefore, project management is different from the daily activity of operation management. Most books and references that discuss project management define the project as a number of tasks and duties to be implemented during a specific period to achieve a specific objective or set of specific targets.

To clarify the difference between project management and operations management, think about what is going on in the minds of two managers. The project manager's dreams are about finishing the project on time and about where he will relocate to after the completion of the project. This is totally different from the thinking of the operation manager. He does not dream about a stop of daily production, which is contrary to the project manager's goal. Therefore, you can imagine the difference between the thinking of the two mangers.

The first difference in the definition of project management, as opposed to operation management, is that the goal is to finish the project within a certain time frame and simultaneously realize a set of objectives.

2.2 Project Characteristics

One of the most important features of the project is the selection of individuals from different locations in the same company. In some international projects, the team members are from different countries, cultures, educations, and employment backgrounds, and all of these individuals have different skills. With all of those differences, they must still work together to complete the work in a specific time and with a definite target.

The project manager has to coordinate between the members of the project to reach the project goal. As a result of the rapid development in modern technology, the specialty has become important. These days, any project contains many different disciplines. An explicit example is a construction project where there are separate teams for constructing the reinforcing concrete, finishing work, plumbing, and other activities. Every branch of the construction activity has its own technology and skills. Therefore, the project manager has to facilitate cooperation between the different disciplines to achieve the project objective.

The primary goal of the project manager is to complete the project, with high quality, and achieve the objectives.

Any project has a main driver, and it is one of the two driving forces. In other words, there are two philosophies in managing the project: one is cost-driven and the other is time-driven. This driver is considered to be the underlying philosophy in the management of the project, which must be determined by the director of the project with other parties, as well as the official sponsor of the project and the stakeholder. The project-driving philosophy should be known to both the technical and administrative department managers.

To illustrate the effect of the two driving factors, we should think about all types of projects that are running around us. We will find that, in some projects, reducing the cost is the major factor and time is the secondary factor, as the increase in project duration time will not affect the project's operation phase. Put more precisely, it will not affect the owner and his investment. Building houses, mosques, churches, museums, and other projects that have a social aspect are examples of this.

On the other hand, there are some projects where reducing time to completion is the main challenge; this is a time-driven project. A clear example of this is in the petroleum industry. For oil, gas, and petrochemical projects, any day saved will be a gain of many millions of dollars per day as the production is measured by barrels of oil per day (BOPD) or million standard cubic feet per day (MMSCFD). By multiplying that by the price of oil or gas, you can calculate the income. For example, if the gain in production from the project is 50,000 BOPD with an oil price of $40 per barrel, for every day saved, the owner gains $2,000,000.

As illustrated, the main driver of the petroleum projects is time. So, the main target of these projects is the reduction of the time to completion.

It is very important to define the basic driving force of a project, whether it is cost or time, and all the staff working on the project should know this information. This is the responsibility of the project manager. Any groups or teams at work, both in design and execution, should provide proposals, recommendations, and action steps that are in line with the project driver, whether that involves reducing time or cost.

It is necessary for the target to be clear to all involved to avoid time wasted discussing ideas and suggestions that are not feasible. Imagine that you are working on a housing project and one of the proposals from the engineers is to use a rapid-setting concrete to reduce the time of construction but with an increase in cost. Is this proposal acceptable? It certainly is not. On the other side, in the case of the construction of an oil or gas plant or new off-shore platform, one of the proposals is for the use of materials that are the least expensive but that need time to be imported from abroad, which will delay the project for some days. Is this proposal accepted? Of course, this proposal is unacceptable. However, if we were to trade off each of these proposals for the other, we will find that the two proposals are excellent and acceptable.

When communication is lost between the project manager and personnel, there is a great deal of confusion; everyone works hard, but in different directions, resulting in wasted effort and lack of success. Moreover, project managers must also communicate with suppliers and contractors to ensure that their proposals regarding supply materials and construction are within the project-driven criteria. The project characteristics can be summarized as follows:

- The project has a specific target.
- The project is unique and cannot be replicated with the same task and resources giving the same results.
- The focus is on the owner's requirements and expectations from the project.
- It is not routine work, although there are some aspects that are routine.
- The project consists of a number of associated activities contributing to the project as a whole.
- There is a specific amount of time allowed to finish the project.
- The project is complex in that it involves a number of individuals from different departments.
- The project manager must be flexible to accommodate any change that might occur during the project.
- There are factors of uncertainty such as the performance of individuals, how their skills adapt to unfamiliar work, and other unknown external influences.
- The total cost is defined and has a limited budget.
- The project presents unique opportunities to acquire new skills.

- The project gives impetus to the project manager to adapt to working under changing circumstances, as the nature of the project is change.
- There are risks with each step of the project, and the project manager should manage those risks to reach the project goal.

2.3 Project Life Cycle

The project definition is a set of activities with a specific start time and end time. These activities vary from project to project depending on the nature of the project. An example of this might be a cultural or social project, such as a public education endeavor, or a civil project, such as the construction of a residential building, hospital, road, bridge, or other industrial projects. In our scope, we will focus on industrial projects.

The civil projects vary from one project to another, depending on the size and value of the project. They can range from constructing a guard room to constructing a nuclear plant; hence, the quality varies with the size of the project, especially in developing countries.

In a small project, it might be sufficient to apply a quality control only where small contracting companies or engineering offices are involved. When the target involves global competition and increasing the quality will increase the total cost of the project, quality control is often applied to the structural safety of the building only.

In the case of a major project, there are many execution companies and engineering offices working at the same time. Therefore, we must also take into account that implementing quality assurance procedures is necessary and vital as are the quality controls carried out in all phases of the project based on the project specifications. Each stage of a construction project starts with a feasibility study, followed by preliminary studies of the project, detailed studies, and, finally, execution. The operation crew will then receive the project to run.

In all of these stages, there are many types of quality control required to achieve a successful project that has benefits and appropriate return on investment for the owner and all parties and participants. Figure 2.1 shows the life cycle of any project. From this figure, it is clear that 5% of the project resources (time and cost) is expended on the feasibility study, 25% is expended on the engineering designs, and the largest percentage of project resources is expended in the execution phase.

As shown in Figure 2.1, after the feasibility study, a decision is required by senior management on the question, "Will the project continue or be terminated?" Imagine a gate and, if the results of the feasibility study are positive,

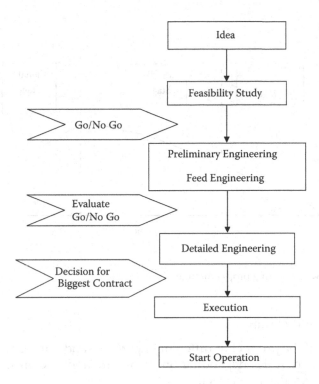

FIGURE 2.1
Project life cycle.

the project passes through to the next stage—the preliminary studies, which will provide a more accurate assessment of the project. After those studies, there is another gate and another decision required. If the answer is positive, this gate will open and the project will move forward to the detailed engineering and construction phase.

At each phase or stage of the project, there are roles for the owner, the contractor, and the consulting engineer, and each system has its own project management approach. Every stage has its own characteristics and circumstances, and each involves change in the scope of work (SOW) for each of the three involved parties and this should be clarified for each stage.

The characteristics of the project life cycle change from time to time. In each period, the number of personnel on the project can change. For example, at the beginning of the project, the number of personnel is very small. It increases with the number of activities being carried out and then gradually decreases until the end of the project. Figure 2.2 shows the changes in the number of personnel in the project and notes that the project manager should have the necessary skill to deal with the changes that occur during the life cycle of the project (Figure 2.3).

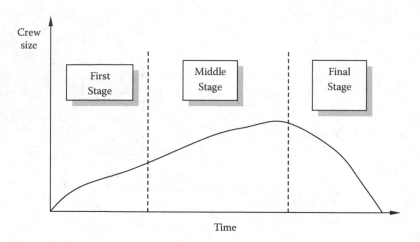

FIGURE 2.2
Change in crew size during project lifetime.

2.3.1 Feasibility Study

Each phase of the project has different importance and impact on the project as a whole and varies depending on the nature and circumstances of the project and its value and target.

The feasibility study is the second phase after the emergence of the owner's idea for the project. The owners of oil and gas projects are the geologist

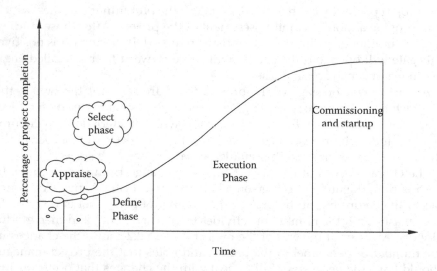

FIGURE 2.3
Project life cycle phases.

and the petroleum engineering team who base their ideas on oil and gas reservoir characteristics.

The economic study for the project will be performed by high-level and highly skilled personnel of the organization, as this study will include expected fluctuations of the price for oil and gas and other petrochemical products during the project life time. Their experience is based on what they have done on similar projects before, as well as records kept and lessons learned from previous projects.

In this initial phase, the selection of team members for the consultant's office is very important as they will perform the feasibility study for the project. In some cases, there is input from an engineering firm that performs a generic engineering study about the project and estimates the cost, based on their experience.

The feasibility study phase, which is also called the *appraise* phase, is followed by the preliminary (FEED) study phase. These two phases are essential as they set the objective of the project and identify engineering ideas through the initial studies. It is preferred to apply the Japanese proverb, "Think slowly and execute quickly," especially in the feasibility study stage. At this stage, the goal of the project is defined and the economic feasibility of any move is determined, as is the move's direction.

For these reasons, we must take full advantage of this phase and its time, effort, study, research, and discussions, with more attention to the economic data. The economic aspect is important at this stage and the engineering input is very limited.

2.3.2 FEED (Preliminary) Engineering

This stage is the second phase after the completion of the feasibility study for the project.

This phase of preliminary engineering studies, which is known as FEED engineering, is no less important than the first phase. It is one of the most important and most critical stages in the engineering of the project because the success of the project as a whole depends on the engineering study in this phase. Therefore, as this stage is vital, the engineering consultancy firm that will perform this study should have an extensive experience in these types of projects.

Specifically, liquefied natural gas (LNG) is a type of project that requires experienced personnel in this field. Another example is a project that uses floating production, storage, and offloading (FPSO). It also requires a special consulting office that has worked on this type of project before. In the case of small projects such as a residential or administrative building or a small factory, the purpose of the FEED phase of engineering is to determine the type of structure, whether it will be steel or concrete. If it is decided to use a concrete structure, the engineer should define whether it will be precast, prestress, or normal concrete and then determine if the type of slab structure system will be solid slab, flat slab, hollow blocks, or another type. This phase also defines the location of the columns and the structure system and whether the project will use a frame or shear wall for a high-rise building.

In summary, the purpose of preliminary engineering is to provide a comparison between the alternatives that vary depending on the size of the building and the requirements of the owner so that a reasonable structure system and appropriate mechanical and electrical systems may be selected. For this reason, this stage has recently been called the *select* phase.

In the case of major projects such as a petrochemical plant or new platforms, there will be other studies at this stage such as geotechnical studies, metocean studies, seismic studies, and environmental studies. The main purpose of this phase is to provide the layout depending on the road design, location of the building, and hazard area classification in the petroleum projects. Moreover, it must select the foundation type, whether it is to be a shallow foundation, or driven or rotary piles, based on the geotechnical studies. In case of oil and gas projects, we need to carefully study the mode of transfer and trade-offs of the product and select the appropriate method of transfer between the available options.

Now, it is clear that, because of the seriousness of this stage and the need for high-level experience, for large projects, the owner should have competent engineers and an administrative organization with the ability to follow-up on initial studies to achieve the goal of the project and coordination between the various project disciplines such as civil, mechanical, electrical, and chemical engineering, as all the disciplines usually intersect at this stage.

In general, regardless of the size of the project, the owner must be presented with the engineering requirements for the project through a statement of requirement (SOR) document, which must be highly accurate and contain the objective of the project and the requirement from the owner. It will also precisely identify the SOW. This document is the starting phase of the mission document quality assurance system and must contain all information requested by the owner. The preparer must be experienced because this document is relied upon to determine the outline of the whole project and to contain all particulars of the project, its objectives, proposals, and the required specifications of the owner.

This document also contains the available technical information from the owner such as the location of the land, its coordinates, and its specifications. This document will be a part of the contract document between the owner and the engineering firm, and the engineering firm will provide cost, time, and resources (CTR) sheets based on it.

In the case of projects such as gas or LNG, it is important to determine the amount of gas, type, and other specifications needed to process and transfer the gas with clarification of temperature, pressure, and all other technical data that allows for the final product to be shipped or transported. Among the most important data to be mentioned in the document is the project lifetime. Again, specifications required by the owner in the project should be defined clearly and precisely in this document.

It should be noted that there will be regular meetings among the owner, the technical team, and the consulting engineers responsible for the preparation of initial studies, and through these meetings, the SOR may be amended several times. Each time, the document must contain the date and revision number as well as all of the requirements—civil, architectural, electrical, mechanical, and others—of the project. We should note here that, for quality assurance, all parties to the project must have a current version of the document and everyone must work according to this document. It is also important to determine the number of meetings and the exact schedule of meetings required to reach the target.

An SOR document is not only required for new projects, but also for modifications to buildings or in the plant. Upon receipt of the SOR document in the engineering office, another document is prepared to respond to the SOR. This document is called the *basis of design*, and through it, the engineering firm clarifies the code and engineering specifications that will operate in the design as well as the calculation methods, theory, and computer software that will be used. This document may also state the required number of copies of the drawings that will be sent to the owner and the sizes of those drawings. The engineering firm should also request any missing data and request that a third party supplement any necessary information such as weather and environmental factors. This document will be carefully reviewed by the owner and can be amended many times until both parties are satisfied.

At this stage, it is important to make sure that both the owner and the engineering firm have the same concept and that there is complete agreement among all parties about the technical aspects. Any drawings prepared during the FEED studies should be delivered to the owner for review and comment. The owner and the engineering firm should agree on the time allotted for review of the document by the owner. If more time is taken than allowed, it indicates owner acceptance. This is very important to control the project time.

This phase may take a number of months in the case of large projects, and therefore, the technical office of the owner must have a qualified engineer with experience in controlling costs and ensuring that follow-up time conforms to the schedule agreed upon in advance. We may need an engineer specialized in planning, called the *planner engineer*. This engineer should be specialized in cost control, ensuring that the estimated cost of the project is comparable to that in the feasibility study.

By monitoring the cost at each stage precisely, at the end of the project the whole cost will be within the estimated value in the beginning of the project.

In petroleum projects, where the return of income or expense is calculated by the day, it is worth noting that any savings in time is a big return for the owner.

The project site itself and the surrounding environment must be considered to determine ways to protect it from weather and reduce the cost of maintenance over time by selecting different methods of maintenance. For example, one way to protect the reinforced concrete foundation from corrosion is by protecting the reinforcing steel with a cathodic protection system, which is expensive at the beginning of the construction but allows periodic, low-cost maintenance later. On the other hand, if we do not want to use an external protection system, we can use a low-cost alternative during construction, although this necessitates high-cost regular maintenance thereafter.

The structure, mode of operation, and maintenance plan all have an impact on the preliminary design. For example, in power stations, we must ask whether the water tank can be repaired, maintained, or cleaned. The answer to this question includes a decision on whether the station needs additional tanks as standbys for maintenance purposes.

In this phase, many other initial design decisions must be decided and, therefore, as previously noted, the parties involved must have extensive experience. An error at this stage could lead to a major problem in the future during operation, when it will cost a lot of money to resolve. Situations like that can be prevented at this stage by a low-cost solution.

2.3.3 Detail Engineering

At the end of this phase, the engineering office will deliver the full construction drawings and specifications for the whole project, which contain all the details that enable the contractor to execute his function. In this phase, there will be a huge number of engineering hours, so there must be good coordination between the different disciplines. This will happen if there is good organization in the engineering office and if the client provides a free mode of communication between the different parties through a system channel with continuous coordination.

The complexity of this phase is such that it needs a quality system. Imagine that you work in an ideal office where everyone's duties are understood, no person comes to work late, the work comes to you in an appropriate manner, and no one ever lets you down. Do you work in an atmosphere like that right now? I doubt it.

As engineers, we always believe that, in an ideal case, all of our lives depend on accuracy. Teamwork does not have this accuracy and is not precise, as all of our experiences at work tell us. Discouraging and disheartening events happen daily, which often means that the work arrives late, in a bad manner, or requires correction before you can complete your work in an efficient manner.

Sometimes there will be changes made in your company or your office without any prior notification to you. This presents obstacles and wastes time. This is the basis of a quality assurance system. When people change, some change might occur in the cooperation between departments. This varies depending on the performance of the managers and the impact they have on work. A system of quality assurance is beneficial because it ensures the basic functioning of all departments, regardless of personnel changes. These problems often occur at a stage in the studies that requires extensive and vital cooperation among the various departments of engineering (civil, architectural, mechanical, and electrical departments).

When the managers of the departments of civil and mechanical engineering have a strong relationship, the work goes well, regular meetings are held, and meetings and correspondence will be fruitful. If one of the department heads is replaced and the new relationship between the two departments is not good, you will find that the final product is also not good. There will likely be no regular meetings. We find that many of these problems do occur; we do not live in a perfect world. The main player who can solve this crisis is the project manager.

You can easily determine whether your business might benefit from a quality assurance system by taking a closer look if you have a bad experience. Does your work suffer because it depends on the work of colleagues who do not complete their tasks or who perform their work poorly? If the answer is yes, then you need a system of quality assurance.

A system of quality assurance is important at this stage because it organizes the work. The target of the project and each team member's responsibility is clear. The concept of quality is defined by supporting documents. The documents are regarded as the executive arm of the quality assurance process. For example, any amendment or correction in the drawings should be made through the agreed procedure and system. Moreover, the drawings should be sent in a specified time to the client for review and discussion, with an official transmittal letter to control the process time. If there are any comments or inquiries, they should be done through agreement between the two technical parties and then the modification should be done by the engineering firm and resent to the client through the same communication procedure.

The development of a system to avoid older copies of the drawings becoming confused with the current copies can prevent human error. The most current set of drawings may be assured by the establishment of a system for continuous amendment of the date and number of the drawings and engineering reviews until the final stage of the project and the approval of the final set of drawings is sealed with a stamp ("Approved for Construction") indicating that these are the final drawings approved for the construction.

After the completion of the detailed engineering phase, the specifications and drawings are ready to be used in the execution phase. You can imagine that in some projects the documents may reach hundreds of volumes, especially the specifications, operation manual, and volumes of maintenance and repairs.

2.3.4 Design Management

The goal of design management is to control the design stage to provide high quality at a better price.

- The design input comprises all technical information necessary for the design process. The basis of this information is the owner through the SOR document, so the engineering firm should review this document carefully, and if there is any confusion or misunderstanding, it should be clarified in the document and through meetings.
- Instructions for control of the design, whether the client controls the whole process or requests some specific action (e.g., a representative from the audit during the design phase) are often provide in the contract.
- The designer must take into account the available materials in the local market in relation to the project and its location and match these with the capabilities of the owner. The designers must have a realistic view and full and up-to-date knowledge of the best equipment, machinery, and available materials.
- The design must conform with the project specifications and the permissible deviation and tolerance should be in accordance with the specifications and requirements of the owner.
- Health, safety, and environment are the most critical subjects these days, so every design should conform to health, safety, and environmental regulations.
- The computer is one of the basic tools used now in the design process as well as in the recording and storage of information and offers the ability to change the design of the work easily. It is now easy to modify the drawings by using Computer Aided Design (CAD) software in order to obtain more precise information with the access to various tables and diagrams.
- The design output must be compatible with all design requirements, and the design should be reviewed through internal audit. The design must be compared with old designs that have been approved before for similar projects. This is a simple procedure for checking designs. Any engineering firm should have a procedural checklist for reviewing designs.

The audits of the design review are intended to be conducted on a regular basis at important stages in the design. The audit must require complete documentation and can take analytical forms such as the analysis of collapse with an assessment of the risk of failure.

2.3.5 Execution Phase

Now that everything is ready for the execution stage, it requires both quality assurance and quality control, especially in the reinforced concrete works where the concrete itself is composed of many materials such as cement, sand and coarse aggregate, water, additives, and steel bars. Therefore, it is essential to control the quality of the received materials as well as the whole mixture. The quality control should follow strict guidelines during all of the construction phase.

It is clear here that the contractor should have a strong, capable organization to achieve good quality control as well as to confirm the existence of documents that define the time and the date on which the work was carried out, who received the materials, who determined the number of samples of concrete, what has been tested by compression, and the exact time, date, and result of each test.

Often during the execution, there occurs some change in the construction drawings of the project as a result of some problems at the site during the construction or the introduction of some ideas or suggestions to reduce the time of the project. It is important that the change of work be done through the documents to manage the change. This called *management of the change document* and requires the approval of the discipline concerned and also approval from the engineering firm. Finally, all of these changes should be reflected in the final drawings.

The supervisor and the owner must both have special organizations. The owner organization in most cases has two scenarios:

- The owner establishes an internal team from the organization to manage the project.
- The owner chooses a consultant office to manage the supervision on-site; in most cases, the design office will do the supervision.

The construction phase shows the contractor's capability for local and international competition if, and only if, the concept of quality assurance for the contractor's project team is very clear and they have experience in a comprehensive quality system. All competitors on the international scene work through an integrated system whose aim is to confirm the quality of the work and control the quality in all stages of execution to achieve full customer satisfaction.

2.3.6 Commissioning and Start-Up

The importance of this stage varies depending on the nature and size of the project itself. In industrial projects such as the construction of pipelines, pumps, turbine engines, or a new plant, a new team will be assembled consisting of project members, operating personnel, and the head of the team.

This team should be competent and have previous experience in commissioning and start-up. The team will have a specific target of starting the operation. The reception is not performed until after the primary operation start-up and commissioning, which at this stage is to make sure that all the mechanical systems work efficiently and safely without any leakage or error in the operation. The length of this stage depends on the size of the project and may extend to months. It is important to have competent personnel at this stage, as it has special characteristics. The cooperation between the operation and project team is essential; communication skills must be at the highest level. The operation starts according to a specific schedule, which can be measured in hours, and all parties should agree to this schedule to provide a smooth transition zone in a safe manner. Increases in temperature or pressure to any equipment without previous study can lead to disaster.

2.4 Is This Project Successful?

When you see a huge oil or gas plant, do you ask yourself, "Is this a successful project?" Although profit and money may come to mind, the focus should be on the management of the project to determine whether it is successful. You see a high-rise building and you say that the project is successful, but is the project manager successful, too? To answer this question, you need to answer the following three questions:

1. What is the plan and actual execution time?
2. What are the actual costs and the budget?
3. Is the project performing according to required specifications?

For the last question, we can answer yes for the big project. We cannot agree or approve anything with low quality or that is beyond the specification. You may assume that quality is a red line that we cannot cross or negotiate. Therefore, the successful project manager has to achieve the goal of the project and satisfy all stakeholders while completing the project on time and within the approved budget.

2.5 Project Management Tasks

Every manager in the project organization is responsible for planning and monitoring the plan and assuring that the execution matches the plan.

To achieve this, the project manager should coordinate with and provide a report to other managers. The main items involved in the planning and monitoring process are:

- Define project objective
- Define work
- Define work period
- Define the available and required resources
- Define cost
- Review and evaluate the master plan
- Accept the master plan
- Follow-up execution
- Follow-up cost
- Compare actual work, cost, and master plan
- Evaluate performance
- Predict and change strategy

The definition of project management is illustrated in Figure 2.4. The first task of project management is to identify the target of the project management process in the planning, execution, and follow-up. There are three key factors involved: resources, time, and funding.

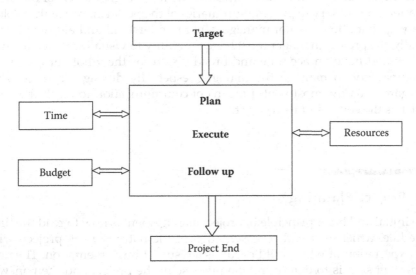

FIGURE 2.4
Project management.

2.6 Project Manager Skill

From the previous discussion, it is obvious that the main player in any project is the project manager, as he carries the biggest load. He is responsible for assembling the team.

When thinking about the selection of the project manager, we must ensure that his role is different from the operational and routine work, as the project is unique and not repetitive.

Therefore, the project manager must be flexible, because the project changes from time to time, for example, at the beginning of the project, there is a small number of individuals and the number increases with time. Then, near the end of the project, the number decreases. The behavior of the project manager must be flexible according to the variables of the project.

The project manager deals with different individuals of all levels from various departments and also with other parties dealing with the project. It is important that the project manager be a good listener and be able to offer guidance and persuasion. Previous experience on the same type of projects is important and provides the necessary technical information to manage the project in an efficient manner. The project manager must also be skilled at managing time, costs, resources, communications, and contracts.

The skill of communication is very important, as the facts do not express themselves; the best idea in the world would not be known without communication. At the lower levels, it is important to provide specific and clear instructions because a misunderstanding causes loss of time and increased costs. On the other hand, communication with the higher levels of the organization is necessary to provide summaries of the performance of the project in a way that allows senior management to understand and aid the project. Finally, the project manager must have the ability to visualize the completed project and have an accurate and broad vision for the whole project, with the capability to manage the dialogue, especially during meetings. This is regarded as the most useful means of communication to reach the goal, which is the success of the project.

2.7 Project Planning

The initial and basic principle of project management is how to read the time schedule, which is a real representation of the nature of the project, with the expectation of what could occur as a result of implementation. The next important step is to determine the purpose of the project and then answer the following question, "Is the driving force time or costs?" Then, the project manager must draw up a timetable with the project represented as a whole range of self-contained but interrelated activities.

For more than 20 years, schedules have been prepared manually. Since the developments in the field of computers, programs are now used to deliver the time schedule. In general, there is more than one method to draw up a time schedule for the project and this happens according to the nature of the project and the required presentation that will be provided to senior management.

We must recognize that the preparation of the schedule is the cornerstone of the management of projects and will follow the work schedule and the allocation of human resources and equipment, as well as the distribution of costs along the project period with the identification of ways to control costs.

In the beginning of 1900 and during World War I, Henry L. Gantt used the first method for preparing a project schedule. This is considered the first scientific method for the preparation of schedules. This simple method of representing activities with rectangles is used in project planning and work schedules at the time of production. The Gantt chart is set up by putting the plan on the magnetic blackboard and using rectangles made of iron and with length as a time unit. This was developed to be an (S) curve and is considered the first method to follow up the project with different activities by distributing the resources on the activity and monitoring the performance.

Until the mid 1950s, there had been no mention of development in project planning. In 1957, there were two different teams working on project planning using networks. The first team was prepared by using the Program Evaluation and Review Technique (PERT); this method depends on probability theory. The second team used a network and depended on the Critical Path Method (CPM). The methods of those networks have the same methodology; the difference is in the objectives. The development is done through operations research.

The first use of the PERT method was when the U.S. Navy was faced with a challenge in the POLARIS system, as they wanted to make rocket launchers in record time in 1958. The basics of the PERT method involved overcoming the lack of definition of activity duration time and used statistical methods to calculate it. This was done by defining the maximum, minimum, and most likely time for each activity, and obtaining the likely time of completion of the project, important parts of the project, and the minimum and maximum probable times to complete the project.

CPM was introduced in 1957 by two companies, Du Pont and Remington Rand Univac. The objective of the working group was to reduce the time for maintenance, overhaul of the rotating machines, and construction work.

It is noted that the calculation of the time required for different activities can easily be applied to projects other than POLARIS activities, as we need to identify one expected period only for each activity and the longer timetable for the course of the series of activities has been defined as the critical path.

CPM is the most common way of networking activities in project planning. It is used along with other methods by computer software. The process of planning is simply planning what will be done on the project in accordance with the order and manner of the execution of the project. Often there will be some changes and they should be followed up and the time schedule adjusted should be in accordance with the changes in the project as well.

To do the work with good planning, you must answer these questions clearly:

What are the activities that you want to do?

When will these activities be executed?

Who will execute the activities?

What are the equipment and tools required?

What are the activities that cannot be executed?

The answer to the previous questions is the key to arranging the work in an appropriate way.

So the project will be understandable, your goal is to transform information in a way that it can be presented to all parties in the project and be clearly understood by all. Your planning team target is to implement the project in a timely manner and in accordance with the specific cost requirements and, at the same time, achieve the required level of quality. Project planning requires the following goals:

- Reduce the risks of the project to the lowest level possible.
- Achieve the performance specifications for the project by establishing an organization for the implementation of business.
- Develop procedures to control the project.
- Achieve the best results in the shortest possible time.

The planner cannot plan for every minute of the project in detail because of the nonavailability of all information. As the work goes on, new details will require an increase in the effort and time to adjust the timetable in accordance with the new information and details.

2.7.1 Who Will Make the Plan?

In simple terms, this should be done by the project manager with the planning team. At the outset, though, it will be necessary to determine if a team will work and if you have sufficient information about their potential and their relation to the size of the project. You will also need to determine if you want to use an experienced planner from another project or contract someone to work with you for some time at the beginning of the project.

You should also know, through the collection of information, if the working group has worked in similar projects or projects that had similar activities. For example, if you are working on an oil and gas project, has the working group worked on the same type of projects before? The project will differ from the construction of housing projects, hotels, road projects, or industrial projects.

Each type of project has its own characteristics, but the working group should have worked on projects similar to your project. The draw is that you must have the human element of efficient planning and the ability to plan the project well. At the beginning of the work, the meeting of the planning group is attended by the official sponsor of the project, the director of the project, and the owner and his representative. In this meeting, the emphasis is on clarification of the main objectives of the project and identification of priorities in the implementation of the driving forces with the presentation of the project. Is it a time- or cost-driven project? What is desired from the project as a whole?

2.7.2 Where Do You Start the Plan?

Some basic definitions are as follows:

- A task is a small job done by one person.
- An activity is a part of the project that is composed of a set of tasks and performed by different individuals.
- Concurrent activities are activities developed in planning to be implemented in parallel.
- Series activities are activities put in the plan, that needs to be executed one after the other so that the second activity does not begin until the first activity is finished.

To be clear, the difference between an activity and a task is that the activity is a set of tasks or work. For example, the preparation of the report is an activity but it is carried out through a series of tasks or functions that are as follows:

1. Information collection
2. Data analysis
3. Initial report preparation
4. Graph generation
5. Final draft preparation
6. Printing of final version

You are now in a phase where it is difficult to know how to start planning. There are many ways and you have to choose among them. The first way to start is to identify the key stages of the project. The key stages of the project will be identified by holding a meeting with the experienced persons in the project team from different disciplines, the stakeholders, and the sponsor. In this meeting, use a brainstorming technique among the attendees. Every person will write his or her suggestions on paper, which will be posted on the wall, to be seen and commented on by all members at the meeting.

The collected papers will contain all the ideas and contributions, regardless of whether they are logical or illogical. An electronic blackboard may be used in some cases. It is important for the specific aim of this exercise that we follow these rules during the meeting:

- Be concerned about quantity and not quality, even if it turns out that some of the tasks and activities have been replicated.
- Avoid any idea of critical observations that would bother participants.

The list that is obtained now is not arranged in any type of order, importance, or time. When you find that all individuals have put forth all the ideas they have, stop the brainstorming. The next step of the action team is to filter the activities. This is done by removing tasks that are repetitive or duplicates and then compiling the tasks there, including the interdependence of both straight and parallel tasks. After reduction, the small number of tasks and activities is often within the range of 30–60, according to the size of the activity and the project. Compile the activities into key stages of the project. By using this method, you will reach precision in planning of up to about 90%, and this is considered the beginning in the planning of the project as a whole.

Now that you have the main stages of the project and all the key stages have been agreed on by the members of the project, you will be arranging them in a logical order. You should avoid the following:

- Avoid defining time or dates.
- Avoid the allocation of employment to those stages.

It is not easy to avoid doing these things, but they will cause problems. To avoid mistakes in the project planning logic control, the key stages must be written on the blackboard on the main wall of the office. Figure 2.5a and b shows an example of the main stages of the draft.

The advantages of this method are that everyone's ideas and opinions on the project is included, which makes them invested in the success of the project. Note that in Figure 2.5, the design phase has been divided into two stages (Figure 2.5a and b) to allow the sending of purchase orders before the end of the first phase of the design. You now have information that can be

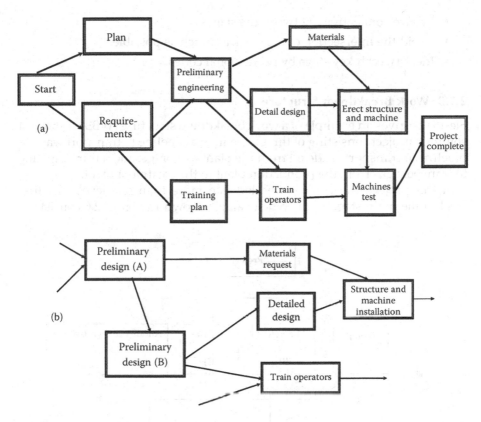

FIGURE 2.5
(a, b) Project key stages.

used in the computer software to prepare a time schedule for the agreed plan for the project.

The basic rules that must be adhered to and strictly followed in the preparation of a project schedule are as follows:

- Movement activities are from the left to the right.
- There is no measure of time.
- Start with the biggest and topmost square. Make sure there is an empty place on the page for each major stage in the project.
- Each phase is described in the present tense.
- The pages are developed in accordance with the logical arrangement.
- Communicate between the different stages of the project.
- Identify responsibilities.

- Ensure connectivity between the stages.
- Avoid the intersection of the stock as much as possible.
- Identify each key stage by professional codes.

2.7.3 Work Breakdown Structure

Figure 2.6 gives an example of a work breakdown structure (WBS). It shows a common project consisting of three pipelines: a pipeline to transport water, a pipeline to transport crude oil from the plant to storage, and a third pipeline to transport gas from the production plant to the treatment plant.

In this project, the work is divided into more than one level. The first level of the main stage in the pipe example shown in Figure 2.6 consists of

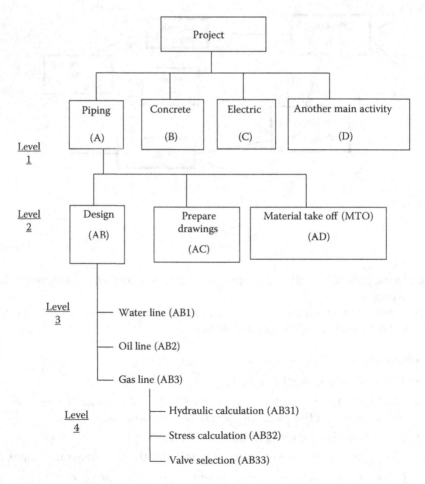

FIGURE 2.6
Work breakdown structure.

the concrete and electrical work. The second level involves more than one stage as the drawing of the designs and the calculation of materials begin. Level 3, in this example, focuses on the design phase and will be divided into stages for the design of each pipeline: gas, oil, and water. At the fourth level, we design a gas pipeline by performing a hydraulic design and pipe stress analysis to choose the thickness of the pipe and the base locations. The subsequent stage involves the selection of the required valves. There may be several stages, depending on the nature of each project.

We note each stage using the appropriate level code for ease of use later. WBS can be completed at any level of description. It does not explain the relations between the activities or show the time or length of time for any activity.

After you have selected the main stages and WBS, the next step is to develop a rough schedule of time for the project. However, you must first define the responsibilities.

2.8 Responsibilities of the Planning Team

The planning team is responsible for preparing the schedule at this stage and has the vital role of distributing the main stages of the project to the team members. Every main stage has a key stage owner (KSO) and it is that person's responsibility to reach the required target in a reasonable amount of time. This involves:

- Defining the work down to the smallest task level
- Ensuring that relations between activities and tasks are clearly defined
- Estimating time with high accuracy
- Ensuring that business is done in a timely manner in accordance with the required quality expectations
- Ensuring that work is proceeding in accordance with the procedures and requirements for quality assurance
- Maintaining the ongoing follow-up
- Ensuring that there are periodic accurate reports

There are some problems that the KSO will face during the project and they should be resolved in a timely manner. The KSO must have:

- The necessary authority to complete the work
- The necessary tools to complete the work
- The right atmosphere to achieve the required quality of work

- Direct support from the project manager or the official sponsor of the project
- Clearly understand performance expectations

When the project manager chooses the KSO for each state, the following should be considered:

- Skills
- Depth of information and knowledge
- Previous experience in the same area
- Rapid completion of work
- Accuracy in the completion of previous work
- Ability to solve problems
- Ability to manage time
- Ability to work as an individual and with a team
- Volume of work
- Ability to offer advice, be supportive, and work under stressful conditions
- Amount of current or future training required

Now that the key stages plan is available and every key stage has a defined person responsible, it is time to begin estimating the time required for each activity of the key stages.

2.9 Estimating Time Required for an Activity

To determine the time required to finish any task, you must know the resources available to perform the work at the required level of quality. To do that, you must know the following:

- The task volume, if it can be measured. For example, first, calculate the time required to prepare the wood form to pour 100 m^3 of concrete for the foundation and, from that, the task volume can be known.
- The work required to finish this activity (by hours, days, or weeks)— noting that the number of workers should be identified and the capability of each worker to perform the task alone should be considered.

The working capacity of a worker will usually be measured in days. You should take care to avoid the pitfalls that usually trap young engineers. For example, you should decrease the capacity of work per day by about 50%, as all of the hours per day do not focus on project activities. Much time is wasted in meetings or special discussions and also by routine events such as bathroom breaks, eating, etc. Moreover, there is some delay involved in the work itself.

After you define the period for each task, keep in mind the other traps when you put it on the calendar. The total time length will be different due to the following factors:

- Weekends
- Official vacations and holidays
- Annual leave for employees
- Work stoppages

It is worth noting that defining the performance rate for each activity depends on the normal rate that is found in textbooks or standard guides for contractors, but it is essential to obtain information from others with experience in the same country, the same location, or from previous projects.

From here, we cannot disregard past experiences. If an activity is the same as in similar projects—pouring concrete foundations, for example—we should use materials and procedures of the same type, preferably from the same location. This is particularly true of remote areas such as the desert because they are more challenging than cities, where there is available labor and higher efficiency of employment.

When you ask individuals and experts for information, you should keep in mind that not everyone has accurate information. It is human nature for good things to remain in memory and bad experiences are erased from the mind with time. When someone says that the expected time for an activity was 18 days, but it was finished in 10 days, that person clearly considers this to be a success story and will keep it in mind for a long time. However, as a rule of thumb, no one can work 100% of the time and that 20% to 50% of time is wasted in the following activities:

- Unnecessary meetings
- Interoffice visits
- Newspapers and mail
- Research
- Assisting or advising others
- Office equipment failure
- Regular daily activities
- Misunderstandings between team members

- Lack of specifications or SOW
- Lack of quality specifications
- Training

2.9.1 Calculating Time Required for an Activity

There is a method of calculating the time required for an activity through the use of performance rates. For example, to calculate the time required to excavate 15,000 m^3 of soil, you must first determine the method of excavation. In this example, that method has been identified as equipment, specifically two bulldozers, two loaders, and four trucks to transport the soil. The performance rate of one bulldozer is 120 m^3/h and the performance rate of one loader and two trucks is 75 m^3/h.

$$\text{Rate of excavation} = 2 \times 120 = 240 \text{ m}^3/\text{h}$$

$$\text{Time required for excavation} = 15{,}000/240 = 62.5 \text{ h} = 9 \text{ days}$$

$$\text{Rate of soil removal and disposal} = 2 \times 75 = 150 \text{ m}^3/\text{h}$$

$$\text{Time required for removal and disposal} = 15{,}000/150 = 100 \text{ h} = 15 \text{ days}$$

Therefore, the excavation time is about 9 days and the disposal of the soil takes about 15 days. Taking into consideration the number of working hours in the day, which is only 7 hours for the purposes of the timetable. For the excavation activity, take the largest period, which is 15 days. However, the excavation may be immediately followed by the pouring of plain concrete. That may begin after 9 days because there is no need to wait until the waste has been removed from the site.

2.9.2 Time Schedule Preparation

We have already discussed the most common ways to draw up a timetable and also how to map along the CPM and PERT. The most important factors in the preparation of the time schedule are to identify the activity, determine the time required for its implementation, and identify relations between the activities. We discussed how to determine the activities and the amount of time required for each activity; we must now determine how to arrange the activities around each other. The different, common relations between activities are shown in Figure 2.7.

1. Activity B cannot start until Activity A is through.
2. Activity A and Activity B start at the same time.
3. Activity A and Activity B finish at the same time.

Order of activities

Types of activity relationships

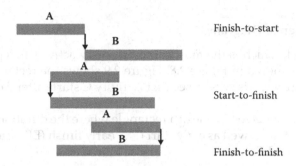

Finish-to-start

Start-to-finish

Finish-to-finish

FIGURE 2.7
Order of activities.

In some cases, Activity A may start and, after a certain period, Activity B will start. This period is called "lag time." A computer can do this task easily. For manual preparation, there are two main methods: the arrow diagram and the precedence diagram, as shown in Figure 2.8.

2.9.3 Arrow Diagram

This diagram depends on defining each activity by an arrow, as in Figure 2.8. The point of connecting arrows is called a node and drawn as a circle. As in Figure 2.8, Activity A depends on Activity B. Activity C depends on Activity

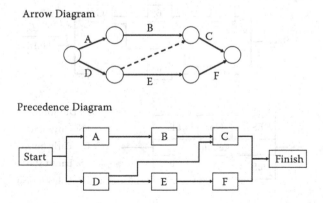

FIGURE 2.8
Tools for arranging the activities.

B and Activity D, but there is no activity line between Activity D and Activity C. So, we use a dummy arrow with time zero to solve this problem. However, using dummy arrows for large activities will present a problem, so the use of a precedence diagram is now more popular.

2.9.4 Precedence Diagram

In this method, which is the most common, every activity is represented by a rectangle, as shown in Figure 2.8. Figure 2.9 shows the rectangle in detail. From these figures, it is easy to see that Activity C starts after Activity D and Activity B.

Figure 2.9 presents the inside of a rectangle where the duration of the activity will be written, as well as early start (ES), early finish (EF), latest start (LS), and latest finish (LF) times.

2.9.5 Gantt Chart

In general, relations between activities are in series or parallel. The relation depends on the nature of the activity and this should be considered in preparing the time schedule.

The Gantt chart is an old, traditional method used until recently to present project schedules (Figure 2.10). The relations between all the activities are not well presented by this method and you cannot go through detailed activity accurately, so it is not a good presentation method for high-level management. It is detailed enough for one level of the schedule, but for more detail, there needs to be another method for scheduling (Figure 2.11).

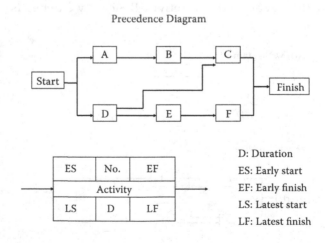

FIGURE 2.9
Method of creating a precedence diagram.

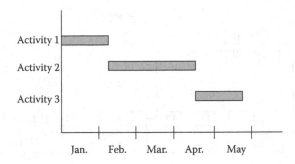

FIGURE 2.10
Example of a Gantt chart.

2.9.6 Critical Path Method

The essential technique for using CPM is to construct a model of the project that includes a list of all activities required to complete the project, the time required for each activity to be completed, and the dependencies between the activities.

Using these values, CPM calculates the longest path of planned activities to the end of the project and the earliest and latest times that each activity can start and finish without making the project longer (Figures 2.12 and 13). This process determines which activities are critical (i.e., on the longest path) and which have "total float" (i.e., can be delayed without making the project longer). Any delay of an activity on the critical path directly impacts the planned project completion date (i.e., there is no float on the critical path). A

Example

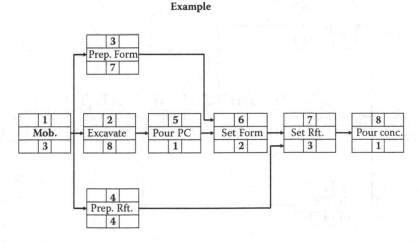

FIGURE 2.11
Precedence diagram example.

Example

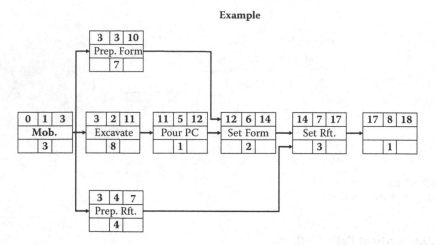

FIGURE 2.12
Calculate the early times.

project can have several critical paths. An additional parallel path through the network with total durations shorter than the critical path is called a subcritical or noncritical path.

2.9.7 Program Evaluation and Review Technique

PERT is required for the execution of special projects or large-scale deliveries with a variety of activities. These activities need to be accurately identified, as do all activities involved in the project that must be completed. The

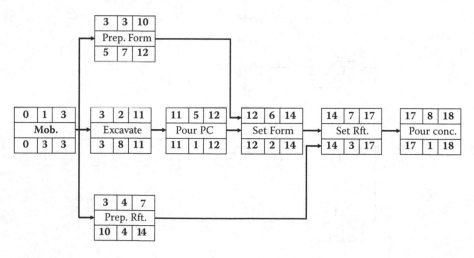

FIGURE 2.13
Calculate the latest times.

completion of these activities must be initiated and carried out in a specific sequence, particularly any activity that must be completed before the start or completion of other activities or that may occur in relay of parallel activities, meaning two or more activities could be completed at the same time.

In addition, there are specific achievements that indicate the completion of key sectors of the project. The successful implementation and management of such projects must be carefully planned to speed up the delivery during a specified period, to coordinate the multiactivity of the project, to monitor the use of various resources necessary for its implementation, and to achieve the project on time and within budget cost.

Although PERT was designed mainly for military purposes, it has been used successfully since 1959 in most large-scale projects. PERT analysis is used in a number of areas of the computer and construction industries in planning shutdowns for maintenance in refineries.

This analysis has confirmed its applicability and importance in different projects. An example of its success for contracting companies was its role in solving the problems of coordination between various activities in a large project of a considerable degree of complexity and in planning the time required for the implementation of each activity in Montreal, Canada, in 1967. This method made it possible to complete the project within the scheduled time.

PERT was developed to simplify the planning and scheduling of large and complex projects. It allowed the incorporation of uncertainty by making it possible to schedule a project even without knowing the precise details or durations of the activities.

2.9.8 Example

Table 2.1 illustrates the relation between activities and how you can create a schedule through the precedence diagram. It shows the activities required for a cast concrete foundation, which is a simple example that shows how to arrange activities and how we can account for the overall time of the project and identify the critical path.

TABLE 2.1

Example of a Foundation Project

Item	Activity	Time (days)	Precedence Activity
1	Mobilization	3	–
2	Excavation	8	1
3	Preparation of wooden form	7	1
4	Preparation of steel reinforcement	4	1
5	Pouring of plain concrete	1	2
6	Setting wooden form	2	3, 5
7	Setting reinforcement	3	4, 6
8	Pouring concrete	1	7

In Table 2.1, the first column contains the item number or the code for that item and its activities and subactivities. The main activity code may be 100 and a subactivity 110, for example. Table 2.1 gives an example of a simple case. The second column contains the name of the activity and the third column contains the time length for each activity in days. The fourth column specifies the relations between activities. Activity 2, which is excavation, is constrained by Activity 1. Activity 6, which is setting the wooden form, depends on Activity 3 (finishing the preparation of the wooden form) and Activity 5 (pouring the plain concrete).

Figure 2.11 shows a precedence diagram. In each diagram, there is a number representing its duration. Figure 2.12 shows the early start and finish times for each activity by using the following equation:

$$EF = ES + D$$

where EF denotes early finish, EF is early start, and D is duration.

Start with Activity 1. The EF of this activity is 3 days. Transfer this value to Activities 2 through 4 in the ES rectangle zone. Activity 6 depends on Activities 3 and 5. Take the higher value for EF, which is 12 days, and use it as the ES for Activity 6. In the same sequence, Activity 7 takes its ES value (14) from the higher value of EF from Activities 6 and 4. For the last activity, take the EF value (18) and transfer its value to the LF rectangular zone. Figure 2.13 shows the latest start and finish for each activity by applying the equation

$$LS = LF - D$$

to the last activity and calculating back. By applying this equation, we determine that the LS of the last activity will be 17 days.

As shown in Figure 2.13, subtracting LS from LF and subtracting ES from EF gives a value of 0, so this means that this activity is on the critical path. If their difference has a value, this means that this activity can be delayed by this period without affecting the total project time.

As an example, let us say that the period for steel bar preparation is 4 days and we assume that the bar preparation is 3 days and 1 day is required for coating the steel bars to protect from corrosion. By performing a simple calculation, we find 7 days to be the total float (TF) for the steel bar preparation and coating. The free float (FF) of the steel bar preparation is 0, as any delay will affect the coating activity. The coating activity has an FF of 7 days, as there is no activity after that.

2.9.9 Applications for the PERT Method

We know that implementation of any activity requires time and resources. The PERT team faces the problem of estimating the time required for each activity. Reliance on a single time estimate is poor planning and incompatible with the conditions of uncertainty, even under the best circum-stances, as the external environmental factors surrounding the project may cause deviations in the time required for activities.

The PERT method overcomes this uncertainty by estimating the time of each activity using three time values that will be compiled to reach a statisti-cal probability for estimating the time required for project completion. This method is applied as follows:

- *Optimistic time.* This is the lowest possible estimation of the time required to complete the activity. It involves all of the factors affect-ing this activity being positive. This usually predicts a short period for completion.
- *Pessimistic time.* This is the estimate of the maximum possible time for implementation of the activity. It presents the worst-case scenario for a delay of this activity. This usually provides a longer period with a low probability.
- *Most likely time.* The most likely time presents the time required to complete this activity under normal conditions for all factors affecting this activity. This value has the highest probability of occurrence.

The PERT method correlates the relation between the above three times by using a beta distribution as illustrated in Figure 2.14. The optimistic and pessimistic values predict the limits of the distribution, and the most likely value presents the time required under normal or expected conditions. To define the activity period, calculate the activity mean time by using Equation (2.1). The summation of all periods on the critical path presents the mean period for the project.

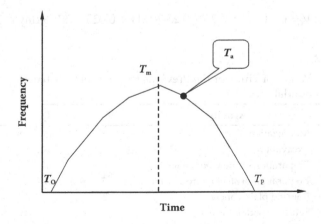

FIGURE 2.14
Probability distribution for every activity.

2.9.9.1 Statistical Calculation of Activity Time

From the beta distribution, the expected, or average, time for each activity is calculated according to Equation (2.1).

$$T_a = \frac{T_o + 4T_m + T_p}{6},$$
(2.1)

where T_a is the average time, T_o is the optimistic time, T_p is the pessimistic time, and T_m is the most likely time.

2.9.9.2 Example

For the example of constructing a concrete foundation, there are three values of time for each activity, as shown in Table 2.2.

After defining the critical path from the previous example, calculate the expected time for each activity as in Equation (2.1). Find the standard deviation using Equation (2.2).

$$S = \frac{T_p - T_o}{6}$$
(2.2)

Variance $(V) = S^2$

The minimum time required to finish the project is 12 days and the maximum time required to finish it is 29 days (Table 2.3). The average time required to finish the project is 19.14 days. What is the probability of the time required to finish the project increasing to more than 21 days? This value can be calculated as:

$$V = 0.44 + 1 + 0.28 + 0.25 + 0.11 + 0.028 = 2.108 \text{ days}$$

TABLE 2.2

The Three Values of Time for the Previous Example of Constructing a Concrete Foundation

Item	Activity	T_p	T_m	T_o
1	Mobilization	5	3	1
2	Excavation	11	8	6
3	Preparation of the wooden form	9	7	6
4	Preparation of steel reinforcement	5	4	3
5	Pouring plain concrete	2	1	1
6	Setting wooden form	4	2	1
7	Setting reinforcement	4	3	2
8	Pouring concrete	2	1	1

TABLE 2.3

Apply the PERT Method

Item	Activity	T_p	T_m	T_o	T_a	CP	S	V
1	Mobilization	5	3	1	3	a	0.67	0.44
2	Excavation	12	8	6	8.3	a	1	1
3	Preparation of the wooden form	9	7	6	7.17		0.5	0.25
4	Preparation of steel reinforcement	5	4	3	4		0.33	0.11
5	Pouring plain concrete	2	1	1	1.14	a	0.167	0.028
6	Setting wooden form	4	2	1	2.17	a	0.5	0.25
7	Setting reinforcement	4	3	2	3	a	0.333	0.11
8	Pouring concrete	2	1	1	1.5	a	0.167	0.028

[a] Activity in the critical path.

Standard deviation, $S = 1.45$ days

$$Z = \frac{21 - 19.14}{1.45} = 1.28$$

From the probability distribution tables, the probability that the project completion period will be more than 21 days is 10%. The probability that the execution time for the project will be 21 days or fewer is 90%, as shown in Figure 2.15.

2.10 Cost Management

2.10.1 Cost Estimate

The cost estimate is the calculation of the estimated cost of performing more than one phase during the project. In the initial phase of studies in the

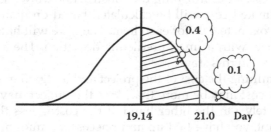

FIGURE 2.15
Probability distribution.

appraise phase, the cost is calculated by a simple principle and the borders of right and wrong are very large. The accuracy of the cost-estimate calculation increases with the forthcoming phases of the project so that the possibility of the actual cost of the project increasing decreases with time until the end of the project and, therefore, 100% of the cost of the project is reached.

Feasibility studies must be performed by persons with extensive experience in similar projects so that accurate cost values can be assigned. All projects have different characteristics and requirements. For example, residential building projects are different from factories and petroleum projects. Let us say that, using a traditional example, your friend is going to purchase a land for a new villa and he requires a rough cost estimate. It will be very complicated as he has no drawings, calculations, or concrete data to base the calculations on. The estimated value will have a high degree of uncertainty because the architecture, exact location, soil type, and other factors are unknown. From previous experience, your friend can obtain an estimate if at least the number of floors and the general location of the land are known. That will provide a cost of US$ per m^2 for that location and number of floors.

The steps in this example are at a very preliminary stage and are commensurate with a small project, such as building a house. In the case of large projects, budgeting at this stage requires making surveys and taking soil samples using boreholes and then calculating the cost of the project. The entire project will then be studied again to determine the effect on profit of this minor clarification.

For large-scale and high-cost industrial projects, especially where some of the equipment and machinery required are unique, if the estimated cost in the initial phase of studies is within ±50%, the proportion of the error and the difference between the expected and real values is very large. In a later similar project, the difference will be smaller because, in industrial facilities, a large proportion of the cost depends on the equipment and machinery used in manufacturing. On the other hand, there are some projects, such as the replacement and renovation of residential or industrial facilities, with very low degrees of accuracy and their initial estimated costs will also be about ±50%.

After finishing the FEED engineering, the accuracy of the cost estimate will be about ±30%. After finishing the construction drawings for the whole project, the estimated cost will be calculated based on quantities and will provide an approximate cost for each item. Then, we will have an estimated cost for the project with a small predicting deviation. The acceptable accuracy is about ±15% in this stage.

If, after determining the cost of the project and budgeting, you have a scenario where the cost may be more than 15%, the project may be halted partially or completely. On the other hand, if the cost is less than 15% of the budgeted amount, you have tied up an unnecessary amount of the owner's investment that could have been invested in other projects. Therefore, the cost estimate is vital to the project's success.

2.10.2 Cost Types

For the owner, there are different types of costs involved in construction projects. These include the costs of assets and capital, including the initial composition of the facilities of the project such as:

- Cost of land and property registration procedures
- Planning and feasibility studies
- Engineering activities and studies
- On-site construction materials, equipment, and supervision
- Insurance and taxes
- Cost of the owner's office
- Cost of equipment not used in construction, such as private cars to transport the owner engineers
- Examination and tests
- Cost of maintenance and operation in each year of the project life cycle include the following:
  - Leasing land
  - Employment and wages
  - Materials required for annual maintenance and repairs
  - Taxes and insurance
  - Other owner costs

The cost values of each of the preceding items vary according to the nature, size, and location of the project and also for other structural organization considerations. We should not forget that the owner's goal is to reduce the total cost of the project to be consistent with the objective of the investors. The highest value in a real estate project is the cost of construction of the structure. In the case of industrial buildings and the petrochemical industry, the cost of civil and structural work is almost small relative to the cost of mechanical and electrical equipment. For example, the cost of a concrete foundation for a power turbine may cost $30,000 and the power turbine may cost more $5 than million. Nuclear and power generation projects are other examples.

When we calculate cost from the viewpoint of the owner, it is very important to calculate the cost of operations and maintenance in each year of the life of the project for each of the alternatives available in the design and to calculate the cost of the life cycle of the project as a whole.

To calculate the estimated costs of a project and to develop a budget, we must point out the limits of deviation and the cost of risk or unexpected events during the execution of the project. The percentage of risk has to be calculated for each item or as a proportion of the total final cost. The calculation of the cost of risks depends on past experience and anticipated problems during the implementation of the project. Increased costs of emergency often occur as a result of each of the following:

- Changes in design
- Differences in the schedule and increased duration time of the project
- Administrative changes such as increased salaries
- Special circumstances at the site, such as unexpected obstacles or defects in the soil
- Special permits required

2.10.3 Construction Cost Estimate

The cost of construction and installation is the biggest part of the total cost of the project. This cost is controlled by the project manager and the project construction manager. The accuracy of calculating the estimated cost of construction is different from one stage to another. Accurate cost calculations require accurate data: the more accurate the data, the more accurate the calculation of the cost.

The estimated cost of construction is calculated from more than one point of view. The cost is estimated for the owner to go through the design and construction drawings. The contractor calculates the cost of construction to tender his bid and obviously tries to be as competitive as possible. The third way is the control-cost estimate, which is used by the owner to control costs. This will be prepared for the owner by professional engineers experienced in the calculation of the cost estimate. These calculations are performed at more than one stage of the project. The estimated cost in general is usually carried out before the design of machinery and equipment in the case of industrial projects.

The initial cost estimate is made after preliminary studies of the project, with an initial identification of industrial facilities such as the number of pumps, compressors, and the size of pipes and lines in diameter. The cost is determined at this stage based on previous experience with similar projects. An example of cost calculation is as follows:

Example

Calculate the cost for a project consisting of two pumps with 1000 hp with a pipe-line 12 inches in diameter and 10 miles in length
The pump cost estimate = $650 per hp
Pump cost = 2 (650) (1000) = $1,300,000
Cost estimate for onshore pipeline = $22,000 per inch per mile
Pipeline cost = 22,000 (12) (100) = $26,400,000
Total cost = 1300000 + 265400000 = $27700000

Example

To calculate the cost of one floor for a building with an area of 300 m² from reinforced concrete:
In this case, assume slab thickness of 250 mm for a slab beam and column as a practical estimate.
Approximate quantity of concrete for a slab and floor = 300 × 0.25 = 75 m³
Assume the cost of concrete is $200 per m³
The concrete cost for one floor = $15,000

Note that the price of concrete is calculated after determining the type of concrete used. There are two methods for the processing of concrete: reinforced concrete and ready-mix concrete.

The cost of reinforced concrete can be calculated by obtaining this information:

1. Quantity of steel per cubic meter of concrete and the price of steel per ton
2. Quantity of cement in the concrete mix and the price of cement per ton
3. Quantity of coarse and fine aggregate and the price per cubic meter
4. Cost of shattering, bending bars, pouring and curing per cubic meter of concrete

To calculate the cost of ready-mix concrete, you must know:

1. Quantity of steel per cubic meter of concrete and the price of steel per ton
2. Price of transporting concrete from the nearest ready-mix location to the site and the price of the concrete pump
3. Cost of a wooden form, usually a special form for the pump concrete
4. Cost of steel fabrication
5. Cost of pouring process and curing

The approximate price for 1 m³ of reinforced concrete prepared on-site, noting that the approximate quantity of coarse aggregate is 0.8 m³ and for fine aggregate is 0.4 m³ and using these value to provide a characteristic cube strength of 25 N/mm².
Steel cost = $1000 per ton
Cement cost = $80 per ton
Coarse aggregate cost = $5 per m³
Sand cost = $1 per m³
Steel quantity = 0.1 t/m³ (see Table 2.4)
Steel cost = 0.1 × 1000 = $100

TABLE 2.4

Guide for Estimating Steel Quantity

Structure Element	Quantity of Steel Reinforcement in Concrete (kg m³)
Slab with beam	90–100
Flat slab	250
Hollow block slab	150–180
Columns	90–120
Isolated footing	100–120
Raft foundation	200–300

Cement cost = 7 / 20 × 80 = $28
Coarse aggregate cost = 0.8 × 5 = $4
Sand cost = 0.4 × 1 = $1
Total material cost = $133 per m³ (material only)
Cost of fabrication of wood and steel and pouring concrete = $30 per m³
Total cost = $163 per m³

Note that in calculating the reinforced concrete cost estimate, the main item is the quantity of steel in concrete and it is different according to the structural elements. Table 2.4 is a guide to estimating the quantity of the steel reinforcement.

The data in Table 2.4 are to be considered as a guideline and depend on the concrete characteristics, strength, and the member design. It is an indicator of the quantity of steel reinforcement. The percentage cost of a domestic or administration building is shown in Table 2.5.

2.10.4   Steel Structure Cost Estimate

The cost estimates for steel structures are significantly different from the calculation of the estimated cost of reinforced concrete structures, as they need special designs. The most important part of the design and construction of steel structures are the connections. It is well known that connection costs are about 50% of the whole building. Table 2.6 defines the cost percentage of each part in a steel structure project.

TABLE 2.5

Percentage of Reinforced Concrete Building Cost

Activity Item	Percentage from the Total Cost (%)
Design and site supervision	3
Concrete works	36
Masonry work	6
Sanitary and plumbing	10
Internal and external finishing	45

TABLE 2.6

Percentage of Steel Structure Costs

Item	Percentage from the Total Cost (%)	Percentage of Cost for Connection (%)	Percentage of the Connection Cost (%)
Preliminary design	2	33	0.7
Final design	3	55	1.7
Detail drawings	8	77	6.2
Total design cost	13		8.6
Material	38	40	15.2
Fabrication	27	63	17.0
Painting	10	35	3.5
Erection	12	45	5.4
Total percentage	100		49.7

Design	13%
Materials	38%
Fabrication	27%
Coating	10%
Erection	12%

From Figure 2.16, one can determine the proportions of the costs of main elements for constructing a steel structure; every item is represented as a percentage of the total cost.

Note that in Table 2.6, the connections, either through the use of welds or high-stress bolts, have the largest share in the process of preparing detailed drawings, where the most important and most critical phase is the accuracy of the details of the connection. There is no doubt that they are important in the cost of manufacturing and in the installation phase as reflected in the connection cost percentages of 63% and 45%, respectively. The connections are of great importance in the calculation of costs, as well as in the preparation of schedules.

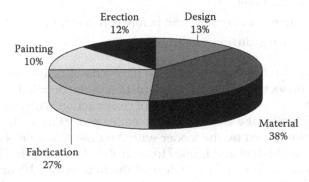

FIGURE 2.16
Percentage of cost for steel structures.

2.10.5 Detailed Cost

The detailed cost estimate will be determined by the detailed construction drawings of the project and by project specification. The engineering office calculates the cost estimates through the experience of the contractors who carry out the construction of those projects. The owner should also add the cost of his or her supervision.

2.10.6 Tendering Cost Estimate

This calculation is performed by the contractor and will be presented to the owner in the form of a tender or offer to negotiate prices. The contractor is often estimating the price with high accuracy to ensure that the work is carried out under the prescribed financial terms and uses every method of calculating the estimated cost for construction. Most construction companies follow a scientific procedure to calculate the construction cost of a project and to develop appropriate pricing of the tender.

There are some work, facilities, and equipment provided by subcontractors and the subcontractors send the bids for the project to the general contractor, who in turn studies the offers, chooses the best price, and adds the percentage of indirect cost and profit. This is included in the cost of supervision and nonprofit expenses, in addition to the direct cost of other factors and is conducted by:

- Subcontractors offers
- Expense amounts
- Construction procedures and steps

2.10.7 Cost Estimate to Project Control

The cost estimate will be calculated to follow up during project execution. The cost estimate will be obtained from the following information:

- Approximate budget
- Budget value after contract and before construction
- Estimated costs during construction

Both the owner and the contractor must have a baseline by which to control costs during project execution. The calculation of the detailed cost estimates is often used to determine an estimate of construction where the balance is sufficient to identify the elements of the entire project as a whole. The contractor price submitted by the tender with the time schedule is determined by the estimated budget and is used to control costs during the implementation of the project. The estimated cost of the budget will be updated periodically during the execution of the project to follow up the cash flow in an appropriate manner along the project period time.

2.10.8 Economic Analysis of Project Cost

The project cost consists of two parts: direct and indirect costs.

Direct costs are the costs of implementing the activities of the project as set forth in the drawings and project documents. The resources to be provided for the implementation of these activities are:

1. Raw materials
2. Employment
3. Equipment
4. Contractor errors
5. Other costs

Direct costs do not include expenses related to the supervision of the implementation work.

Indirect costs are expenses that cannot be avoided such as supervising work and facilitating the implementation. They can also involve terms such as *administrative* and *general* expenses and are usually fixed during the implementation period of the project.

2.10.8.1 Work Breakdown Structure

Most of the projects aim to achieve a certain degree of specific quality. This objective can be divided into a set of components that reflect the entire project example; if the desired goal is to build a building, it can be identified by five main components:

- Public works site
- Concrete structure
- Mechanical
- Electrical
- Finishing works

However, if the project's aim is to create a missile defense system, the key elements will be:

- Rocket
- Control
- Communications
- Firing elements

After identification of the first level of the project, elements can be divided into a subset of components and can continue to be divided down to successive levels of detail to achieve the level of the basic elements of project implementation. Therefore, great importance is attached to the breakdown structure in making sure that all elements of the project have been taken into account.

The time spent in preparing and processing the WBS is important, as it will ensure that one of the important elements of project implementation is not forgotten or deleted. After determining the WBS, work must be put into a particular method of identifying the elements with a coding system.

2.10.8.2 Organization Breakdown Structure

In the same way that determining the levels of components of the project is done, systems can be divided, as well as project management and the organizational structure, so that management can understand and facilitate the identification of responsibilities within the project.

Organization breakdown structure (OBS) assists in surveying work activities. The organization should be known to everyone involved in the project facilitation of communication and definition of the responsibility. In the organization, structure is usually followed by the project manager and a group of managers, each of whom oversees a portion of the project. For example, note that the electrical work supervisor will be responsible only for the activities required to implement the electrical work for the project. It is clear to us how important it is to develop a project management structure that is conducive to the easy flow of information to and from the project manager to support decision-making.

2.10.8.3 OBS and WBS Matrix

The OBS and WBS matrix is shown in Table 2.7. The OBS in the horizontal and vertical presents the organization. In this table, every person in the organization has a number of cells that presents the WBS. The project manager is responsible for all the stages, so that cost will be for all activities. The mechanical supervisor is allocated mechanical work only and the civil supervisor is allocated the civil work and foundation work.

2.10.8.4 Work Packages

Cost will be divided into work packages. This is the last level for WBS that presents the project activities and the time schedule and cost that depend on them. For every activity, it is required to define the following to calculate cost:

- Execution time from start to finish
- Resources required to execute this activity
- Cost to execute the activity

TABLE 2.7

WBS and OBS Matrix

	Planning Work	Design Work	Electrical Work	Mechanical Work	Foundation Work	Civil Work
Project manager	×	×	×	×	×	×
Planning supervisor	×					
Design supervisor		×				
Electrical work supervisor			×			
Mechanical work supervisor				×		
Civil work supervisor					×	×
Concrete works supervisor					×	

Note: OBS, organization breakdown structure.

Cost account (CA) is the cost budget for the activities executed on the project. It is a budget for every responsible person in the organization and shows the direct cost that every person is responsible for executing. For example, the concrete supervisor will be responsible for executing the following activities:

- The foundation with cost $100,000
- The first floor with cost $50,000
- The other floors $300,000

So, in this case, the concrete supervisor will be responsible for three cost allocations with a total cost of $450,000.

Indirect cost is, so far, not taken into account, as the indirect expenses do not figure into the work expenses or, to be more precise, in the execution of the WBS structure and, therefore, such costs are charged to the administrative management of the project or elements of project management.

As an example, the concrete foreman in the previous example has a salary and uses equipment and other expenses; these expenses cannot be added or charged to the specific activities, since he supervises the foundation, columns, slabs, etc. Those expenses must be paid on a regular basis during the execution of these activities along with any other expenses of the supervisor, other administrative expenses such as offices for the engineers, technicians, computers, cars to transport engineers, and expenses and salaries of engineers and senior management.

For example, if concrete work will continue for a period of 6 months and indirect expenses for each month amount to $8000, the total budget or estimated indirect costs is 6 × 8000 = $48,000. Thus, the total reinforcing concrete budget will be as follows:

- Direct cost = $450,000
- Indirect cost = $48,000
- Total cost = $498,000

2.10.8.5 Cost Control

Cost control is very important in the management of projects as they relate to the economics of the project as a whole, which is a key element in the success of any project.

The objectives of cost control are to follow-up what has been spent compared to what was planned and to identify deviations. Therefore, cost control and intervention are ongoing processes in the domain of the project manager, who is directly responsible for supervision.

In calculating the actual costs, different costs (employment, materials, equipment, subcontractors, etc.) must be considered for each calculation depending on the specific document that was agreed on at the start of the project. If the actual cost is more than the planned cost, it will be due to the following reasons:

- Low cost estimate
- Circumstances of the project not well studied
- Increases in the prices of raw materials and labor
- Climatic conditions
- Poor selection of equipment
- Inefficient supervision

While it is difficult to correct the impact of the first four factors, selection of equipment can be improved and the department should always be aware and choose competent supervisors or at least help increase their capability.

The cost-control process should be more than simply collecting data on the cost. The codification of collected data as well as a copy of gains and losses after implementation of project cost control can be given to the project manager. They will be useful in the analysis of the performance rate of equipment productivities and manpower. A review of the total amount spent on the project since the beginning of work until the date of the audit will present one of the following three cases:

- Which was exactly equal to the spending planned according to the plan of implementation and estimated budget?
- Which spent more than had been planned according to the plan of implementation? This represents an overexpenditure or cost overrun.
- Which spent less than planned according to the plan of implementation? This represents savings in spending or a cost underrun.

In general, overexpenditure is not desirable and must be prevented. If the analysis can detect the causes so that it can be avoided in the future, that is desirable. Researching the causes of costs overruns to reduce them in the future is one of the characteristics of successful project management.

The following parameters are the main tools to control the cost:

ACWP is actual cost work performed

BCWP is budget cost work performed

BCWS is budget cost work scheduled

BAC is budget at completion

EAC is estimation at completion

To illustrate the above factors in a simple way, assume that an activity is scheduled to be completed in 40 h. The actual work will be 50 h and the work already completed has taken 40 h. Assume that the cost of 1 h is $100:

$$ACWP = \$5000$$

$$BCWP = \$4000$$

$$BCWS = \$4000$$

$$Cost\ variation = ACWP - BCWP$$

In this example, the cost variation equals $1000.

$$Percentage\ of\ cost\ deviation = (ACWP - BCWP)/BCWP$$

$$Schedule\ Index\ (SI) = BCWP/BCWS$$

When the SI value is higher than 1, it indicates an acceptable performance. A value less than 1 indicates an unacceptable performance.

$$\text{Cost Index (CI)} = \text{BCWP/ACWP}$$

$$\text{EAC} = \text{BAC/CI}$$

As we have stated, these factors must be calculated at regular intervals during project implementation and should preferably be compatible with the monthly accounting date of the company. The monitoring of these factors on a monthly basis assists in evaluating the project and the final cost of the project.

2.10.8.6 The Cost Curve

The cost curve, called the *S curve*, takes the form of the letter "S" in the calculation of costs and distributions to the timetable. Using the previous example of pouring the concrete foundations, in Figure 2.17, the first column represents the various activities on the other side, a Gantt chart shows the cost of each item per the day, and we can easily assume that the cost of every activity cost is $1000 per day.

This is done through the compilation of costs in the bottom row. In the first case, each activity is started and finished early, and in the second case, each activity is started and finished late.

When we assign the cumulative cost curve and make the relationship between the cost and time, we get the S curve, as shown in Figure 2.18. We

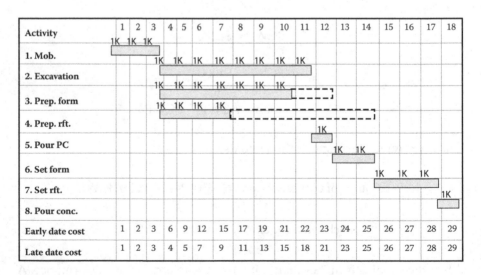

FIGURE 2.17
Distribution of cost of the activity.

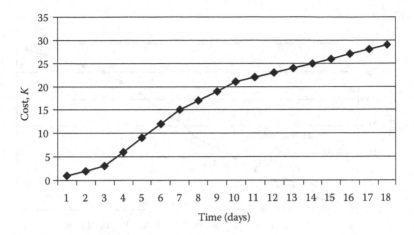

FIGURE 2.18
Cash flow in the case of early dates.

will find the first curve in the case of an early implementation of activities and the second curve after the activities have been implemented (Figure 2.19).

In Figure 2.20, we find that if the cost curve is higher than planned and we are in shape to do the work well, it is called "ahead of schedule." When the cost curve is within the envelope, this indicates that you are in the scheme called "within schedule." Where the curve is less than planned, as in Figure 2.20, the situation is critical, as the project is late and called "behind schedule."

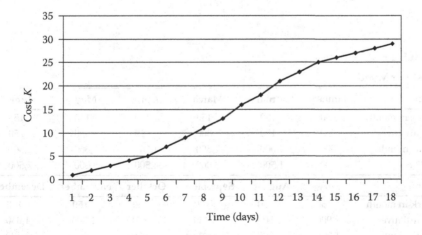

FIGURE 2.19
Cash flow in the case of late dates.

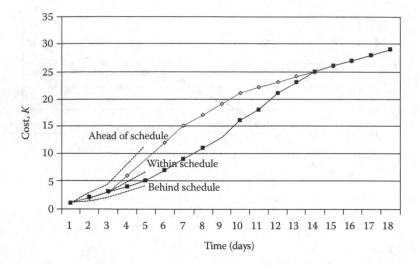

FIGURE 2.20
Cash flow envelope.

Example

The following table gives the values of the cost of planning a project for a period of 12 months in accordance with the timetable explained earlier. In the projected 6 months after the transaction, we will calculate the position of the crisis to see the project as presented in Table 2.8.

From Table 2.9, calculate the cost control parameters 6 months from the start of the project.

TABLE 2.8

Cost per Month

Plan	January	February	March	April	May	June
Work in month	50	100	150	150	150	150
Cumulative	50	150	300	450	600	750
Cost/month	500	1,000	1,500	1,500	1,500	1,500
BCWS	500	1,500	3,000	4,500	6,000	7,500

Plan	July	August	September	October	November	December
Work in month	150	50	100	150	150	150
Cumulative	900	1,500	1,450	1,350	1,200	1,050
Cost/month	1,500	500	1,000	1,500	1,500	1,500
BCWS	9,000	15,000	14,500	13,500	12,000	10,050

TABLE 2.9

Cost Control Parameters Calculation

	January	February	March	April	May	June
Cost per month	500	1,000	1,500	1,500	1,500	1,500
BCWS	500	1,500	3,000	4,500	6,000	7,500
BCWP (EV)	300	1,000	1,600	1,600	1,500	1,500
ACWP	300	1,000	1,800	1,800	1,600	1,500
EV cumulative	300	1,300	2,900	4,500	6,000	7,500
ACWP	300	1,300	3,100	4,900	6,500	8,000
CV	0	0	−200	−400	−500	−500
SV	200	200	−100	0	0	0
Cost index	1	1	0.936	0.918	0.923	0.937
Schedule index	0.6	0.87	0.967	1	1	1
BAC	15,000	15,000	15,000	15,000	15,000	15,000
EAC	15,000	15,000	16,034	16,333	26,250	16,000

The parameters in Table 2.9 provide us with tools to evaluate the project every month as follows:

January
The cost of execution is $300 less than ACWP, so the work is slow, but the cost is acceptable. The reason for this may be the late hiring of new labor.

February
The work is progressing as planned, but still slow. Deal with this situation by letting work continue on weekends to achieve the required time schedule.

March
The work is close to reaching the plan. The time schedule is not the only problem; the cost increase is a problem as well.

April
More work was done than planned. Now, the work is going according to schedule and will return to a normal work mode and end weekend work.

May
The activity increases slightly more than planned, but we are in line with the time schedule.

June
The work is progressing according to plan. After 6 months, work is proceeding according to schedule, but there has been an increase in cost. It is expected that the budget at the end of the project will be about $16,000.

2.10.9 Cash Flow Calculation

Cash flow is the real movement of money from and to the project, also called the *flow of money*. The cash flow is positive if money is received by the company. The cash flow is negative if the money is paid by the company, and the difference between the positive and negative cash flows is called the *net cash flow*.

In the case of a contractor, the positive cash flow is the money that he receives from his invoices or monthly payments. The negative cash flow is the money paid for labor salaries, equipment, subcontractors, and other items during the construction. In any project, the contractor, at some point, will have a negative cash flow, so money must be provided as an investment. In the case of increasing the net cash flow, this contract is self-financed.

In general, construction companies work by terms of contractual reservation in case of a lack of funding during a certain period of the project.

2.10.9.1 Cash Flow during the Project

Cash flows are calculated according to your role, whether you are the owner or a contractor. If you are the owner, you will pay the contractor and engineering offices throughout the project execution time. At the start of the operation of the project, the owner will gain money from the products; after a certain period, the owner will obtain profit from the operation along the facilities' lifetime.

The process of cash flow during the project has more than one phase, as shown in Figure 2.21, and is associated with the project plan, overall performance, and cost of the project through the life cycle of the project overall. During the first period of the project, the preparation of feasibility studies and preliminary engineering studies, the owner will pay the invoices. When the project is designed in detail during the *define* phase, the number of team members on the project increases as well as the cost.

During the construction phase itself, there is a significant cost increase, as the rate of spending increases at this stage. There will be large payments made to purchase materials and equipment and pay contractor invoices. The maximum negative value during the project comes at the end of the project and the start of the operational phase.

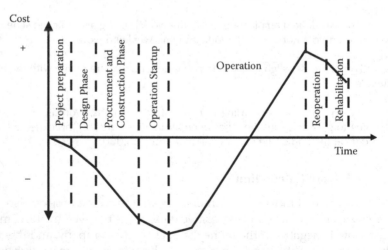

FIGURE 2.21
Cash flow along the project lifetime.

After the start of the operation and the sale of the product, which increases with time, the negative cash flow will decrease until it becomes positive. It is worth noting that, in the industrial project, after time has elapsed and depending on the project design, lifetime overhauls will be required as well as maintenance and rehabilitation, which will decrease profit as shown in Figure 2.21.

2.10.9.2 Impact on Increasing Cost

Sometimes costs increase during the execution. There are many reasons for that and the most common is a greater increase in the price of materials than was estimated. There may be an increase in the foreign exchange rate, which often occurs and affects the machines that are imported from abroad. Another factor that increases costs is difference in the quantities from what was calculated from the drawings. Yet another common example occurs during excavation where different soil characteristics are found than what was in the soil report or the existence of problems in the soil were not taken into account. This requires restudies of the soil and foundations, with a change in the foundations design. This will often cause an increase in costs. All of the above are project risks and, as they occur, will eventually lead to an increase in project costs.

Figure 2.22 presents the impact of increases in cost to the project investment with respect to the owner. From this figure, you can find that the time period to start a positive cash flow increases. Moreover, the total income from the project at the end of its lifetime is less than in the case of a project executed within its budget.

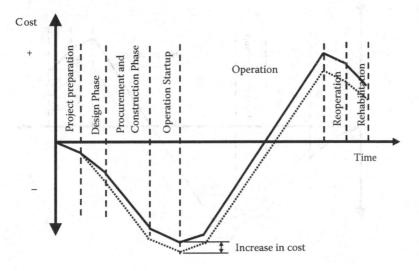

FIGURE 2.22
Effect of increase in costs on the whole project.

2.10.9.3 Project Late Impact

Project delays can occur because of either poor project management or a problem that cannot be controlled. A clear example is a change in the state of the country where the project is located that occurs because of political turmoil and causes the suspension of work for long periods. The planner and the owner are often experienced in these matters and therefore, those considerations are usually taken into account. A delay may also occur on the part of the suppliers, whether through a poor selection or a delay that is beyond their control.

In any case, the result is the same and it leads to delay in cash flow, as shown in Figure 2.23, resulting in an increased period before the return of investment. Projects such as this often have drawn on loans from banks, and as a result of the delay, we may find that the impact on profitability at the end of the project lifetime may be more than in the case of increased costs only, as in Figure 2.23.

2.10.9.4 Impact of Operation Efficiency

Operation efficiency is the responsibility of the owner. When an error occurs in the operation, either as a result of faulty design or incompatibility with the requirements of the owner or required reliability, the error usually lies between the owner and the engineering office.

The most common cause of errors in industrial projects is the lack of choice regarding high-quality equipment, which affects operation performance and, thereby, the overall revenue of the project. This is evident in Figure 2.24.

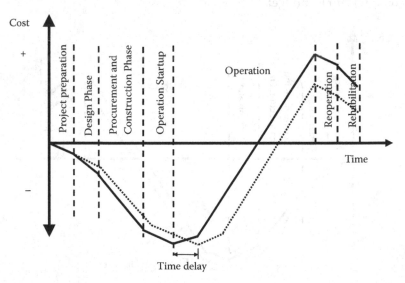

FIGURE 2.23
Effect of project delay on the whole project.

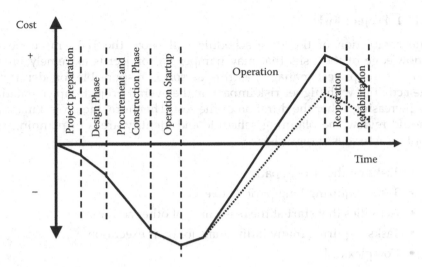

FIGURE 2.24
Effect of operation performance on the whole investment.

Another reason and an important element is that the operations need to have experienced, well-trained staff. If there is a lack of competent leaders and highly skilled professional operators, this will have a negative impact on production and reduce the overall revenue of the project.

2.11 Project Risk Management

Risk management of a project is the process of identifying risks that could affect the project and determining how to control them. We must trust that we can overcome these risks. In general, risks are the events that, if they occur, will affect the outcome of the project and result in the dissatisfaction of the stakeholder. To be more specific, they are the events or activities that can affect performance negatively and increase project costs or time.

Risk management is an ongoing process, from the implementation of the project to the end of the project, which has substantially fewer pontential risk. Risks can be classified into two categories:

- Project risks—risks that can happen because of technical mistakes that occur during construction.
- Process risks—risks that can occur because of procedural mistakes, lack of communication between the project team, or bad team performance.

2.11.1 Project Risks

The completion of the time schedule will show the risks more clearly. Knowledge of the risks that may hamper the project is extremely important for the project manager, as this person is responsible for identifying the activities with higher risk impact on the overall project implementation, by increasing either the duration or the cost. Therefore, the project manager should review the planning schedule and identify areas of planning that contain activities at high risk:

- Tasks on the critical path
- Tasks requiring long periods to execute
- Activities that start at the beginning of other activities
- Tasks requiring many individuals for their execution
- Complex tasks
- Activities and tasks requiring condensed training
- Tasks requiring advanced new technology

After you select the tasks that pose a risk to the project and determine their position relative to other high-risk tasks, you will need to identify and plan the necessary steps to implement those tasks, follow up on them from day to day, and be aware of the person responsible for that stage of the project. To get a sense of high-risk activities, let us think together of the case of the pouring concrete example and answer the question, "What are the risk activities?" We find that the more serious ones are the excavation activities, which require a long time to complete as they are in the critical path, and there is a high probability of delay, in addition to their potential impact on the entire project.

In the case of the second project, the construction of the pipeline, one finds that the excavation activity is critical as well. It takes 60 days, the longest period for any activity and is located on the critical path, but when you study the project from another point of view, the activity of the arrival of materials is more risky, as most of the materials are from abroad and, therefore, outside of full control. Therefore, the probability is high for their delay and that will have a direct impact on the project completion time.

The project's success is in achieving the objectives of the project on time and budget. It is known that there is nothing specific in nature, for example, specific costs can increase and decrease and the time length and objectives of the project can change as well. It is worth noting that these three elements affect each other so the success of the project requires a good control of each element of the project as presented in Figure 2.25.

There are many areas in the project that are not specific and these are sources of risks, which include the following:

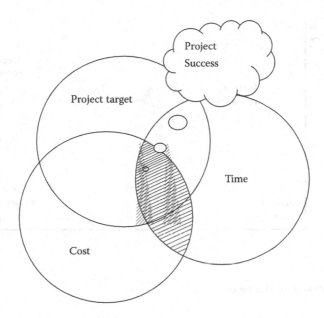

FIGURE 2.25
Point of project success.

- Activities on the critical path that require long periods to complete
- Lack of identification of the project objectives
- Incompetent project management
- Inaccurate cost estimate
- General lack of a positive atmosphere
- Achieving customer satisfaction
- Rapid change in a resource (see Figure 2.26)

2.11.2 Risk Assessment

Each project has risks, no matter what the project is. Here we focus on the risks affecting the management of the project, and by knowing the risks, we can set priorities to develop solutions or mitigations. To assess the risks, we must answer the following questions accurately and impartially:

- What is the exact risk?
- How does the risk affect the project?
- What can be done to reduce the impact of the risk?

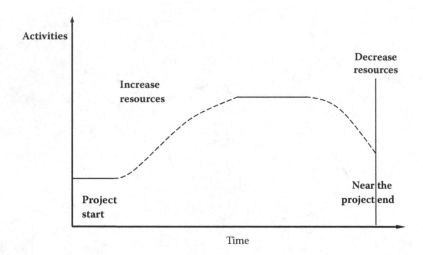

FIGURE 2.26
Change in staff volume during the project.

2.11.3 Defining Risk Using Quantitative Risk Assessment

There is another method for determining risks. First, as before, assess the risk of the project as a whole by a qualitative risk assessment methodology as part of the feasibility study (Figure 2.27).

At this stage, risks are assessed based on their effect on objectives, time, and cost. We need an easy way to assess the risks practically. This method is called *quantitative risk assessment*.

Initially, you want to determine the risks and you can do that by inviting the working group in the project to a special meeting to determine the potential risks to the project. At this meeting, use the brainstorming method

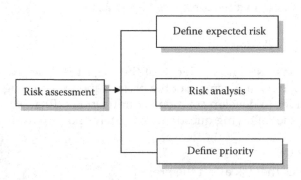

FIGURE 2.27
Risk assessment tools.

so that each individual can share his/her thoughts regarding project risks. The second step is to collect all the ideas and then write them in the order shown in Table 2.10, which is an example of the basic document for listing project risks.

Now that all possible risks to the project have been listed, we want to determine the priorities of those risks. We will again use the experience of the team in risk assessment. This will be a two-stage evaluation process.

For the first phase, determine the likelihood of a risk on a scale from 1 to 9, with 1 being the least likelihood of the event and 9 being the greatest probability of the event.

The second stage will require the team to determine the outcome of an event occurring on the project and the expected losses due to this event. This is divided into the following categories:

- High: The event has a significant impact on the time schedule or cost of the project.

TABLE 2.10

List of Project Risks

Project name:

Project manager:

Client name:

Item No.	Key Stage Code	Risk Description	Probability 1–9	Impact (High/ Medium/Low)	Date	Risk Degree (High/ Medium/Low)	Responsible

Project manager
  signature: Date:

TABLE 2.11

Risk Assessment Matrix

		Impact on the Project		
		Low	Medium	High
Probability of	7–9	Medium	High	Not acceptable
occurring event	4–6	Low	High	Not acceptable
	1–3	Low	Medium	High

- Medium: The event has a medium amount of impact on the time schedule or cost of the project.
- Low: The event has a little impact on the time schedule and cost of the project.

Table 2.11 may be used to determine the risks priorities.

The team must know the probability of the event occurring and what its impact would be. From the previous matrix shown in Table 2.11, you can determine the risk category. In a case where there is an unacceptable risk, that event should be analyzed with high accuracy and there should be a thorough examination, focusing on finding solutions to the risks, as they will cause the failure of the whole project if they occur. This means that the project cannot be carried out as planned because the risks are unacceptable and must be resolved.

Risks are designated as high in a situation when the event would have a significant impact on the project schedule and cost. Therefore, it is necessary to proceed with sustained monitoring and great care.

Medium-risk events, if they occur, will have a medium-level impact on the project but are not expected to affect the key points. Therefore, they must be reviewed at each meeting of the project team and with periodic follow-up. Low-risk events are not expected to have a significant impact on the project, if they occur, but should be followed up from time to time (Figure 2.28).

The risks that have been assessed as high risk during follow-up may change due to some work being done that reduces or increases the probability of the event's occurrence.

After identifying the risks, order them from high- to low-risk levels. For every item, find a solution and identify accurately who has the responsibility for the activities that reduce the risk of those events. This is illustrated in Table 2.12.

2.11.4 Qualitative Risk Assessment

A recognized method for calculating risks is the probability of the event occurring with knowledge of the impact of that event on the project as a whole. The importance is calculated by means of monetary value and, from

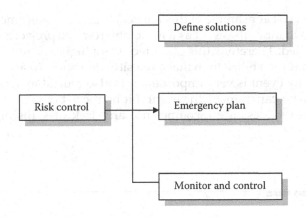

FIGURE 2.28
Steps in controlling risks.

that, the degree of the event impact on the project will be defined. Some modern views inevitably take into account the ability to manage the event and understand that there are some pitfalls that can be solved easily managerially. For example, discrepancies between departments can be solved through the convening of a meeting between the departments from time to time. Therefore, because this event is very easy to manage, it receives a lower score (to decrease the whole risk-score value) for the manageability of the event. On the other hand, if the problem is related to contact with an outside

TABLE 2.12

Risk Management

Project:	
Project manager:	
Risk title:	Risk number:
Risk description:	
Probability: 1 2 3 4 5 6 7 8 9	Risk degree:
Impact: High Medium Low	
Project areas that will be affected by the event:	
Ordering of the event impact on the project:	
Solution step:	Who is responsible:
Prepared by:	
Project manager signature:	Date:

organization (a clear example of this is a deal with the government), it would be difficult to manage the event as it is remote from the project and cannot be easily controlled. Therefore, this event receives a higher score.

We note that the ability to manage the situation so as to avoid the occurrence of a risky event is very important. It is also crucial to appoint competent person who can manage the event if it happens because in some cases, the barriers to success are incompetence and lack of skills of the person responsible.

References

PMPGuide Book. 2000.

Tricker, R. 1997. *ISO 9000 for Small Business: A Guide to Cost-Effective Compliance.* Butterworth-Heinemann.

3

Loads on Industrial Structures

3.1 Introduction

The philosophy behind designing and constructing any load-bearing structure will affect the structure along its lifetime. The structure's probability of safety depends on the fluctuations of its load and this increases the importance of its design. The types of loads and their values vary according to geographic location. In European countries, the ice load is the most critical factor in the design of structures, whereas in Middle Eastern countries with very hot climates, especially in summer, the thermal load caused by increasing temperature is the most critical factor.

To define loads, we must think about an answer to a very important question, "What is the building's designed lifetime?" The answer to this question is easy in the case of residential buildings, but for commercial, industrial, and special-purpose buildings, it is not easy. The answer depends on the project economy as a whole. To know the structural lifetime is very important for structures that are affected by loads such as wind, earthquake, or waves in the case of offshore structures.

3.2 Loads

The main factors that must be considered in structural design are the definition of the load on the structure and the calculation of the structural resistance to that load. Defining the load values is the first step in the structural design process. The load values that the designer will consider are defined according to codes and specifications. There are different types of loads and we will look at them in the following sections.

3.2.1 Dead Load

A dead load is a constant load caused by the weight of the structure itself. It can be determined if the dimensions of the structural elements and the density of the materials are known. The structural members will be made from concrete, steel, wood, and other materials. For the first step, it is necessary to know the preliminary dimensions of the structure and the materials from which it will be constructed. The dead load also includes the weight of finishing materials such as plaster, tiles, brick walls, and materials not considered part of the main structure.

It is important during design to remember that dead load is a large portion of the total load and, therefore, the weight of the building members must be calculated precisely, especially in the case of a reinforced concrete structure.

The density for different materials usually used in buildings is listed in Table 3.1 to be considered in calculating a dead load.

3.2.1.1 General Design Loads

The main concept of the design of new facilities, buildings, and other structures—including floor slabs and foundations—is the ability to resist loads throughout the structures' lifetimes.

The minimum live load shall be given on two documents (the statement of requirements and basis of design) and in the standard code of the project. It shall refer to the local building codes for the country of your facilities location as defined by the client. In general, the traditional code for minimum loads in industrial buildings is SEI/ASCE 7-05.

In addition to the loads we discuss in this chapter, other loads shall be considered as appropriate. These loads shall include, but are not limited to, snow, ice, rain, hydrostatic, dynamic, upset conditions, earth pressure, vehicles, buoyancy, and erection. The rain and snow load shall be defined according to the environmental conditions of the location of the project.

It is worth noting that the prediction of future loads shall be considered in the design when specified by the owner or by the engineering companies, based on their past experience. The eccentric loads for the piping, platforms, and other elements of the structure, particularly on horizontal and vertical vessels and heat exchangers, shall be considered. These loads are illustrated by examples in Chapter 6.

The dead load is the actual weight of the materials forming the building, structure, foundation, and all permanently attached appurtenances that are calculated from the dimensions and the material density as in Table 3.1. The dead load includes the weights of fixed process equipment and machinery, piping, valves, electrical cable trays, and also their contents. Throughout this book, the dead load nomenclature will be D_s, D_f, D_e, D_o, and D_t. D_s is the structure dead load, which is the weight of the structural materials (not the empty weight of process equipment, vessels, tanks, piping, nor cable trays), the foundation,

TABLE 3.1

Weight of Different Materials

Material Type	kg/m³
Plain concrete	2200
Reinforced concrete	2500
Light weight concrete	1000–2000
Air entrained concrete	600–900
Heavy concrete	2500–5500
Cement (loose)	1100–1200
Clinker	1500–1800
Big aggregate	1700
Sand	1500
Foamed aerated slag	1700
Granulated	1200
Expanded clay	200–900
Pumice stone	350–650
Exfoliated vermiculite	60–200
Fly ash	600–1100
Water	1000
Liquid or powder	1000–1200
Masonry Stones	
A. Igneous rocks	
Granite	2800
Basalt	3000
Basalt lava	2400
Trechzte	2600
B. Sedimentary rocks	
Limestone	2700
Marble	2800
Sandstone	2700
C. Transformed rocks	
Slate	2800
Gneiss	3000
Serpentine	2700
Marble	2700
Masonry bricks	
Red brick	1600–1800
Solid brick	1850
Hollow brick	1400
Light brick	700–800
Refractory Brick for General Purposes	
Fire clay	1850
Silica	1800

(continued)

TABLE 3.1 Continued

Weight of Different Materials

Material Type	kg/m³
Magnesite	2800
Chrome-magnesite	3000
Chorundum	2600
Brick anti acid	1900
Glass brick	870
Masonry Blocks	
Concrete blocks	1400–1900
Hollow concrete blocks	1150
Leca concrete blocks	600–800
Gypsum blocks	950
Lime	
Limestone powder	1300
Calcined, in lumps	850–1300
Calcined	600–1300
Calcined slaked	110
Gypsum	800–1000
Mortar	
Cement mortar	2100
Lime mortar	1800
Lime cement mortar	750–1800
Gypsum mortar	1400–1800
Bitumin mortar with sand	1700
Wood and Substitutes	
A. Hardwood	
Beech	680
Oak	790
B. Softwood	
Pitch pine	570
White wood	400
C. Fiberboard	
Hard	900–1100
Medium–hard	600–850
Porous insulating	250–400
Plywood	750–850
Core board	450–650
Other Building Materials	
Asbestos	800
Asbestos boarders	1600
Asbestos cement pipe	1800

(continued)

TABLE 3.1 Continued

Weight of Different Materials

Material Type	kg/m³
Celton	120
Dry earth	1700
Wet earth	2000
Rubber floor	1800
Asphalt	3200
Bitumen	1000–14,000
Tar	1100–1400
Cement tile	2400
Mosaic tile	2200
Epoxy resin	
Without fills	1150
With mineral materials	2000
With fiberglass	1800
Plastic tile	1100
Polyester resin	1350
Polyethelene	930
PVC hardboard	1400
PVC flooring board	1600
PVC flooring tile	1700
Fiberglass	160–180
Glass wool	100–110
Slag wool	200–300
Cork	60
Plaster	1100–1500
Glass, in sheets	2500
Wired glass	2600
Acrylic glass	1200
Linen baled	600
Leather, in piles	900–1000
Paper	
In stocks	1200
In rolls	1100
Rubber	
Rolled-up flooring materials	1300
Raw balled	1100
Wool	
Bales	700
Pressed baled	1300
Metallic Materials	
Steel	7850

(continued)

TABLE 3.1 Continued

Weight of Different Materials

Material Type	kg/m³
Wrought iron	7850
Cast iron	7250
Iron ore	3000
Aluminum	2700
Aluminum alloy	2800
Lead	
White lead (powder)	9000
Red lead (powder)	8000
Copper	8700–8900
Brass	8300–8500
Bronze	8900
Nickel	7900
Zinc	8200
Zinc rolled	7200
Tin rolled	7200–7400
Magnesium	1850
Antimony	6620
Barium	3500
Cadmium	8650
Cobalt	8700
Gold	19,300
Silver	10,500
Manganese	7200
Molybdenum	10,200
Platinum	21,300
Titanium	4500
Tungsten	19,000
Uranium	5700
Vanadium	5600
Zirconium	6530
Fuels	
Mineral coal	900–1200
Coke	450–650
Charcoal	250
Coal dust	700
Oils	
Diesel oil	800–1000
Crude oil	950
Petrol (gasoline)	800–850
	(continued)

TABLE 3.1 Continued

Weight of Different Materials

Material Type	kg/m³
Petroleum	
Liquid gas	
Propane	500
Butane	580
Wood	
Hardwood chopped	400–600
Hardwood logs	500
Softwood chopped	250
Softwood logs	300
Firewood	400
Liquids	
Glycerin	1250
Oil paint, canned, or boxed	1100
Milk, in tanks	950–1000
Milk, in boxes	1000
Milk, in cans	600
Honey, in tanks	1300
Honey, in boxes	1000
Honey, in tanks	600
Nitric acid	1500
Hydrochloric acid	1200
Sulfuric acid	1400
Food Stuffs and Agricultural Products	
Butter, in barrel	550
Butter, in boxes	500–800
Sugar, in paper	600
Sugar, in big boxes	800
Hump sugar, in paper sacks	600
Sugar, boxed	700
Tea packet	400
Alcohol	800
Butter, in tanks	1000
Butter, in barrel	900
Cacao, in boxes	550
Eggs, in carton carries	550
Fat boxed	800
Fish, in barrel	600
Fish in	800
Fruits, in boxes	350–400

(continued)

TABLE 3.1 Continued

Weight of Different Materials

Material Type	kg/m³
Fruits storage	500–700
Hay, baled	150–200
Maize corn	450
Margarine, in barrels	550
Margarine, in boxes	700
Meat, refrigerated	400–700
Onion, in bags	550
Pickled, bottled in sacs	700
Drinks bottled	800
Rise	500
Rise, in bags	560
Salt, in piles	10,000
Salt, in bags	1120
Starch flour, in bags	800
Straw baled	170
Tobacco baled	300–500
Wheat	800–900
Wine, in tanks	1000
Wine, in barrels	850
Coffee, in bags	700
Soap powder, in sacks	610
Other Materials	
Books and files	1000–1100
Ice, in blocks	850–900
Textiles	1100
Cellulose, baled	800
Cloth, baled	700–1300
Cotton, baled	500
Hemp, baled	400
Jute, baled	700

the soil above the foundation resisting uplift, and all permanently attached appurtenances. Examples of this are the lighting, instrumentation, HVAC (heating, ventilating, and air conditioning), sprinkler and deluge systems, fireproofing, and insulation. D_f is the erection dead load, which is the fabricated weight of the process equipment or vessels. D_e is the empty dead load, which is the empty weight of process equipment, vessels, tanks, piping, and cable trays. D_o is the operating dead load, which is the weight of process equipment, vessels, tanks, piping, and cable trays filled with fluid under normal operating conditions. D_t is the test dead load and comprises the weight of process equipment, vessels, tanks, and piping filled with water for test procedures. It should be noted that, in some cases, they may be filled with nitrogen.

To consider all the possible situations that could occur during the structure's lifetime, the structural engineer shall obtain all the available data through cooperation with static equipment or pressure vessel engineers as well as the operations engineers as these parameters depend largely on the mode of operation.

3.2.1.2 Pipe Rack

A famous structure in any industrial project is the pipe rack that carries the piping and the cable tray. At the beginning of the design process, the actual piping load and that of the cable tray will not be precisely defined. The structural engineer will have the precise actual load data at the end of the project and the first activity on-site is the civil and structure work. Instead, we will use the following estimate loads for the piping and cable tray to begin the design. These nearly match the actual loads and, in the final phase, will be checked against the actual load.

- *Piping loads*
  1. Dead loads for the piping carried on pipe racks shall be estimated using the following measurements, unless the actual load information is available and requires otherwise.
     (a) Operating dead load (D_o): a uniformly distributed load of 40 psf (1.9 kPa) for piping, product, and insulation. This value is equivalent to schedule 40 pipes, 8 in. (203 mm) in diameter, full of water, at 15-in. (381-mm) spacing.
     (b) Empty dead load (D_e): for checking uplift and components controlled by minimum loading, 60% of the estimated piping operating loads shall be used if combined with a wind or earthquake load, unless the actual conditions require a different percentage.
     (c) Test dead load (D_t): the empty weight of the pipe plus the weight of a test medium contained in a set of simultaneously tested piping systems. Unless otherwise specified in the contract documents or by the owner, a minimum specific gravity of 1.0 shall be used for the test medium.
  2. For any pipe larger than 12 in. (304 mm) in nominal diameter, a concentrated load, including the weight of piping, product, valves, fittings, and insulation, shall be used instead of the 40-psf (1.9-kPa) load used for 8-in. pipes. This load shall be uniformly distributed over the pipe's associated area.
  3. Pipe racks and their foundations shall be designed to support loads associated with full utilization of the available rack space

and any specified future expansions. Among all the disciplines represented on an industrial project, structural engineers have the hardest luck. Unlike other disciplines, they must perform activities before the actual load data can be obtained. Therefore, it is necessary to use load values obtained from past experiences and projects. They will have the actual data about the load after procurement and before running the final simulation of the piping stress analysis. The guidelines below are derived from the experience of engineering offices and shall be used in conjunction with the structural engineer's judgment, good industrial practice, and owner specifications.

- *Pipe rack cable tray loads*

  Dead loads for cable trays on pipe racks shall be estimated as follows (unless actual load information is available and requires otherwise):

  1. Operating dead load (D_o): a uniformly distributed dead load of 20 psf (1.0 kPa) for a single level of cable trays and 40 psf (1.9 kPa) for a double level of cable trays. These values estimate the full (maximum) level of cables in the trays.

  2. Empty dead load (D_e): to check uplift and components controlled by minimum loading, a reduced level of cable tray load (i.e., the actual configuration) shall be considered the empty dead load. The engineer's judgment shall be exercised in defining the dead load for uplift conditions.

3.2.1.3 Ground-Supported Storage Tank Loads

Dead loads for ground-supported storage tanks use the same nomenclature as the dead loads in the previous sections. Here we will discuss the individual load components that make up the dead loads. They are separated here by their actual use and design.

1. Operating dead load (D_o): The operating dead load for a ground-supported storage tank is made up of the metal load from the tank shell and roof, vertically applied through the wall of the tank, as well as the fluid load from the stored product. The fluid load acts through the bottom of the tank, not vertically through the wall of the tank. Therefore, the metal dead load and the fluid load must be used separately by design.

2. Empty dead load (D_e): To check uplift and components controlled by minimum loading, the corroded metal weight (when a corrosion allowance is specified) shall be considered the empty dead load.

3. Test dead load (D_t): The test dead load for a ground-supported storage tank is made up of the metal load from the tank shell and roof,

vertically applied through the wall of the tank, as well as the fluid load from the test medium. The fluid load acts through the bottom of the tank, not vertically through the wall of the tank. Therefore, the metal dead load and the fluid load must be used separately by design. The test medium shall be as specified in the contract documents or as specified by the owner. Unless otherwise specified, a minimum specific gravity of 1.0 shall be used for the test medium.

The load and design principle of storage tanks, with examples, is illustrated in Chapter 5.

3.2.2 Live Loads

Live loads (L) are gravity loads produced by the use and occupancy of buildings and structures. These include the weight of all movable loads such as personnel, tools, miscellaneous equipment, movable partitions, wheel loads, parts of dismantled equipment, stored material, and other related materials. Areas specified for maintenance (e.g., the servicing of heat exchanger tube bundles) shall be designed to support the live loads.

The minimum live loads shall be in accordance with SEI/ASCE 7-05, applicable codes and standards, and, unless otherwise specified, the values illustrated in Table 3.2 can be used.

The uniform and concentrated live loads listed in Table 3.2 shall not be applied simultaneously.

According to SEI/ASCE 7-05, concentrated loads of 1000 lb (4.5 kN) or more may be assumed to be uniformly distributed over an area of 2.5 × 2.5 ft (750 × 750 mm) and shall be located to produce the maximum load effects in the structural members. Stair treads shall be designed according to Occupational Safety and Health Administration (OSHA) regulations or building codes, as applicable. Live load reductions shall be in accordance with SEI/ASCE 7-05.

TABLE 3.2

Minimum Live Loads

Structure Type	Uniform[a]	Concentrated[a]
Stairs and exit ways	100 psf (4.8 kN/m²)	1000 lb (4.5 kN)
Operating, access platforms, and walkways	75 psf (3.6 kN/m²)	1000 lb (4.5 kN)
Control, I/O, HVAC room/floors	100 psf (4.8 kN/m²)	1000 lb (4.5 kN)
Manufacturing floors and storage areas		
Light	125 psf (6.0 kN/m²)	2000 lb (9.0 kN)
Heavy	250 psf (12.0 kN/m²)[b]	3000 lb (13.5 kN)
Ground-supported storage tank roof	25 psf (1.2 kN/m²)	NA

[a] Loads provided in this table are to be used unless noted otherwise on the owner's data sheet.

[b] This 250 psf (12.0 kN/m²) live load includes small equipment.

For manufacturing floor areas not used for storage, it is acceptable to use the live load reduction specified by SEI/ASCE 7-05. The loadings on handrails and guardrails for process equipment structures shall be in accordance with OSHA regulations (29 CFR 1910). The loadings on handrails and guardrails for buildings and structures under the jurisdiction of a building code shall be in accordance with that code.

3.2.3 Wind Loads

Wind load (*W*) is the most critical influence on some projects because of location. In industrial, nonbuilding structures, wind loads shall be computed and applied in accordance with SEI/ASCE 7-05. Oil and gas projects shall also follow the recommended guidelines for open-frame structures, pressure vessels, and pipe racks based on the ASCE guidelines for wind loads and anchor bolt design for petrochemical facilities unless otherwise specified by the client's requirements (ASCE 1997). The owner shall be consulted on the determination of the classification category, which will be based on the location of the project.

3.2.3.1 Basic Wind Load Formula

Recently, it became possible to use computer software designed for structural analysis that calculates the self-weight of a structural member after you define its materials and dimensions. The finishing materials and the dead load of the structure must be input manually.

Wind loads shall be computed and applied in accordance with SEI/ASCE 7 and the recommended guidelines for pipe racks and open-frame structures contained in *Wind Loads and Anchor Bolt Design for Petrochemical Facilities* (ASCE, 1997). Using this method, wind load calculations shall be based on the basic wind speed, *V*, of a 3-s gust at 33 ft (10 m) above the ground at exposure C and the annual probability of 0.02 being equaled or exceeded (50-year mean recurrence interval). The basic wind speed, *V*, for each site shall be obtained from the owner or from the code of the country in which these facilities will be built. Based on experience, the importance factor (*I*) shall be category IV.

In most cases, facilities for industrial businesses, such as the oil and gas business, shall be located away from the buildings and exposure C shall be used. In some cases, the facilities will be near the shoreline so exposure D is more appropriate. As defined by ASCE 7-02, exposure D is defined as "flat, unobstructed areas exposed to wind flowing over open water for a distance of at least 1 mi (1.61 km)."

The exposure categories designated by ASCE7-02 are as follows:

- Exposure B: Urban and suburban areas, wooded areas, or other terrain with numerous, closely spaced obstructions having the size of single-family dwellings or larger.

- Exposure C: Open terrain with scattered obstructions having heights generally less than 30 ft (9.1 m). This category includes flat open country, grasslands, and all water surfaces in hurricane-prone regions.
- Exposure D: Flat, unobstructed areas and water surfaces outside hurricane-prone regions. This category includes smooth mud flats, salt flats, and unbroken ice.

The applied wind force, F, shall be determined by the basic equation from ASCE 7-02:

$$F = q_z G C_f A_e \tag{3.1}$$

where
q_z = Velocity pressure at height z above ground
G = Gust effect factor
C_f = Net force coefficient
A_e = Projected area

Velocity pressure q_z is determined in accordance with the provisions of Section 6.5.10 of ASCE 7-02.

$$q_z = 0.613 \, K_z K_{zt} K_d V^2 I \ (\text{N/m}^2)$$

where
K_z = Velocity pressure exposure coefficient defined in SEI/ASCE 7-05. This is shown in Table 3.3.
K_{zt} = Topographic factor as in SEI/ASCE 7-05. K_{zt} is equal to 1.0 for pipe racks and open-frame structures with normal topography.
K_d = Wind directionality factor (see Table 3.4), which is adapted from ASCE 7-02. When used with load combinations, K_d is equal to 0.85 for pipe racks and open-frame structures.
V = Basic wind speed of a 3-s gust at 11 m (33 ft) above the ground.
I = Importance factor, depending on the occupation category shown in Table 3.5 and the occupancy category shown in Table 3.6. For most category IV structures, $I = 1.15$. In most cases, pipe racks and open-frame structures are considered category IV structures.

It shall be noted that, for process industry facilities, SEI/ASCE 7-05 category III is the most likely classification because of the presence of hazardous materials. Category II may be used if the owner can demonstrate that release of the hazardous material does not pose a threat to the public.

- *Gust effect factor*

  The gust effect factor, G, is determined in accordance with ASCE 7-02, which contains an equation for calculating G. For rigid structures,

TABLE 3.3

Velocity Pressure Exposure Coefficients, K_h and K_z

Height above Ground Level, z (m)	B		C	D
	Case 1	Case 2	All	All
			Exposure (Note I)	
0–4.6	0.70	0.57	0.85	1.03
6.1	0.70	0.62	0.90	1.08
7.6	0.70	0.66	0.94	1.12
9.1	0.70	0.70	0.98	1.16
12.2	0.76	0.76	1.04	1.22
15.2	0.81	0.81	1.09	1.27
18	0.85	0.85	1.13	1.31
21.3	0.89	0.89	1.17	1.34
24.4	0.93	0.93	1.21	1.38
27.4	0.96	0.96	1.24	1.40
30.5	0.99	0.99	1.26	1.43
36.6	1.04	1.04	1.31	1.48
42.7	1.09	1.09	1.36	1.52
48.8	1.13	1.13	1.39	1.55
54.9	1.17	1.17	1.43	1.58
61.0	1.20	1.20	1.46	1.61
76.2	1.28	1.28	1.53	1.68
91.4	1.35	1.35	1.59	1.73
106.7	1.41	1.41	1.64	1.78
121.9	1.47	1.47	1.69	1.82
137.2	1.52	1.52	1.73	1.86
152.4	1.56	1.56	1.77	1.89

Note: Where
    Case 1:
        (a) All components and cladding
        (b) Main wind force resistance system in low-rise buildings
    Case 2:
        (a) All main wind force resisting systems in buildings except those in low-rise buildings designed
        (b) All main wind force resisting systems in other structures

the gust effect factor shall be taken as $G = 0.85$, as illustrated in ASCE 7-02. This value shall be used for all pipe racks and open-frame structures.

- *Force coefficient*

  The force coefficient, C_f, for various structures shall be as listed as discussed in the following sections, based on guidelines regarding the shape of the equipment.

TABLE 3.4

Wind Directionality Factor, K_d

Structure Type	Directionality Factor, K_d
Buildings	
Main wind force resisting system	0.85
Components and cladding	0.85
Arched roofs	0.85
Chimneys, tanks, and similar structures	
Square	0.90
Hexagonal	0.95
Round	0.95
Solid signs	0.85
Open signs and lattice framework	0.85
Trussed towers	
Triangular, square, rectangular	0.85
All other cross sections	0.95

- *Projected area*

  The projected area, A_o, the normal wind direction for the various structures in question, shall be as defined in this guideline.

3.2.3.2 Wind Loads on Pipe Racks and Open-Frame Structures

3.2.3.2.1 Pipe Racks

Wind on the pipe rack structure itself should be calculated assuming no shielding.

For all structural members except columns, $C_f = 1.8$ shall be used. For columns, $C_f = 2.0$ shall be used.

- *Tributary area for piping*

  The tributary area for piping shall be based on the diameter of the largest pipe plus 10% of the width of the pipe rack. This sum is multiplied by the length of the pipes (bent spacing) to determine the tributary area.

- *Tributary area for cable trays*

  The tributary area for cable trays shall be based on the height of the largest tray plus 10% of the width of the pipe rack. This sum is multiplied by the length of the pipes (bent spacing) to determine the tributary area.

TABLE 3.5

Occupancy Category of Buildings and Other Structures for Flood, Wind, Snow, Earthquake, and Ice Loads

Occupancy Category	Description
I	Buildings and other structures that represent a low hazard to human life in the event of failure, including, but not limited to • Agricultural facilities • Certain temporary facilities • Minor storage facilities
II	All buildings and other structures except those listed in Occupancy Categories I, III, and IV
III	Buildings and structures that represent a substantial hazard to human life in the event of failure, including, but not limited to • Buildings and other structures where more than 300 people congregate in one area • Buildings and other structures with daycare facilities with a capacity greater than 150 • Building and other structures with elementary school or secondary school facilities with a capacity of more than 250 • Buildings and other structures with a capacity of more than 500 for colleges or adult education facilities, health care facilities with a capacity of 50 or more resident patients, but not having surgery or emergency treatment facilities, jails, and detention facilities Buildings and other structures, not included in occupancy category IV, with potential to cause a substantial economic impact and/or mass disruption of day-to-day civilian life in the event of failure, including, but not limited to: • Power generating stations • Water treatment facilities • Sewage treatment facilities • Telecommunication centers Buildings and other structures not included in occupancy category IV (including, but not limited to, facilities that manufacture, process, handle, store, use, or dispose of such substances as hazardous fuels, hazardous chemicals, hazardous waste, or explosives) containing sufficient quantities of toxic or explosive substances to be dangerous to the public if released Buildings and other structures containing toxic or explosive substances
IV	Buildings and other structures designated as essential facilities, including, but not limited to: • Hospitals and other health care facilities having surgery or emergency treatment facilities. • Fire, rescue, ambulance, and police stations, and emergency vehicle garages. • Designated earthquake, hurricane, or other emergency shelters. • Designated emergency preparedness, communication, and operation centers, and other facilities required for emergency response. • Power generating stations and other public utility facilities required in an emergency.

(continued)

TABLE 3.5 Continued

Occupancy Category of Buildings and Other Structures for Flood, Wind, Snow, Earthquake, and Ice Loads

Occupancy Category	Description
	• Ancillary structures (including, but not limited to, communication towers, fuel storage tanks, cooling towers, electrical substation structures, fire water storage tanks or other structures housing or supporting water, or other fire-suppression material or equipment) required for operation of occupancy category IV structures during an emergency. • Aviation control tower, air traffic control center, and emergency aircraft hangars. • Water storage facilities and pump structures required to maintain water pressure for fire suppression. • Buildings and other structures having critical national defense functions. • Buildings and other structures (including, but not limited to, facilities that manufacture, process, handle, store, use, or dispose of such substances as hazardous fuels, hazardous chemicals, or hazardous waste) containing highly toxic substances where the quantity of the material exceeds a threshold quantity established by the authority having jurisdiction. (Buildings and other structures containing highly toxic substances are eligible for classification as category II structures.)

- *Force coefficients for pipes*

  The force coefficient, $C_f = 0.7$, shall be used as a minimum. The force coefficient, C_f, is taken from ASCE 7-02, as shown in Table 3.7 for a round shape, with $h/D = 25$, $D\sqrt{q_z} > 2.5$, and a moderately smooth surface, that is, $C_f = 0.7$. If the largest pipe is insulated, then consider using a C_f for a rough pipe, depending on the roughness coefficient of the insulation, (D'/d).

TABLE 3.6

Importance Factor, *I* (Wind Loads)

Category	Nonhurricane-Prone Regions and Hurricane-Prone Regions, Where $V = 85$–100 mi/h, and Alaska	Hurricane-Prone Regions, Where $V > 100$ mi/h
I	0.87	0.77
II	1.00	1.00
III	1.15	1.15
IV	1.15	1.15

TABLE 3.7

Force Coefficient Factor (C_f) for Chimneys, Tanks, Rooftops, and Similar Structures

Cross Section	Type of Surface	h/D		
		1	7	25
Square (wind normal to face)	All	1.3	1.4	2.0
Square (wind along diagonal)	All	1.0	1.1	1.5
Hexagonal or octagonal	All	1.0	1.2	1.4
Round ($D'\sqrt{q_z} > 2.5$)	Moderately smooth	0.5	0.6	0.7
($D'\sqrt{q_z} > 5.3$, D in m, q_z in N/m²)	Rough ($D'/D = 0.02$)	0.7	0.8	0.9
	Very rough ($D'/D = 0.08$)	0.8	1.0	1.2
Round ($D'\sqrt{q_z} > 2.5$)	All	0.7	0.8	1.2
($D'\sqrt{q_z} \leq 5.3$, D in m, q_z in N/m²)				

Note: The design wind force shall be calculated based on the area of the structure projected on a plane normal to the wind direction. The force shall be assumed to act parallel to the wind direction.

   Linear interpolation is permitted for h/D values other than shown.

   D, diameter of circular cross section and least horizontal dimension of square, hexagonal, or octagonal cross sections at elevation under consideration, in feet (meters); D', depth of protruding elements such as ribs and spoilers, in feet (meters); h, height of structure, in feet (meters); q_z, velocity pressure evaluated at height z above ground, in pounds per square foot (N/m2).

- *Force coefficients for cable trays*

   For cable trays, the force coefficient $C_f = 2.0$ shall be used. The force coefficient, C_f, for cable trays is taken from ASCE 7-02. Table 3.7 gives the force coefficient for a square shape with the face normal to the wind and with $h/D = 25$, that is, $C_f = 2.0$.

3.2.3.2.2 Open-Frame Structures

Wind loads shall be calculated in accordance with the general procedures and provisions of ASCE 7-02 for wind loads on "other structures." The force coefficients of components for wind loads for the design of individual

TABLE 3.8

Force Coefficient (C_f) for Miscellaneous Structures

Item	C_f	Projected Area
Handrail	2.0	0.24 m²/m
Ladder without cage	2.0	0.15 m²/m
Ladder with cage	2.0	0.23 m²/m
Solid rectangles and flat plates	2.0	
Stair with handrail		Handrail area plus channel depth
Side elevation	2.0	50% gross area
End elevation	2.0	

Source: ASCE, 40262, *Wind Load and Anchor Bolt Design for Petrochemical Facilities*, ASCE, New York, 1997.

components, cladding, and appurtenances (excluding vessels, piping, and cable trays) shall be calculated according to the provisions of ASCE 7-02. Table 3.8 presents the force coefficients for several items as a guideline.

3.2.3.2.2.1 Open-Frame Load The wind forces used in the design of the main wind force resisting system for an open-frame structure should be determined by Equation (3.1). The structure is idealized as two sets of orthogonal frames. The maximum wind force on each set of frames is calculated independently, noting that C_f accounts for the entire structure in the direction of the wind.

Wind forces are calculated for the structure as a whole. The method is described for structures that are rectangular in plan and elevation. The solidity ratio, ε, for pipe rocks is

$$\varepsilon = A_s/A_g \qquad (3.2)$$

where

A_g = Gross area of the windward frame

A_s = Effective solid area of the windward frame defined by the following:

1. The solid area of a frame is defined as the solid area of each element in the plane of the frame projected normal to the nominal wind direction. Elements considered as part of the solid area of a frame include beams, columns, bracing, cladding, stairs, ladders, handrails, etc. Items such as vessels, tanks, piping, and cable trays are not included in calculations of the solid area of the frame; the wind loads on these items are calculated separately.

2. The presence of flooring or decking does not cause an increase in the solid area beyond the inclusion of the thickness of the deck.

3. For structures with frames of equal solidity, the effective solid area should be taken as the solid area of the windward frame.

4. For structures where the solid area of the windward frame exceeds the solid area of the other frames, the effective solid area should be taken as the solid area of the windward frame.

5. For structures where the solid area of the windward frame is less than the solid area of the other frames, the effective solid area should be taken as the average of all the frames.

The calculation of force coefficients, C_f, depends on the spacing between the frame and its width and the ratio between the area of the steel frame and the total applied wind load.

The force coefficient for a set of frames shall be calculated by

$$C_f = C_{Dg}/\varepsilon \qquad (3.3)$$

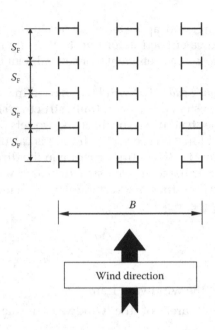

FIGURE 3.1
Force coefficient, C_{Dg}, for an open-frame structure. (Adapted from ASCE, 40262, *Wind Load and Anchor Bolt Design for Petrochemical Facilities*, ASCE, New York, 1997.)

where C_{Dg} is the force coefficient for the set of frames given in Figure 3.1 and ε is the solidity ratio calculated according to Equation (3.3).

Force coefficients are defined for wind forces acting normally to the wind frames, irrespective of the actual wind direction.

The force coefficient can be calculated for an open-frame configuration as shown in Figures 3.2–3.5 and taking the following points into consideration:

1. The frame spacing (SF) is measured from centerline to centerline.
2. The frame width (*B*) is measured from the outside edge to the outside edge.
3. The frame spacing ratio is defined as SF/*B*.
4. The number of frames (*N*) is the number of framing lines normal to the nominal wind direction (*N* = 5 as shown in Figure 3.1).

The area of application of force (A_e) shall be calculated in the same manner as the effective solid area, except that it is for the portion of the structure height consistent with the velocity pressure, q_z.

3.2.3.2.2.2 Design Load Cases   The total wind force acting on a structure in a given direction, F_T, is equal to the sum of the wind load acting on the structure and appurtenances (F_S), plus the wind load on the equipment and vessels, plus the wind load on piping. See Figure 3.6 for complete definitions of F_T and F_S.

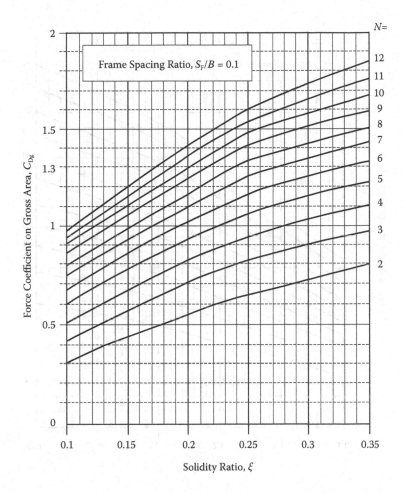

FIGURE 3.2
Force coefficient, C_{Dg}, for an open-frame structure with $S_F/B = 0.1$. (Adapted from ASCE, 40262, *Wind Load and Anchor Bolt Design for Petrochemical Facilities*, ASCE, New York, 1997.)

If piping arrangements are not known, the engineer may assume the piping area to be 10% of the gross area of the face of the structure for each principal axis. A force coefficient of 0.7 should be used for this piping area. The following two load cases must be considered as a minimum:

Frame load + equipment load + piping load (F_T) for one axis, acting simultaneously with 50% of the frame load (F_S) along the other axis, for each direction. These two combinations are indicated in Figure 3.6.

3.2.3.2.2.3 Partial Wind Load A solid width of 1.5 ft (350 mm) shall be assumed when calculating the wind load on ladder cages. The partial wind

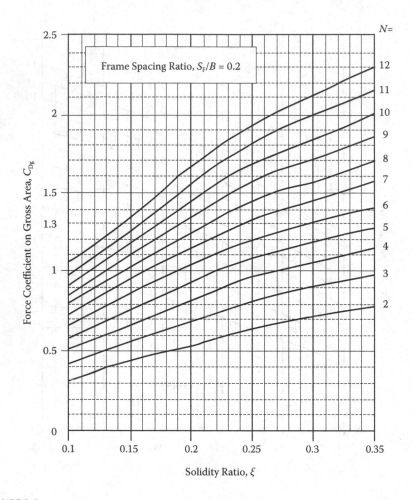

FIGURE 3.3
Force coefficient, C_{Dg}, for an open-frame structure with $S_F/B = 0.2$. (Adapted from ASCE, 40262, *Wind Load and Anchor Bolt Design for Petrochemical Facilities*, ASCE, New York, 1997.)

load (WP) shall be based on the requirements of SEI/ASCE 37-02, Section 6.2.1, for the specified test or erection duration. The design wind speed shall be 109 km/h, which is 0.75 × 145 km/h, according to SEI/ASCE 37-02, for test or erection periods of less than 6 weeks.

For tests or erection periods of 6 weeks or longer, or if the test or erection is in a hurricane-prone area and is planned during the peak hurricane season (from August 1 to October 31 in the United States), refer to SEI/ASCE 37-02.

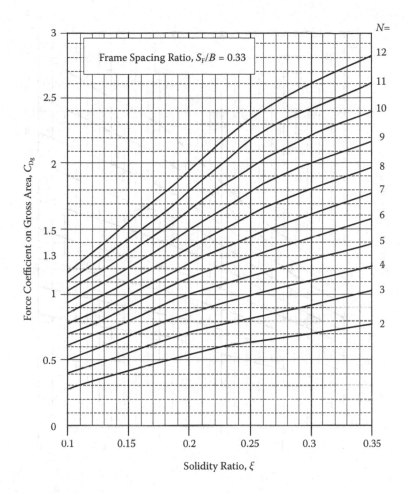

FIGURE 3.4
Force coefficient, C_{Dg}, for an open-frame structure with $S_F/B = 0.33$. (Adapted from ASCE, 40262, *Wind Load and Anchor Bolt Design for Petrochemical Facilities*, ASCE, New York, 1997.)

Example 1: Wind Load on Pipe Rack

Design a pipe rack in an oil plant. The pipe rack case shall be as shown in Figure 3.7 with bent spacing = 6 m, with a 3-s gust wind speed of 155 km/h.

The design wind forces are determined using Equation (3.1), where F is the force per unit length of the piping or cable tray.

The design wind pressure for a 9-m elevation:

$$q_z = 1.269 \text{ kN/m}^2 \tag{3.4}$$

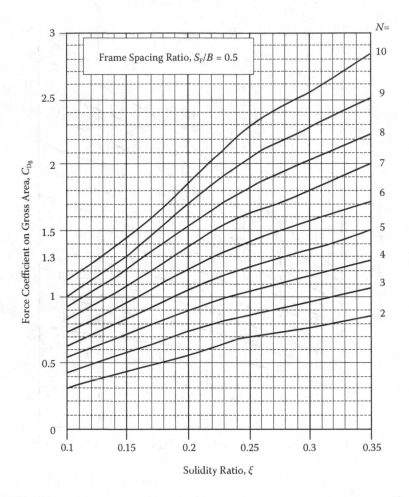

FIGURE 3.5
Force coefficient, C_{Dg}, for an open-frame structure with $S_F/B = 0.5$. (Adapted from ASCE, 40262, *Wind Load and Anchor Bolt Design for Petrochemical Facilities*, ASCE, New York, 1997.)

with the gust effect factor, $G = 0.85$.

Force Coefficients

For structural members	$C_f = 1.8$
For columns	$C_f = 2.0$
For pipes	$C_f = 0.7$
For cable trays	$C_f = 2.0$

Projected Area

The projected area per meter of pipe rack, A_e, is equal to the largest pipe diameter or cable tray height + 10% of the pipe rack width.

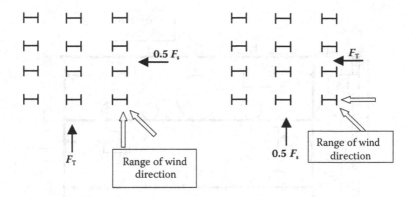

F_s denotes the wind force on the structural frame and appurtenances in the indicated direction and excludes wind load on equipment, piping, and cable tray.

F_T denotes the total wind force on the structure in the direction indicated, which is the sum of forces on the structural frame and appurtenances, equipment, and piping.

FIGURE 3.6
Design load cases from wind load according to ASCE.

A. PIPING AND CABLE TRAY

The guidelines require the consideration of the piping or cable trays separately from the structural members. The following calculations are only for piping and cable trays without the structural support members:

Force Calculation

- For cable tray, 150 mm deep (at an elevation of 9 m)

$$C_f = 2.0$$

$$A_e = 0.15 + (10\% \times 7.5) = 0.9 \text{ m}^2/\text{m}$$

$$F_1 = q_z G C_f A_e$$

$$F_1 = [(1.27)(0.85)(2.0)(0.9)] \times 6.0 \text{ m frame spacing} \quad F_1 = 11.6 \text{ kN}$$

- For F_2 pipe level 7.5 m–24 in. maximum outside diameter (OD)

$$C_f = 0.7$$

$$A_e = 0.61 + (10\% \times 7.5) = 1.36 \text{ m}^2$$

$$F_2 = q_z G C_f A_e$$

$$F_2 = [(1.276 \text{ kN/m}^2)(0.85)(0.7)(1.36)] \times 6.0 \text{ m bent spacing} \quad F_2 = 6.18 \text{ kN}$$

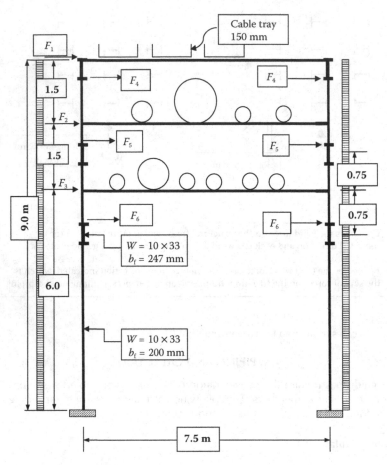

FIGURE 3.7
Example 1—sketch of configuration.

- For pipe level 6.0 m–18 in. maximum OD

$$C_f = 0.7$$

$$A_e = 0.45 + (10\% \times 7.5) = 1.2 \text{ m}^2$$

$$F_3 = q_z GC_f A_e$$

$$F_3 = [(1.27)(0.85)(0.7)(1.2)] \times 6.0 \text{ m bent spacing} \qquad F_3 = 5.45 \text{ kN}$$

B. STRUCTURAL MEMBERS

For structural members, assume a 7.5-m wide rack with bent spacing of 6.0-m centers and all stringers unshielded. Stringers are at elevations 9, 6.75, and 5.25 m (as shown in Figure 3.8).

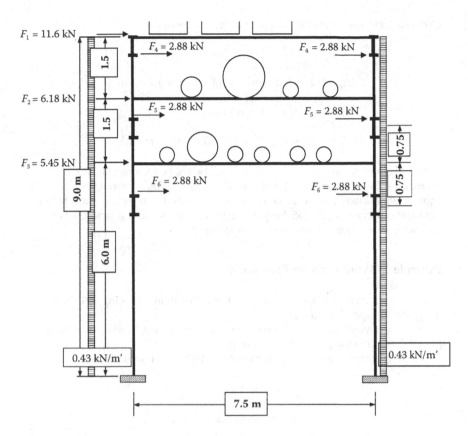

FIGURE 3.8
Example 1—sketch of loads on pipe rack.

Assume $q_z = 1.27$ kN/m² psf for all three levels of stringers (conservative):

$$C_f = 1.8$$

$$A_e = 0.247 \text{ (beam depth)} \times 6.0 \text{ m (beam length)} = 1.482 \text{ m}^2$$

$$F = q_z G C_f A_e$$

$$F_4 = F_5 = F_6 = 1.269 \times 0.85 \times 1.8 \times 1.482 = 2.88 \text{ kN}$$

For columns:

$$q_z = 1.269 \text{ kN/m}^2 \text{ at an elevation of } 9.0 \text{ m}$$

$$q_z = 1.217 \text{ kN/m}^2 \text{ at an elevation of } 7.5 \text{ m}$$

$$q_z = 1.166 \text{ kN/m}^2 \text{ at an elevation of } 6.0 \text{ m}$$

Use $q_z = 1.217$ kN/m² for the whole column (conservative)

$$C_f = 2.0$$

$$A_e = 0.2\text{m}^2 \text{ (column width)} \times 1 \text{ m} = 0.2 \text{ m}^2/\text{m}$$

$$F = q_z G C_f A_e$$

$$F = (1.269) \times 0.85 \times 2.0 \times 0.2 = 0.43 \text{ kN/m}^2$$

A wind directionality factor, $K_d = 0.85$, shall be applied to calculate the q_z values above or, otherwise, multiply the above-calculated loads F_1 through F_6 including column wind loads with a factor of 0.85, when used with service and factored load combinations. $K_d = 0.85$ for pipe racks and open-frame structures. For pipe rack wind load values on the bent, refer to Figure 3.8.

Example 2: Wind Load on Pipe Rack

The pipe rack case shall be as shown in Figure 3.9 (frame spacing = 5 m), with a 3-s gust wind speed of 46 m/s.

The design wind forces are determined using Equation (3.1), where F is the force per unit length of the piping or cable tray.

The design wind pressure, for an elevation of 11 m, is calculated as:

$$q_z = 4.9 \text{ kN/m}^2$$

Gust effect factor, $G = 0.85$

Force Coefficients
In accordance with Section 3.2.3.2, the following are C_f values for different members:

Structural members, $C_f = 1.8$

Columns, $C_f = 2.0$

Pipes, $C_f = 0.7$

Cable trays, $C_f = 2.0$

Projected Area
The projected area per foot of pipe rack, A_e, equals the largest pipe diameter or the cable tray height + 10% of the pipe rack width.

A. PIPING AND CABLE TRAY

The guidelines require the consideration of the piping or cable trays separately from the structural members. The following calculations are only for piping and cable trays without the structural support members.

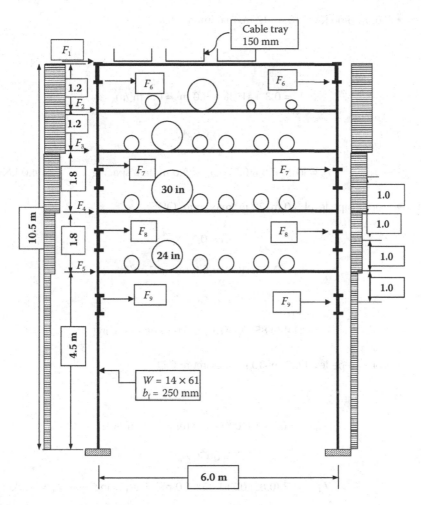

FIGURE 3.9
Example 2—sketch of pipe rack case.

Force Calculation

- For cable tray, 150 deep (at an elevation of 11.0 m)

$$C_f = 2.0$$

$$A_e = 0.15 + (10\% \times 6\ m) = 0.75\ m^2/m$$

$$F_1 = q_z G C_f A_e$$

$$F_1 = [(1.51)(0.85)(2.0)(0.75)] \times 5.0\ m\ bent\ spacing \qquad F_1 = 10.2\ kN$$

- For F_2 pipe level 10 m–12 in. maximum OD

$$C_f = 0.7$$

$$A_e = 0.3 + (10\% \times 6.0 \text{ m}) = 0.9 \text{ m}^2/\text{m}$$

$$F_2 = q_z G C_f A_e$$

$$F_2 = [(1.5)(0.85)(0.7)(0.9)] \times 5.0 \text{ m bent spacing} \qquad F_2 = 4.0 \text{ kN}$$

- For F_3 pipe level 8.0 m–20 in. maximum OD

$$C_f = 0.7$$

$$A_e = 0.5 + (10\% \times 6.0 \text{ m}) = 1.1 \text{ m}^2/\text{m}$$

$$F_3 = q_z G C_f A_e$$

$$F_3 = [(1.4)(0.85)(0.7)(1.1)] \times 5.0 \text{ m bent spacing} \qquad F_3 = 4.6 \text{ kN}$$

- For F_4 pipe level 6.3 m–30 in. maximum OD

$$C_f = 0.7$$

$$A_e = 0.75 + (10\% \times 6.0 \text{ m}) = 1.35 \text{ m}^2/\text{m}$$

$$F_4 = q_z G C_f A_e$$

$$F_4 = [(1.3)(0.85)(0.7)(1.35)] \, 5.0 \text{ m bent spacing} \qquad F_4 = 5.2 \text{ kN}$$

- For F_5 pipe level 4.5 m–24 in. maximum OD

$$C_f = 0.7$$

$$A_e = 0.6 + (10\% \times 6.0 \text{ m}) = 1.2 \text{ m}^2/\text{m}$$

$$F_5 = q_z G C_f A_e$$

$$F_5 = [(1.1)(0.85)(0.7)(1.2)] \, 5.0 \text{ m bent spacing} \qquad F_5 = 3.93 \text{ kN}$$

B. STRUCTURAL MEMBERS

For structural members, assume a 6-m wide rack with bent spacing of 5-m centers, with all stringers unshielded. The stringers are at elevations 11, 7.2, 5.4, and 4 m.

$$C_f = 1.8$$

$$A_e = 0.24 \text{ (beam depth)} \times 6.0 \text{ m (beam length)} = 1.5 \text{ m}^2$$

$$F = q_z GC_f A_e = q_z \times 0.85 \times 1.5 = q_z \times 1.27$$

$$F_6 = 1.6 \times 1.27 = 2 \text{ kN}$$

$$F_7 = 1.4 \times 1.27 = 1.8 \text{ kN}$$

$$F_8 = 1.3 \times 1.27 = 1.65 \text{ kN}$$

$$F_9 = 1.1 \times 1.27 = 1.4 \text{ kN}$$

Columns

$$q_z = 1.6 \text{ kN at an elevation of 11.0 m}$$

Use $q_z = 1.6$ kN for all columns (conservative)

$$C_f = 2.0$$

$$A_e = 0.25 \text{ (column width)} \times 1.0 \text{ m} = 0.25 \text{ m}^2/\text{m}$$

$$F = q_z GC_f A_e$$

Force per column, $F = (1.6) \times 0.85 \times 2.0 \times 0.25 = 0.68$ kN/m

A wind directionality factor, $K_d = 0.85$, shall be applied to the calculated q_z values above; otherwise, multiply the above-calculated loads F_1 through F_9, including column wind loads, with a factor of 0.85 when used with service and factored load combinations.

Example 3: Wind Load on Open Frame Structure

The plan and elevation views of the structure are shown in Figures 3.10, 3.11, 3.12, and 3.13. The structure considered is $10 \times 10 \times 18$ m high, with three open frames in the direction of the wind. The basic wind speed, V, equals 119 mi/h. This is a 3-s gust wind speed with an annual probability exceeding 0.02. Member sizes are assumed to be as follows:

$$\text{Columns} = 0.31 \times 0.31 \text{ m}$$

$$\text{Beams at an elevation of 6.0 m} = W_{36}$$

$$\text{Beams at an elevation of 18.0 m} = W_{18}$$

$$\text{Braces} = W_8$$

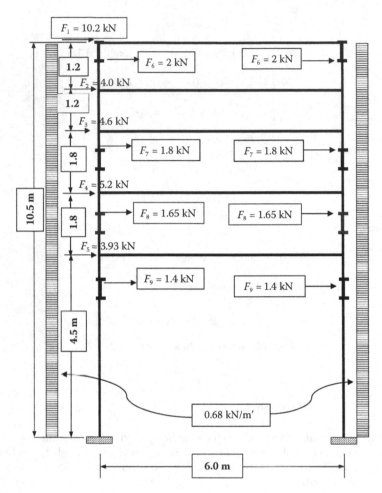

FIGURE 3.10
Example 2—configuration sketch.

$$\text{Intermediate Beams} = W_{12}$$

$$F_S' = q_z G C_f A_e$$

where F_S is the wind force on the structural frame and appurtenance.

It is convenient to determine the velocity pressures at the mid-floor heights and at the top of the structure. The following are the q_z values at different levels of velocity pressure, q_z.

Height above Ground, z (m)	q_z (kN/m²)
3.0	36.0
15.0	42.4

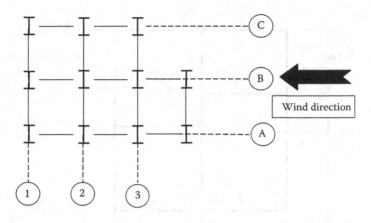

FIGURE 3.11
Example 3—sketch of plan for open-frame structure.

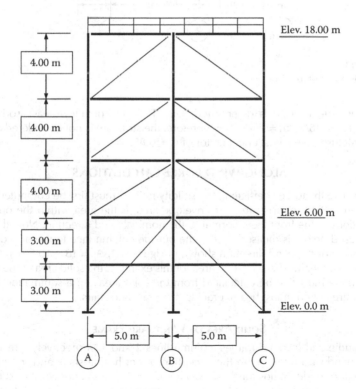

FIGURE 3.12
Example 3—view of line 3.

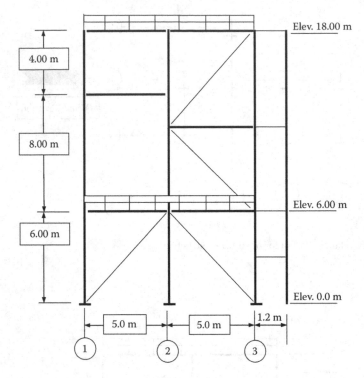

FIGURE 3.13
Example 3—view of axis A.

The gust effect factor is determined next. The ratio of height/least horizontal dimension = 18/10 m = 1.8 < 4.0; therefore, the structure is not considered a flexible structure. Use a gust effect factor of $G = 0.85$.

ALONG-WIND FORCE CALCULATIONS

To calculate the force coefficient, the solidity ratio, ε, must first be computed from Equation (3.3). The gross area (or envelope area) is the area within the outmost projections of the front face, normal to the nominal wind direction. Note that the width used below is measured from the outside column face to the outside column face. For the wind direction shown in Figure 3.11, $A_g = 18 \times 10.5 = 189$ m^2.

To determine the effective solid area for this example, the solid area of the windward frame must first be calculated from the solid area, A_s, a summation of the surface area of columns, beam, bracing, handrail, and stairs.

SOLID AREA OF WINDWARD FRAME

The middle and leeward frames (column lines 2 and 1, respectively) are similar to the windward frame, with the exception of not having stairs and solid areas. Hence, the solidity ratios for these two frames will be less than windward frame, and so A_s is equal to the solid area of windward frame, $\varepsilon = A_s/A_g = 53.8/189 = 0.285$.

TABLE 3.9

Total Force—Structural Frame and Appurtenances, F_S

Floor Level	q_z (psf)	G	C_f	K_d	A_e (m²)	F (kN)
0	1.72	0.85	3.33	0.85	10	41.4
1	2.0	0.85	3.33	0.85	20	96.2
2	2.34	0.85	3.33	0.85	24	135.1
	$F_S = \sum F =$					272.7

Next, the coefficient C_{Dg} is obtained from curves given in Figure 3.1 as a function of the solidity ratio, ε, the number of frames, N, and the frame spacing ratio, S_F/B. As defined in Figure 3.1, $N = 3$ and $S_F/B = 5/10.5 = 0.48$. From Figure 3.1c and d, $N = 3$ and, extrapolating slightly, $\varepsilon = 0.285$, and $C_{Dg} = 0.95$. Next, the gross area force coefficient, C_{Dg}, is converted into a force coefficient compatible with ASCE 7-02:

$$C_f = C_{Dg}/\varepsilon = 0.95/0.285 = 3.33.$$

The area of application of force, A_e, has already been determined per floor level during the calculation of the solidity ratio. The wind force transmitted to each floor level may now be found. The total force on the structural frame and appurtenances, F_S, is 272.7 kN, found by summing the forces at all levels in Table 3.9.

These forces are due to wind acting on the frames only. The wind loads on vessels, equipment, and piping are not considered in this example.

CROSSWIND FORCE CALCULATIONS

The next step is to repeat the analysis for the nominal wind direction normal to column line A, as shown in Figure 3.13, a "nonwindward" frame. The member sizes are the same at this elevation, except that:

Intermediate beams $= W_{10s}$

Beams at an elevation of 6.0 m $= W_{14}$

Beams at an elevation of 18.0 m $= W_{16}$

TABLE 3.10

Solid Area, A_s

Floor Level	Tributary Height (m)	Columns	Beams	Intermediate Beams	Bracing	Handrails	Stairs	Total
0	3	2.7	0	0	1.7	0	2.2	10.6
1	3–12	6.5	4.0	0.72	3.2	3.6	3.8	22
2	12–18	8.4	4.77	3.7	3.2	3.6	3.24	27
Total solid area of windward frame (m²)								59.6

(Solid Area (m²) spans Columns through Total)

TABLE 3.11

Total Force—Structural Frame and Appurtenance, F_S

Floor Level	q_z (psf)	G	C_f	A_e (ft²)	K_d	F (kN)
0	1.72	0.85	3.2	10.6	0.85	42
1	2.0	0.85	3.2	22	0.85	101.7
2	2.34	0.85	3.2	27	0.85	146.1
	$F_S = \sum F =$					289.9

The gross area of the windward face includes the stair tower on the right-hand side of the structure, $A_g = (18 \times 10.5) + (1.2 \times 18) = 210.6$ m². The solid areas for the windward frame are given in Table 3.10. The *stairs* column in this table includes areas of the stair column, struts, and handrails.

Since the solidity of neither the middle nor leeward frames (column lines B and C, respectively) exceeds that of the windward frame, A_s is equal to the solid area of the windward frame, yielding

$$\varepsilon = A_s/A_g = 59.6/210.6 = 0.28.$$

The frame spacing ratio in this direction is $S_F/B = 6/11.2 = 0.54$. Because the width is not uniform (the stair tower stops at the second floor level), an average value of B is used. From Figures 3.12 and 3.13, $N = 3$ and $\varepsilon = 0.28$; therefore, we use $C_{Dg} = 0.9$, $C_f = C_{Dg}/\varepsilon = 0.9/0.28 = 3.21$. The wind forces per floor level are shown in Table 3.11.

3.2.3.2.2.4 Open-Frame Example: Summary and Conclusion   Table 3.12 summarizes the results. The load combinations for design are applications of F_T in one direction, simultaneously with $0.5\ F_S$ in the other direction. These combinations are shown in Figure 3.14.

In normal cases, there are a piping and a vessel, so its wind load must be included in the example, showing the load on the equipment and piping for illustration. A wind directionality factor, $K_d = 0.85$, shall be applied to calculate q_z values or, otherwise, multiply the calculated loads (including column wind loads) by 0.85 when using service and factored load combinations.

TABLE 3.12

Summary of Load Direction

	Load in Wind Direction 1 (kN)	Load in Wind Direction 2 (kN)
Wind load on structure frame, F_S	273	290
Wind load on piping, F_E	100	90
Total wind load, F_T, on the structure	373	380

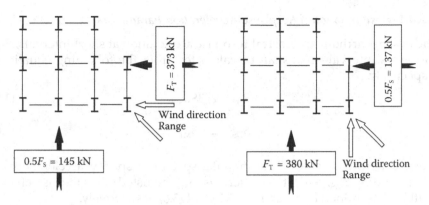

FIGURE 3.14
Sketch of design load cases in open-frame structures.

3.2.4 Earthquake Loads

An earthquake is a sudden and rapid shock that occurs when layers of the earth's crust slide over each other. Rocks are bent to adapt to the new situation and, when the rocks return to their original position in a sudden motion, the earth's surface is shaken and quake waves spread through the earth's crust from the quake center.

Earthquake loads (F) are very important. Serious structural damage can occur because this load has a special nature—it occurs at intervals of up to several years, last several seconds when it occurs, and can have a significant impact on buildings. In the past, people held the belief that an animal held the land and would shake the earth when it moved. The scholar Ibn Sina developed one of the first scientific theories about the cause of earthquakes, believing that the movement originated from activity underneath the land.

It is worth noting that about 300,000 earthquakes occur annually throughout the world, but most are not strong or occur away from residential areas. In the case of powerful earthquakes that occur near areas where facilities are located, landslides may also occur and shall be factored into the design of the building. Earthquake codes require that a structure shall be designed to resist a minimum total lateral load, V, which shall be assumed to act nonconcurrently in orthogonal directions parallel to the main axes of the structure.

The procedures and limitations for the design of structures shall be determined by considering seismic zoning, site characteristics, occupancy, configuration, structural system, and height in accordance with the information presented in this section. Structures shall be designed with adequate strength to withstand the lateral displacements induced by the design basis ground motion, considering the inelastic response of the structure and the inherent redundancy, over-strength, and ductility of the lateral force–resisting system.

3.2.4.1 Design Spectral Response Acceleration Parameters

The design earthquake spectral response acceleration at short intervals, S_{DS}, and at 1-s intervals, S_{D1}, shall be calculated from the following equations, respectively.

$$S_{DS} = 2/3S_{MS} \qquad (3.5)$$

$$S_{DS} = 2/3S_{M1} \qquad (3.6)$$

The maximum considered earthquake spectral response acceleration for short periods, S_{MS}, and at 1-s intervals, S_{M1}, adjusted for site-class effects, shall be determined by Equations (3.7) and (3.8), respectively.

$$S_{MS} = F_a S_S \qquad (3.7)$$

$$S_{M1} = F_v S_1 \qquad (3.8)$$

where
  S_1 = Mapped maximum considered earthquake spectral response acceleration at a period of 1 s
  S_S = Mapped maximum considered earthquake spectral response acceleration at short periods

Because S_S and S_1 are determined using the seismic map for the country where the project will be located, the seismic zone of the plant location must be included in the planning.

It is worth noting that if the soil properties are not known in sufficient detail to determine the site class, class D shall be used (Table 3.13). Class E shall be used when the authority of jurisdiction determines that site class E is present at the site or in the event that site class E is established by geotechnical data (Tables 3.14 and 3.15).

3.2.4.2 Architectural, Mechanical, and Electrical Components Systems

In industrial buildings and structures, there are always architectural, mechanical, electrical, and nonstructural systems, components, and elements that are permanently attached to structures, including supporting structures and attachments. The design criteria established minimum equivalent static force levels and relative displacement demands for the design of components and their attachments to the structure, recognizing ground motion, structural amplification, component toughness and weight, and performance expectations. In most cases, the components' design shall be considered to have the same seismic design category as that of the structure that they occupy or to which they are attached. Therefore, it is necessary to establish a minimum seismic design force requirement for nonbuilding

TABLE 3.13

Site Classification with Soil Parameters

Site Class Definitions	Soil Parameters
A	Hard rock with measured shear wave velocity, V_s, > 1500 m/s
B	Rock with 760 m/s < V_s, ≤ 1500 m/s
C	Very dense soil and soft rock with 370 m/s ≤ V_s ≤ 760 m/s or N or N_{ch} > 50 or S_u ≥ 100 kPa
D	Stiff soil with 180 ≤ V_s ≤ 370 m/s or with N and N_{ch} > 50 or 50 kPa ≤ S_u ≤ 100 kPa
E	A soil profile with V_s < 180 m/s or any profile with more than 3 m of soft clay[a]
F	Soils requiring site-specific evaluations: soils vulnerable to potential failure or collapse under seismic loading such as liquefiable soils, quick and highly sensitive clays, collapsible weakly cemented soils

Note: S_u, average undrained shear strength in top 100 ft (30 m) according to ASTM D2166-91 or ASTM D2850-87; N, standard penetration resistance according to ASTM D 1568-84; N_{ch}, average field standard penetration resistance for cohesionless soil layer for the top 100 ft (30 m).

[a] Soft clay is defined as soil with plasticity index (PI) > 20, w ≥ 40%, and S_u <25 kPa.

structures that are supported by other structures, where the weight of the nonbuilding structure is less than 25% of the combined weight of the nonbuilding structure and the supporting structure. The seismic design force will be calculated using the following formula:

$$F_p = \frac{0.4a_p S_{DS} W_p}{R_p / I_p} \left(1 + 2\frac{z}{h} \right).$$

TABLE 3.14

Value of F_a as a Function of Site Class and Mapped Short Period Maximum Considered Earthquake Spectral Acceleration

Site Class	Mapped Maximum Considered Earthquake Spectral Response Acceleration at Short Periods				
	S_S ≤ 0.25	S_S = 0.5	S_S = 0.75	S_S = 1.0	S_S ≥ 1.25
A	0.8	0.8	0.8	0.8	0.8
B	1.0	1.0	1.0	1.0	1.0
C	1.2	1.2	1.1	1.0	1.0
D	1.6	1.4	1.2	1.1	1.0
E	2.5	1.7	1.2	0.9	0.9
F	a	a	a	a	a

Note: Use straight-line interpolation for intermediate values of S_S.

[a] Site-specific geotechnical investigation and dynamic site response analyses shall be performed except that for structures with periods of vibration equal to or less than 0.5 s, values of F_a for liquefiable soils may be assumed as equal to the values for the site class determined without regard to liquefaction.

TABLE 3.15

Values of F_v as a Function of Site Class and Mapped 1-s Period
Maximum Considered Earthquake Spectral Acceleration

Site Class	Mapped Maximum Considered Earthquake Spectral Response Acceleration at 1-s Periods				
	$S_1 \leq 0.1$	$S_1 = 0.2$	$S_1 = 0.3$	$S_1 = 0.4$	$S_1 \geq 0.5$
A	0.8	0.8	0.8	0.8	0.8
B	1.0	1.0	1.0	1.0	1.0
C	1.7	1.6	1.5	1.4	1.3
D	2.4	2.0	1.8	1.6	1.5
E	3.5	3.2	2.8	2.4	2.4
F	a	a	a	a	a

Note: Use straight-line interpolation for intermediate values of S_1.

[a] Site-specific geotechnical investigation and dynamic site response analyses shall be performed except that for structures with periods of vibration equal to or less than 0.5 s, values of F_v for liquefiable soils may be assumed as equal to the values for the site class determined without regard to liquefaction.

F_p is not required to be taken as greater than

$$F_p = 1.6 S_{DS} I_p W_p,$$

and F_p shall not be taken as less than

$$F_p = 0.3 S_{DS} I_p W_p,$$

where

F_p = Seismic design force centered at the component's center of gravity and distributed relative to the component's mass distribution

S_{DS} = Spectral acceleration short period

a_p = Component amplification factor that varies from 1.00 to 2.50

I_p = Component importance factor, which varies as shown in Table 3.16

W_p = Component operating weight

R_p = Component response modification factor that varies from 1.50 to 5.00

z = Height of the structure at the point of attachment of the component, with respect to the base. For items at or below the base, z shall be 0. The value of z/h shall not exceed 1.0.

h = Average roof height of the structure with respect to the base

Force, F, shall be independently applied longitudinally and laterally in combination with service loads associated with the component. The component importance factor, I_p, shall be selected as shown in Table 3.16.

TABLE 3.16

The Importunacy Factor, I_p

Equipment Criticality	I_p
Life safety component required to function after an earthquake (e.g., fire protection sprinkler system)	1.5
Components that contain hazardous content	1.5
Storage racks in structures open to the public (e.g., warehouse retail stores)	1.5
All other components	1.0
In addition, for structures in Seismic Use Group III:	
All components needed for continued operation of the facility or whose failure could impair the continued operation of the facility	1.5

3.2.4.2.1 Mechanical and Electrical Component Period

The mechanical and electrical equipment attached to the structure have a natural frequency. The natural frequency period of the mechanical and electrical components, T, shall be determined by the following equation, provided that the component and attachment can be reasonably represented analytically by a simple spring and mass single degree of freedom system:

$$T = 2\pi\sqrt{\frac{W_p}{K_p g}}$$

where

T = Component's fundamental period
W_p = Component's operating weight
g = Gravitational acceleration
K_p = Stiffness of the resilient support system of the component and attachment, determined in terms of load per unit deflection at the center of gravity of the component

Consistent units must be used. Otherwise, determine the fundamental period of the component (in seconds) from experimental test data or by a properly substantiated analysis.

3.2.4.3 HVAC Ductwork

Attachments and supports for HVAC ductwork systems should be designed to meet the force and displacement requirements. In addition to their attachments and supports, ductwork systems designated as having an I_p of 1.5 shall be designed to meet the force and displacement requirements, as the drift will not be more than 1/100 of the height level. Where HVAC ductwork runs between structures that could be displaced relative to one another and for seismically isolated structures where the HVAC ductwork crosses the seismic isolation interface, the HVAC ductwork shall be designed to accommodate the seismic relative displacements.

Based on ASCE 7-02, the seismic restraints are not required for HVAC ducts with an I_p of 1.0 if either of the following conditions are met:

1. HVAC ducts are suspended from hangers 12 in. (305 mm) or less in length from the top of the duct to the supporting structure. The hangers shall be detailed to avoid significant bending of the hangers and their attachments.
2. HVAC ducts have a cross-sectional area of less than 6 ft^2 (0.557 m^2).

HVAC duct systems that are fabricated and installed in accordance with standards approved by the regulating authority will be deemed as having met the lateral bracing requirements. Equipment items installed in line with the duct system—for example, fans, heat exchangers, and humidifiers weighing more than 75 lb (344 N)—shall be supported and laterally braced, independent of the duct system, and shall also meet any force requirements. Furthermore, the appurtenances, such as dampers, louvers, and diffusers, shall be positively attached to the structure with mechanical fasteners. Unbraced piping attached to in-line equipment shall be provided with adequate flexibility to accommodate differential displacements.

3.2.4.4 Piping Systems

The piping systems attached and supporting the structure shall be designed to meet the force and displacement requirements for safe operation. In addition to their attachments and supports, piping systems designated as having an I_p of 1.5 shall be designed to meet any force and displacement requirements. Where piping systems are attached to structures that could be displaced relative to one another and for seismically isolated structures where the piping system crosses the seismic isolation interface, the piping system shall be designed to accommodate the seismic relative displacements requirement to limit the drift to 1/100 of the level height.

Seismic effects that shall be considered in the design of a piping system include the dynamic effects of the piping system, its contents, and, when appropriate, its supports. The interaction between the piping system and the supporting structures, including other mechanical and electrical equipment, shall also be considered. Pressure piping systems designed and constructed in accordance with the ASME B31 Code for Pressure Piping may be considered to also meet the specific force and displacement provisions provided in the ASME B31.

Fire protection sprinkler systems usually designed and constructed in accordance with the NFPA 13 Standard for the Installation of Sprinkler

Systems may be deemed to meet the other requirements and the force will be calculated using the force equation. Other piping systems designated as having an I_p of 1.5 that are not designed and constructed in accordance with the ASME B31 or NFPA 13 should meet the criteria described in Section 3.2.4.6.

3.2.4.5 Boilers and Pressure Vessels

The seismic design of a boiler or pressure vessel shall include analysis of the following: the dynamic effects of the boiler or pressure vessel, its contents, its supports, the sloshing of liquid contents, loads from attached components such as piping, and the interaction between the boiler or pressure vessel and its support. Boilers or pressure vessels designed in accordance with the ASME Boiler and Pressure Vessel Code may be considered to meet the force, displacement, and other requirements of this code.

3.2.4.6 General Precaution

Based on the ASCE 7-02 guidelines for sites in seismic zones, the design strength for seismic loads of general equipment such as boilers, pressure vessels, and mechanical and electrical equipment, in combination with other service loads and appropriate environmental effects, should not exceed the following:

1. For equipment constructed with ductile materials (steel, aluminum, or copper), 90% of the material's minimum specified yield strength.
2. For threaded connections in equipment or their supports constructed with ductile materials, 70% of the material's minimum specified yield strength.
3. For equipment constructed with nonductile materials such as plastic, cast iron, or ceramics, 25% of the material's minimum specified tensile strength.
4. For threaded connections in equipment or their supports constructed with nonductile materials, 20% of the material's minimum specified tensile strength.

3.2.4.7 Building and Nonbuilding Structures

The definition for *nonbuilding structures* includes all self-supporting structures that carry gravity loads and that may be required to resist the effects of earthquake, with the exception of buildings, vehicular and railroad bridges, nuclear power generation plants, offshore platforms, dams, and other structures. Nonbuilding structures supported by the earth or supported by other structures shall be designed and detailed to resist the minimum lateral forces as specified.

For general equipment such as machines, piping, pressure vessels, and tanks, there are two cases during construction, maintenance, or, in some cases, shutdown: in the first case, the equipment is full of liquid, and in the second case, the equipment is empty. The weight of the equipment is different in the two cases, so there will be two calculated value of earthquake loads:

- E_o is the earthquake load considering the unfactored operating dead load and the applicable portion of the unfactored structure dead load.
- E_e is the earthquake load considering the unfactored empty dead load and the applicable portion of the unfactored structure dead load.

The design of nonbuilding structures shall provide stiffness, strength, and ductility consistent with the requirements specified here for buildings to resist the effects of seismic ground motions, as represented by these design forces:

1. Applicable strength and other design criteria shall be obtained from codes and standards that apply to the project.
2. When applicable strength and other design criteria are not contained in project specifications, such criteria shall be obtained from approved national standards. Where approved national standards define acceptance criteria in terms of allowable stresses as opposed to strength, the design seismic forces shall be obtained and used in combination with other loads according to the national standards.

3.2.4.7.1 Design Basis

Nonbuilding structures with specific seismic design criteria established in approved standards shall be designed using those standards as amended here. Nonbuilding structures shall be designed to resist minimum seismic lateral forces that are not less than the requirements for building structures with the following values of the response modification coefficient, R, as shown in Table 3.14. Apply the following equation:

$$C_S = 0.14 \, S_{DS}I.$$

The minimum value specified shall be replaced by:

$$C_S = 0.8 \, S_1I/R.$$

The overstrength factor, Ω_o, and the R factor shall be taken from Table 3.17. The importance factor, I, shall be as given in Table 3.18. The vertical distribution of the lateral seismic forces in nonbuilding structures shall be determined in accordance with an approved standard applicable to the specific nonbuilding structure (Table 3.19). For nonbuilding structural systems containing liquids, gases, and granular solids supported at the base, the minimum seismic design force shall not be less than that required by the approved standard for the specific system. Irregular structures, where S_{DS} is greater than or equal to 0.50, that cannot be modeled as a single mass shall use the modal analysis procedures.

TABLE 3.17

Architectural Component Coefficient

Architectural Component or Element	a_p	R_p
Interior Nonstructural Walls and Partitions		
Plain (unreinforced) masonry walls	1	1.5
All other walls and partitions	1	2.5
Cantilever elements (unbraced or braced to structural frame below its center of mass)		
Parapets and cantilever interior nonstructural walls	2.5	2.5
Chimneys and stacks when laterally braced or supported by the structural frame	2.5	2.5
Cantilever elements (braced to structural frame above its center of mass)		
Parapets	1.0	2.5
Chimneys and stacks	1.0	2.5
Exterior Nonstructural Wall Elements and Connections		
Wall element		
Body of wall panel connections	1	2.5
Fasteners of the connecting system	1	2.5
Storage cabinets and laboratory equipment	1	2.5
Special access floors	1	2.5
All access floors	1	1.5
Signs and billboards	2.5	2.5
Other rigid components		
High deformability elements and attachments	1	3.5
Limited deformability elements and attachments	1	2.5
Low deformability materials and attachments	1	1.5
Other flexible components		
High deformability elements and attachments	2.5	3.5
Limited deformability elements and attachments	2.5	2.5
Low deformability materials and attachments	2.5	1.5

TABLE 3.18

Mechanical and Electrical Components Seismic Coefficients

Mechanical and Electrical Component or Element	a_p	R_p
General mechanical equipment		
Boilers and furnaces	1.0	2.5
Pressure vessels on skirts and free-standing	2.5	2.5
Stack	2.5	2.5
Cantilevered chimneys	2.5	2.5
Other	1.0	2.5
Manufacturing and process machinery		
Genera	1.0	2.5
Conveyors (nonpersonnel)	2.5	2.5
Piping Systems		
High deformability elements and attachment	1.0	3.5
Limited deformability elements and attachments	1.0	2.5
Low deformability elements and attachments	1.0	1.5
HVAC System		
Vibration isolated	2.5	2.5
Nonvibration isolated	1.0	2.5
Mounted in line with ductwork	1.0	2.5
Other	1.0	2.5
Elevator components	1.0	2.5
Escalator components	1.0	2.5
Trussed towers (free-standing or guyed)	2.5	2.5
General electrical		
Distribution systems (bus ducts, conduit, cable tray)	2.5	5.0
Equipment	1.0	2.5
Lighting Fixtures	1.0	1.5

Note: A lower value for a_p shall not be used unless justified by detailed dynamic analyses. The value for a_p shall not be less than 1.00. The value of $a_p = 1$ is for equipment generally regarded as rigid or rigidly attached. The value of $a_p = 2.5$ is for equipment generally regarded as flexible or flexibly attached.

Components mounted on vibration isolation systems shall have a bumper restraint or snubber in each horizontal direction. The design force shall be taken as $2F_p$ if the maximum clearance (air gap) between the equipment support frame and restraint is greater than 6 mm.

If the maximum clearance is specified on the construction documents to be not greater than 6 mm, the design force may be taken as F_p.

3.2.4.7.2 Rigid Nonbuilding Structures

Nonbuilding structures that have a fundamental period, T, less than 0.06 s, including their anchorages, shall be designed for the lateral force obtained from the following equation:

$$V = 0.30 S_{DS} W I$$

TABLE 3.19

Seismic Coefficients for Nonbuilding Structures

Nonbuilding Structure Type	R	Ω	C_d
Nonbuilding frame systems			
Concentric braced frames of steel	6	2.5	5
Moment-resisting frame systems			
Special moment frames of steel	8	3	5.5
Ordinary moment frames of steel	3.5	3	2.5
Special moment frames of concrete	8	3	5.5
Intermediate moment frames of concrete	5	5	4.5
Ordinary moment frames of concrete	3	3	2.5
Steel storage racks	4	2	3.5
Elevated tanks, vessels, bins, or hoppers			
On braced legs	3	2	2.5
On unbraced legs	3	2	2.5
Irregular braced legs, single pedestal, or skirt-supported	2	2	2
Welded steel	2	2	2
Concrete	2	2	2
Horizontal, saddle-supported welded steel vessels	3	2	2.5
Tanks or vessels supported on structural towers similar to building	3	2	2
Flat bottom, ground supported tanks, or vessels			
Anchored (welded or bolted steel	3	2	2.5
Unanchored (welded or bolted steel)	2.5	2	2
Reinforced or prestress concrete			
Tanks with reinforced nonsliding base	2	2	2
Tanks with anchored flexible base	3	2	2
Tanks with unanchored and unconstrained			
Flexible base	1.5	1.5	1.5
Other material	1.5	1.5	1.5
Cast-in-place concrete silos, stacks, and chimneys having walls continuous to the foundation	3	1.75	3
All other reinforced masonry structures not similar to buildings	3	2	2.5
All other nonreinforced masonry structures not similar to buildings	1.25	2	1.5
All other steel and reinforced concrete distributed mass (cantilever structures not covered herein including stacks, chimneys, silos, and skirt-supported vertical vessels that are not similar to buildings)	3	2	2.5
Trussed towers (freestanding or guyed), guyed stacks, and chimneys	3	2	2.5
Cooling towers			
Concrete or steel	3.5	1.75	3

(continued)

TABLE 3.19 Continued

Seismic Coefficients for Nonbuilding Structures

Nonbuilding Structure Type	R	Ω	C_d
Telecommunication towers			
Truss: Steel	3	1.5	3
Pole: Steel	1.5	1.5	1.5
Wood	1.5	1.5	1.5
Concrete	1.5	1.5	1.5
Frame: Steel	3	1.5	1.5
Concrete	2	1.5	1.5
Inverted pendulum-type structures (except elevated tanks, vessels, bins, and hoppers)	2	2	2
Signs and billboards	3.5	1.75	3
All other self-supporting structures, tanks, or vessels not covered above or by approved standards that are similar to buildings	1.25	5	2.5

where

  V = Total design lateral seismic base shear force applied to a nonbuilding structure

  S_{DS} = The site's design response acceleration

  W = Nonbuilding structure's operating weight

  I = Importance factor as determined from Table 3.20

3.2.4.8 Flexibility of Piping Attachments

The piping systems connected to tanks and vessels are sensitive to the potential movement of the connection points during earthquakes and therefore require a sufficient flexibility to avoid release of the product by failure of the

TABLE 3.20

Importance Factor (*I*) and Seismic Use Group Classification for Nonbuilding Structures

Importance factor	$I = 1.0$	$I = 1.25$	$I = 1.5$
Seismic use group	I	II	III
Hazard	H-I	H-II	H-III
Function	F-I	F-II	F-III

Note: H-I, nonbuilding structures not assigned to H-II or H-III; H-II, nonbuilding structures containing hazardous materials and classified as category III structures in Table 3.5; H-IV, nonbuilding structures containing extremely hazardous materials and classified as category IV structures in Table 3.5; F-I, nonbuilding structures not classified as F-II or F-III; F-II, nonbuilding structures classified as category III structures in Table 3.5; F-III, nonbuilding structures classified as category IV structures (essential facilities) in Table 3.5.

TABLE 3.21

Minimum Displacement for Piping Attachments

	Displacements (mm)
Anchored Tanks or Vessels	
Vertical displacement relative to support or foundation	50
Horizontal (radial and tangential) relative to support or foundation	12
Unanchored Tanks or Vessels (at Grade)	
Vertical displacement relative to support or foundation if designed to meet approved standards	150
If designed for seismic loads per these provisions but not covered by an approved standard	300
For tanks and vessels with a diameter <40 ft, horizontal (radial and tangential) relative to support or foundation	200

piping system. Forces and deformations imparted by piping systems or other attachments should not exceed the design limitations at points of attachment to the tank or vessel shell. Mechanical devices that add flexibility such as bellows, expansion joints, and other flexible apparatus may be used when they are designed for seismic loads and displacements. Unless otherwise calculated, the minimum displacements in Table 3.19 shall be assumed. For attachment points located above the support or foundation elevation, the displacements in Table 3.19 shall be increased to account for drift of the tank or vessel relative to the base of support.

When the elastic deformations are calculated, the minimum design displacements for piping attachments shall be the calculated displacements at the point of attachment, increased by the amplification factor C_d. Table 3.21 does not include the influence of relative movements of the foundation and piping anchorage points due to foundation movements (e.g., settlement, seismic displacements). The effects of the foundation movements shall be included in the piping system design, including the determination of the mechanical loading on the tank or vessel and the total displacement capacity of the mechanical devices intended to add flexibility.

3.2.4.9 Design Review for Seismic Loads

In the case of a project in a zone with high seismic activity or a special case such as a liquefied natural gas plant, a design review of the seismic force resisting system and the structural analysis should be performed by an independent team of registered design professionals. This shall involve professionals in the appropriate disciplines and other individuals experienced in seismic analysis methods and the theory and application of nonlinear seismic analysis and structural behavior under extreme cyclic loads. The design review shall include, but not be limited to:

1. Review of any site-specific seismic criteria used in the analysis, including the development of site-specific spectra and ground motion time histories
2. Review of acceptance criteria used to demonstrate the adequacy of structural elements and systems to withstand the calculated force and deformation demands, together with laboratory and other data used to substantiate these criteria
3. Review of the preliminary design, including the selection of the structural system and the configuration of structural elements
4. Review of the final design of the entire structural system and all supporting analyses

3.2.5 Impact Loads

Impact loads shall be in accordance with SEI/ASCE 7-05. Impact loads for davits shall be the same as those for monorail cranes (powered). Lifting lugs, or pad eyes, and internal members (including both end connections) framing into the joint where the lifting lug or pad eye is located shall be designed for 100% impact. All other structural members transmitting lifting forces shall be designed for 15% impact. The allowable stresses shall not be increased when combining impact with dead load.

3.2.6 Thermal Loads

In this book, thermal loads are designated by T_p, T, A_f, and F_f, where:

T_p represents the forces on vertical vessels, horizontal vessels, or heat exchangers caused by the thermal expansion of the pipe attached to the vessel.

T represents the self-straining forces caused by the restrained thermal expansion of structural steel in pipe racks or horizontal vessels and heat exchangers.

A_f represents the pipe anchor and guide forces.

F_f represents pipe rack friction forces caused by the sliding of pipes or by the sliding of horizontal vessels or heat exchangers on their supports in response to thermal expansion.

All support structures and their elements shall be designed to accommodate the loads or effects produced by thermal expansion and the contraction of equipment and piping. Thermal loads shall be included with operating loads

TABLE 3.22

Coefficient of Friction

Steel to steel	0.4
Steel to concrete	0.6
Special sliding surfaces as Teflon	According to manufacturer's instruction

in the appropriate load combinations and shall have the same load factor as dead loads.

Thermal loads and displacements shall be calculated on the basis of the difference between the ambient or equipment design temperature and the installed temperature. To account for the significant increase in temperatures of steel exposed to sunlight, 35°F (20°C) shall be added to the maximum ambient temperature. Friction loads caused by thermal expansion shall be determined using the appropriate static coefficient of friction. Coefficients of friction shall be in accordance with Table 3.22.

To be clear, the friction loads shall be considered temporary and shall not be combined with wind or earthquake loads. However, anchor and guide loads (excluding their friction component) shall be combined with wind or earthquake loads.

For pipe racks supporting multiple pipes, 10% of the total piping weight shall be used as an estimated horizontal friction load and applied only to local supporting beams. However, an estimated friction load equal to 5% of the total piping weight shall be accumulated and carried into pipe rack struts, columns, braced anchor frames, and foundations.

Under normal loading conditions with multiple pipes, tensional effects on the local beam need not be considered because the pipes supported by the beam limit the rotation of the beam to the extent that the torsion stresses are minimal. Under certain circumstances, engineering judgment shall be applied to determine whether a higher friction load or torsion effects should be used.

Pipe anchor and guide loads shall have the same load factor as dead loads. The internal pressure and surge shall be taken into consideration for pipe anchor and guide loads.

During the design stage of the beams, struts, columns, braced anchor frames, and foundations, they should be able to resist actual pipe anchor and guide loads.

For local beam design, only the top flange shall be considered effective for horizontal bending unless the pipe anchor engages both flanges of the beam.

3.2.7 Bundle Pull Load

Structures and foundations supporting heat exchangers subject to bundle pulling shall be designed for a horizontal load equal to 1.0 times the weight of the removable tube bundle, but not less than 2000 lb (9.0 kN). If the total weight of the exchanger is less than 2000 lb (9.0 kN), the bundle pull load (B_p) may be taken as the total weight of the exchanger. Noting that, the bundle pull load shall be applied at the center of the bundle. This is discussed in Chapter 6 in relation to heat exchanger foundation design.

If it can be assured that the bundles will be removed strictly by the use of a bundle extractor attaching directly to the exchanger (such that the bundle pull force is not transferred to the structure or foundation), the structure or foundation need not be designed for the bundle pull force. Such assurance would typically require the addition of a sign posted on the exchanger to indicate bundle removal by an extractor only.

The portion of the bundle pull load at the sliding end support shall equal the friction force or half the total bundle pull load, whichever is less. The remainder of the bundle pull load shall be resisted at the anchor end support.

3.2.8 Ice Loads

Atmospheric ice loads (S) due to freezing rain and snow that include icing shall be considered in the design of ice-sensitive structures. In areas where records or experience indicate that snow or in-cloud icing produces larger loads than freezing rain, site-specific studies shall be used. Structural loads due to hoarfrost are not a design consideration.

3.2.8.1 Site-Specific Studies

Mountainous terrain and gorges shall be examined for unusual icing conditions. Site-specific studies shall be used to determine the 50-year mean recurrence interval ice thickness and concurrent wind speed in:

1. Alaska
2. Areas where records or experience indicate that snow or in-cloud icing produces larger loads than freezing rain
3. Special icing regions, which should be determined using an official map
4. Mountainous terrain and gorges where examination indicates unusual icing conditions exist

Site-specific studies shall be subject to review and approval by the governing authority. In lieu of using the mapped values, it is permitted to determine the ice thickness and the concurrent wind speed for a structure from local meteorological data based on a 50-year mean recurrence interval provided that:

1. The quality of the data for wind and type and amount of precipitation has been taken into account.

2. A robust ice accretion algorithm has been used to estimate uniform ice thicknesses and concurrent wind speeds from the data.

3. Extreme-value statistical analysis procedures acceptable to the governing authority have been employed in analyzing the ice thickness and concurrent wind speed data.

4. The length of the record and sampling error have been taken into account.

3.2.8.2 Loads due to Freezing Rain

The ice load shall be determined using the weight of glaze ice formed on all exposed surfaces of structural members, guys, components, appurtenances, and cable systems. On structural shapes, prismatic members, and other similar shapes, the cross-sectional area of ice shall be determined by:

$$A_i = \pi t_d (D_c + t_d),$$

where D_c is shown for a variety of cross-sectional shapes.

On flat plates and large three-dimensional objects such as domes and spheres, the volume of ice shall be determined by:

$$V_i = \pi t_d A_s,$$

where $A_s = \pi r^2$.

It is acceptable to multiply 0.6 for horizontal plates by 0.8 for vertical plates. The ice density shall not be less than 56 pcf (900 kg/m^3). The weather data of the site location and its maps show the equivalent uniform radial thicknesses, t, of ice due to freezing rain at a height of 33 ft (10 m) over the contiguous United States for a 50-year mean recurrence interval. The data shall also provide concurrent 3-s gust wind speeds. Thicknesses for Alaska and Hawaii, and for ice accretions due to other sources in all regions, shall be obtained from local meteorological studies. The height factor, f_z, used to increase the radial thickness of ice for a height above ground, z, shall be determined by:

$$f_z = \left(\frac{z}{10}\right)^{0.1} \quad \text{for } 0 < z \le 275 \text{ m}$$

$$f_z = 1.4 \quad \text{for } z > 275 \text{ m.}$$

Importance factors to be applied to the radial ice thickness and wind pressure according to the structure classifications are shown in Table 3.5. The importance factor multiplier, I_i, must be on the ice thickness, not the ice weight, because ice weight is not a linear function of thickness.

Both the ice thickness and concurrent wind speed for structures on hills, ridges, and escarpments are higher than those on level terrain because of wind speed-up effects. The topographic factor for the concurrent wind pressure is K_{zt} and the topographic factor for ice thickness, K_{rt}, is 0.35 w. Here, K is obtained from Equation (6.1).

3.2.8.3 Design Ice Thickness for Freezing Rain

The design ice thickness, t_d, shall be calculated using the following equation:

$$t_d = 2.0 t I_i f_z (K_{zt})^{0.35}.$$ (3.9)

Wind on ice-covered chimneys, tanks, and similar structures. Force coefficients for structures with square, hexagonal, and octagonal cross sections shall be as given in Table 3.7. Force coefficients for structures with round cross sections shall be as given in Table 3.7 for round cross sections with $D(q_z)^{0.5} \le 2.5$ for all ice thicknesses, wind speeds, and structure diameters.

3.2.8.4 Wind on Ice-Covered Structures

Ice accreted on structural members, components, and appurtenances increases the projected area of the structure exposed to wind. The projected area shall be increased by adding t_d to all free edges of the projected area. Wind loads on this increased projected area shall be used in the design of ice-sensitive structures. The wind load coefficient area shall be increased due to the ice thickness.

3.3 Load Combinations

Buildings, structures, equipment, vessels, tanks, and foundations shall be designed for the following:

1. SEI/ASCE 7-05 except as otherwise specified by the owner
2. Local building codes
3. Any other applicable design codes and standards
4. Any other probable and realistic combination of loads

TABLE 3.23

Loading Combinations—Allowable Stress Design (Service Loads)

Load Combination No.	Load Combination	Allowable Stress Multiplier	Description
1	$D_o + L$	1.00	Operating weight + live load
2	$D_o + (W \text{ or } 0.7 E_o{}^a)$	1.00	Operating weight + wind or earthquake
3	$D_e + W$	1.00	Empty weight + wind (wind uplift case)
4a	$0.9(D_o) + 0.7E_o{}^a$	1.00	Operating weight + earthquake (earthquake uplift case)
4b	$0.9(D_e) + 0.7E_e{}^a$	1.00	Empty weight + earthquake (earthquake uplift case)
5	$D_f + W_p$	1.00	Erection weight + partial wind[b] (wind uplift case)
6	$D_t + W_p$	1.20	Test weight + partial wind

[a] For skirt-supported vertical vessels and skirt-supported elevated tanks classified as occupancy category IV in accordance with SEI/ASCE 7-05, the critical earthquake provisions and implied load combination of SEI/ASCE 7-05 shall be followed.

[b] Erection weight + partial wind is required only if the erection weight of the vessel is significantly less than the empty weight of the vessel. This case is only applied if the erection weight of the vessel is significantly less than the empty weight of the vessel.

TABLE 3.24

Loading Combinations and Load Factors—Strength Design

Load Combination No.	Load Combination	Description
1	$1.4\,(D_o)$	Operating weight
2	$1.2(D_o) + 1.6L$	Operating weight + live load
3	$1.2(D_o) + (1.6W \text{ or } 1.0E_o{}^a)$	Operating weight + wind or earthquake
4	$0.9(D_e) + 1.6W$	Empty weight + wind (wind uplift case)
5a	$0.9(D_o) + 1.0E_o{}^a$	Operating weight + earthquake (earthquake uplift case)
5b	$0.9(D_e) + 1.0E_e{}^a$	Empty weight + earthquake (earthquake uplift case)
6	$0.9(D_f) + 1.6W_p$	Erection weight + partial wind[b] (wind uplift case)
7	$1.4(D_t)$	Test weight
8	$1.2(D_t) + 1.6W_p$	Test weight + partial wind

[a] For skirt-supported vertical vessels and skirt-supported elevated tanks classified as occupancy category IV in accordance with SEI/ASCE 7-05, the critical earthquake provisions and implied load combination of SEI/ASCE 7-05 shall be followed.

[b] Erection weight + partial wind is required only if the erection weight of the vessel is significantly less than the empty weight of the vessel. This case is only applied if the erection weight of the vessel is significantly less than the empty weight of the vessel.

3.3.1 Load Combinations

Load combinations are provided to cover all cases of probability of combination for any types of loads to the structure at the same time along its lifetime. For the industrial projects, the specific combination should be studied carefully and reviewed in every project, as it is different from one mode of operation to another, as is the way of testing the vessel, the piping, and the construction procedure itself. In this book, the load combination will be according to ASCE 7-05. If the owner has not specified any another code, it is preferable to use this standard and the load combination in this section may be considered as a guideline to the load combinations for specific types of structures in both allowable stress design (ASD) and load resistance.

TABLE 3.25

Loading Combinations—Allowable Stress Design (Service Loads)

Load Combination No.	Load Combination	Allowable Stress Multiplier	Description
1	$D_o + (T \text{ or } F_f)^b$	1.00	Operating weight + thermal expansion or friction force
2	$D_o + L + (T \text{ or } F_f)^b$	1.00	Operating weight + live load + thermal expansion or friction force
3	$D_o + (W \text{ or } 0.7E_o)^a$	1.00	Operating weight + wind or earthquake
4	$D_e + W$	1.00	Empty weight + wind (wind uplift case)
5a	$0.9(D_o) + 0.7E_o$	1.00	Operating weight + earthquake (earthquake uplift case)
5b	$0.9(D_e) + 0.7E_e$	1.00	Empty weight + earthquake (earthquake uplift case)
6	$D_f + W_p{}^c$	1.00	Erection weight + partial wind (wind uplift case)
7	$D_t + W_p$	1.20	Test weight + partial wind (for horizontal vessels only)
8	$D_e{}^d + B_p$	1.00	Empty weight + bundle pull (for heat exchangers only)

[a] Wind and earthquake forces shall be applied in both transverse and longitudinal directions, but shall not necessarily be applied simultaneously.

[b] The design thermal force for horizontal vessels and heat exchangers shall be the lesser of T or F_f.

[c] Erection weight + partial wind is required only if the erection weight of the vessel or exchanger is significantly less than the empty weight of the vessel or exchanger.

[d] Heat exchanger empty dead load will be reduced during bundle pull because of the removal of the exchanger head.

Allowable Stress Design

1. The noncomprehensive list of typical load combinations for each type of structure provided here shall be considered and used as applicable.

2. Engineering judgment shall be used in establishing all appropriate load combinations.

3. The use of a one-third stress increase for load combinations, including wind or earthquake loads, shall not be allowed for designs using the American Institute of Steel Construction (AISC) ASD.

4. Steel structures in seismic design category D or higher shall use factored load combinations as specified in ANSI/AISC 341-02, Part III (Allowable Stress Design Alternative).

TABLE 3.26

Loading Combinations and Load Factors—Strength Design

Load Combination No.	Load Combination	Description
1	$1.4(D_o) + 1.4(T$ or $F_f)^b$	Operating weight + thermal expansion or friction force
2	$1.2(D_o) + 1.6L + 1.2(T$ or $F_f)^b$	Operating weight + live load + thermal expansion or friction force
3	$1.2(D_o) + (1.6W$ or $1.0E_o)^a$	Operating weight + wind or earthquake
4	$0.9(D_e) + 1.6W$	empty weight + wind (wind uplift case)
5a	$0.9(D_o) + 1.0E_o$	Operating weight + earthquake (earthquake uplift case)
5b	$0.9(D_e) + 1.0E_e$	Empty weight + earthquake (earthquake uplift case)
6	$0.9(D_f)^c + 1.6W_p$	Erection weight + partial wind (wind uplift case)
7	$1.4(D_t)$	Test weight (for horizontal vessels only)
8	$1.2(D_t) + 1.6W_p$	Test weight + partial wind (for horizontal vessels only)
9	$1.2(D_e^d) + 1.6B_p$	Empty weight + pundle pull (for heat exchangers only)
10	$0.9(D_e^d) + 1.6B_p$	Empty weight + bundle pull (for heat exchangers only) (bundle pull uplift case)

[a] Wind and earthquake forces shall be applied in both transverse and longitudinal directions, but shall not necessarily be applied simultaneously.

[b] The design thermal force for horizontal vessels and heat exchangers shall be the lesser of T or F.

[c] This case only applies if the erection weight of the vessel is significantly less than the empty weight of the vessel.

[d] Heat exchanger empty dead load will be reduced during bundle pull because of the removal of the exchanger head.

The dead load factor used for the seismic uplift ASD load combinations is generally taken as 0.9. This factor is greater than the 0.6 dead load factor used in the ASD load combinations in Section 2 of ASCE 7-02, because the dead loads of nonbuilding structures are known to a higher degree of accuracy than are the corresponding dead loads of buildings. A dead load factor of 0.9 instead of 1.0 is used to account for the effect of vertical seismic forces.

This reduction is necessary because foundations sized using ASD loads, except for foundations for ground-supported storage tanks, are generally not required to consider the effect of vertical seismic forces. A dead load factor of 1.0 is used for the wind uplift ASD load combinations because of the higher accuracy of dead loads of nonbuilding structures.

Load Resistance Factor Design

1. The noncomprehensive list of typical factored load combinations for each type of structure provided shall be considered and used as applicable.

2. Engineering judgment shall be used in establishing all appropriate load combinations.

TABLE 3.27

Loading Combinations—Allowable Stress Design (Service Loads)

Load Combination No.	Load Combination	Allowable Stress Multiplier	Description
1	$D_s + D_o + F_f + T + A_f$	1.00	Operating weight + friction force + thermal expansion + anchor force
2	$D_s + D_o + A_f + (W^a$ or $0.7E_o{}^b)$	1.00	Operating weight + anchor + wind or earthquake
3	$D_s + D_e{}^c + W^a$	1.00	Empty weight + wind (wind uplift case)
4a	$0.9(D_s) + 0.6(D_o) + A_f + 0.7E_o{}^d$	1.00	Operating weight + earthquake (earthquake uplift case)
4b	$0.9(D_s + D_e{}^c) + 0.7E_e$	1.00	Empty weight + earthquake (earthquake uplift case)
5	$D_s + D_t + W_p$	1.20	Test weight + partial wind[e]

[a] Considerations of wind forces are normally not necessary in the longitudinal direction because friction and anchor loads will normally govern.
[b] Earthquake forces shall be applied in both transverse and longitudinal directions, but shall not necessarily be applied simultaneously.
[c] $0.6D_o$ is used as a close approximation of the empty pipe condition D_e.
[d] Full $D_s + D_o$ value shall be used for the calculation of E_o in load combination 4a.
[e] Test weight + partial wind normally is required only for local member design because test is not typically performed on all pipes simultaneously.

3. The following load combinations are appropriate for use with the strength design provisions of either AISC LRFD (third edition or later).

For all the load combinations, the value of the structure dead load, D_S—the weight of material forming the structure—will be included in other values of the dead load that is most applicable, in case only the operating load and empty load will be available. The foundation weight of uplift should be calculated and added as a dead load to resist the uplift and this will be illustrated in Chapter 6.

3.3.1.1 Vertical Vessels

Load combinations affecting the vertical vessel are shown in Table 3.23 as a guide (see also Table 3.24). The design of the vertical vessel is illustrated in Chapter 6.

3.3.1.2 Horizontal Vessels and Heat Exchangers

Designers of heat exchangers and horizontal vessels, such as the separator knockout drums and others, should use the load combinations for these vessels given in Table 3.25 (see also Table 3.26).

TABLE 3.20

Loading Combinations and Load Factors Strength Design

Load Combination No.	Load Combination	Description
1	$1.4(D_o + F_f + T + A_f)$	Operating weight + friction force + thermal expansion + anchor
2	$1.2(D_o + A_f) + (1.6W^a \text{ or } 1.0E_o{}^b)$	Operating weight + anchor + wind or earthquake
3	$0.9(D_e{}^c) + 1.6W^a$	Empty weight + wind (wind uplift case)
4a	$0.9(D_o) + 1.2(A_f) + 1.0E_o$	Operating weight + earthquake (earthquake uplift case)
4b	$0.9(D_e{}^c) + 1.0E_e$	Empty weight + earthquake (earthquake uplift case)
5	$1.4(D_t)$	Test weight
6	$1.2(D_t) + 1.6W_p$	Test weight + partial wind[d]

[a] Considerations of wind forces are normally not necessary in the longitudinal direction because friction and anchor loads will normally govern.

[b] Earthquake forces shall be applied in both transverse and longitudinal directions, but shall not necessarily be applied simultaneously.

[c] The value of $0.6D_o$ is used as a close approximation of the empty pipe condition D_e.

[d] Test weight + partial wind normally is required only for local member design because the test is not typically performed on all pipes simultaneously.

3.3.1.3 Pipe Rack and Pipe Bridge Design

See Tables 3.27 and 3.28.

3.3.1.4 Ground-Supported Storage Tank Load Combinations

Load combinations for ground-supported storage tanks shall be taken from the API Standard 650 and modified for use with SEI/ASCE 7-05 loads and PIP nomenclature (Table 3.29).

Load combinations for static machinery, skid and modular equipment, filters, and other equipment shall be similar to the load combinations for vertical vessels.

3.3.2 Test Combinations

Engineering judgment shall be used in establishing the appropriate application of test load combinations to adequately address actual test conditions in accordance with project and code requirements, while avoiding overly conservative designs. Consideration shall be given to the sequence and combination of testing for various equipment, vessels, tanks, and piping systems supported on common structures, pipe racks, or foundations. Full wind

TABLE 3.29

Loading Combinations—Allowable Stress Design (Service Loads)

Load Combination No.	Load Combination	Description
1	$D_o + P_i$	Operating weight + internal pressure[a]
2	$D_t + P_t$	Test weight + test pressure
3	$(D_e \text{ or } D_o) + W + 0.4P_i$[b]	Empty or operating weight + wind + internal pressure[a]
4	$(D_e \text{ or } D_o) + W + 0.4P_e$[b]	Empty or operating weight + wind + external pressure
5	$D_o + (L \text{ or } S) + 0.4P_e$[b]	Operating weight + live or snow + external pressure
6	$(D_e \text{ or } D_o) + 0.4(L \text{ or } S) + P_e$	Empty or operating weight + live or snow + external pressure
7	$D_o + 0.1S + E_o$[c] $+ 0.4P_i$[b]	Operating weight + snow + earthquake + internal pressure[a] (earthquake uplift case)
8	$D_o + 0.1S + E_o$[c]	Operating weight + Snow + Earthquake

[a] For internal pressures sufficient to lift the tank shell according to the rules of API Standard 650, tank, anchor bolts, and foundation shall be designed to the additional requirements of API Standard 650 Appendix F.7.

[b] If the ratio of operating pressure to design pressure exceeds 0.4, the owner shall consider specifying a higher factor on design pressure in load combinations 3, 4, 5, and 7 of this table.

[c] Earthquake loads for API Standard 650 tanks taken from ASCE/SEI 7 "bridging equations" or from API Standard 650 already include the 0.7 ASD seismic load factor.

and earthquake loads are typically not combined with test loads unless an unusually long test duration is planned (i.e., if a significant probability exists that the partial wind velocity will be exceeded or an earthquake event may occur). Additional loading shall be included with the test if specified in the contract documents. For ASD, a 20% allowable stress increase shall be permitted for any test load combination. For ultimate strength and limit states design, no load factor reduction shall be permitted for any test load combination.

References

ACI 318-02, 2002. Building code requirements for structural concrete, *ACI Manual of Concrete Practice*. Farmington Hills, MI: ACI.

AISC 341-97. 1997. *Seismic Provisions for Structural Steel Buildings*, including Supplement No. 1. Part I: February 15, 1999. Part II: July 1997. Chicago, IL: AISC.

AISC 316-89. 1989. *Manual of Steel Construction: Allowable Stress Design*. Chicago, IL: AISC.

AISC LRFD 325-01. *Manual of Steel Construction*, 3rd Edition. Chicago, IL: AISC.

ASCE 40262. 1997. *Wind Load and Anchor Bolt Design for Petrochemical Facilities*. Ann Arbor, MI: ASCE.

ASCE. ASCE 7–02. 2002. *Minimum Design Loads for Buildings and Other Structures*. Ann Arbor, MI: ASCE.

ASCE. SEI/ASCE 7 –05, 2005: *Minimum Design Loads for Buildings and Other Structures*. Ann Arbor, MI: ASCE.

ASME Standard B31.8. 1995. *Gas Transmission and Distribution Piping Systems*. New York, NY: ASME.

ASME Standard. 2000. *Boiler and Pressure Vessel Code*, including Addenda. New York, NY: ASME.

ASTM Standard D2166-91. 1991. Test Method for Unconfined Compressive Strength of Cohesive Soil, *ASTM Annual Book of Standards*. West Conshohocken, PA: ASTM International.

ASTM Standard D2850-87. 1987. Test Method for Unconsolidated, Undrained Compressive Strength of Cohesive Soils in Triaxial Compression, *ASTM Annual Book of Standards*. West Conshohocken, PA: ASTM International.

ASTM D1586-84 Standard Test Method for Standard Penetration Test (SPT) and Split-Barrel Sampling of Soils.

SEI/ASCE 3702. Design Loads on Structures during Construction Standard.

4

Design of Foundations for Vibrating Equipment

4.1 Introduction

The design of foundations and structures subjected to vibratory loads is considered to be a complex process presenting problems involving structural and geotechnical engineering and the theory of vibration. The structural form of a machine foundation is generally determined by the information provided by the geotechnical consultant and the machine's manufacturer. However, it is necessary during the design phase to adjust the dimensions of the foundation to meet design criteria or to accommodate other fixed facilities such as pipelines, other foundations, or neighboring buildings.

The foundation may be shallow or deep like the pile, which is the most well-known type of foundation and is used in cases of weak soil. There are two other structural configurations of foundations:

1. A block-type foundation consists of a thick slab of concrete fixed directly to the machine and other fixed auxiliary equipment. This type of foundation is the most traditional in oil and gas processing plants and petrochemical plants.
2. An elevated pedestal foundation consists of a base-slab and vertical columns supporting a grid of beams at the top, on which rests skid-mounted machinery. This type of support is traditionally used in electrical power station projects.

4.2 Machine Requirements

There are two large groups of machines: centrifugal and reciprocating machines. It is essential to obtain a machine's technical data from the manufacturer in order to design a foundation for it. The required machine data, properties, and parameters include the following:

- Outline drawings of the machine assembly
- Function of the machine
- Weight of the machine and the weight of the rotor
- Location of the center of gravity in the space x, y, and z
- Speed ranges of the machine or the free frequency of unbalanced primary and secondary forces
- Magnitude and direction of unbalanced forces, both vertically and horizontally, and their points of application
- Limits imposed on the foundation with respect to differential deflection between points on the planned area
- Foundation requirements

The magnitude and direction of the unbalanced forces are often not available from the machine manufacturer, as many manufacturers maintain that their centrifugal machines are perfectly balanced. For reciprocating machines, the unbalanced forces are generally of considerable magnitude and are provided by manufacturers. The settlement should have a limit to avoid any damage to the pipes that are fixed to the machine. Limits of differential settlement in the case of high-pressure gas 50,000-psi piping should be less than 0.0025 mm (0.0001 in.).

4.3 Foundation Design

The main physical concept in designing foundations for vibrating equipment is based on the logical phenomena that a lightweight body cannot vibrate a heavyweight body. So, one must design a machine foundation that has a higher weight than that of the machine it will support. This chapter presents how to choose the distribution of this weight based on reasonable dimensions.

4.3.1 Trial Foundation Sizing Guidelines

Arya et al. (1984) provide the following guidelines, which are very useful for choosing the dimensions of the foundation and for checking the analysis that is acceptable in most cases.

- A rigid block-type foundation resting on soil should have a weight of 2 to 3 times the weight of a centrifugal-type machine. For a reciprocating machine type, the foundation weight should be from 5 to 7 times the weight of the machine.

- The top of the foundation is usually 300 mm above the finished floor or pavement elevation to avoid damage from surface water runoff due to rain or washing of the facilities.

- In any case, the depth of the foundation should not be less than 600 mm. The depth is generally verified by the required foundation weight. The depth of the foundation is governed by the length of anchor bolts.

- The foundation should be wide enough to increase damping in rocking mode. Therefore, the width should be at least 1 to 1.5 times the vertical dimension from the machine centerline to the foundation base.

- After obtaining the physical dimensions of the machine, you can now define the foundation length, including a clearance of about 300 mm from the machine edge to the foundation edge for maintenance purposes.

- The length and width of the foundation should be adjusted to verify that the center of gravity of the machine plus the equipment coincides with the foundation's center of gravity.

The design of a block foundation resting on piles should include the following precautions:

1. The pile cap weight should be 1.5 to 2.5 times the machine weight for centrifugal machines and from 2.5 to 4 times the machine weight for reciprocating machines.

2. The pile should be arranged so as to ensure that the centroid of the pile group coincides with the center of gravity of the combined structure and machine loads.

Figure 4.1 presents a diesel generator package considered to be a type of reciprocating machine.

Most industries use a centrifugal machine in different applications, such as compressors or pumps. Figure 4.2 shows a gas turbine generator.

In a gas turbine generator, the vibration is usually lower than other applications. Figure 4.3 illustrates the process of fixing the skid to concrete. As you can see, there are two bolts—one is the anchor bolt and the other is located under a steel plate that is fixed during the pouring of the concrete. When rotated, the bolt will rest on this steel plate and the level of the machine may be adjusted during installation by moving the bolt.

FIGURE 4.1
Diesel generator.

4.3.2 Foundation Dynamic Analysis

Machinery foundations shall be designed based on the equipment manufacturer's recommendations, published design procedures, and criteria for dynamic analysis.

The foundation under machine will be considered a single degree of freedom when performing the dynamic analysis and the summary of the derived expression is shown in Table 4.1.

If the equipment manufacturer's vibration criteria are not available, the maximum velocity of movement during steady-state normal operation shall be limited to 0.12 in./s (3.0 mm/s) for centrifugal machines and 0.15 in./s

FIGURE 4.2
Gas turbine generator.

FIGURE 4.3
Fixing the skid to concrete.

(3.8 mm/s) for reciprocating machines. Any support structures or foundations for centrifugal machinery greater than 500 hp shall be designed for the expected dynamic forces using dynamic analysis procedures.

For centrifugal machinery less than 500 hp, in the absence of a detailed dynamic analysis, the foundation weight shall be designed to be at least 3 times the total machinery weight, unless specified otherwise by the equipment manufacturer. When using reciprocating machinery less than 200 hp,

TABLE 4.1

Summary of Derived Expressions for a Single-Degree-of-Freedom System

Expression	Constant Force Excitation F_o Constant	Rotating Mass Type Excitation $F_o = me\omega^2$
Magnification factor	$M = \dfrac{1}{\sqrt{(1-r^2)^2 + (2Dr)^2}}$	$M = \dfrac{r^2}{\sqrt{(1-r^2)^2 + (2Dr)^2}}$
Amplitude at frequency, f	$Y = M(F_o/k)$	$Y = M_r(m_1 e/m)$
Resonant frequency, f_{mr}	$f_{mr} = f_n\sqrt{1-2D^2}$	$f_{mr} = \dfrac{f_n}{\sqrt{1-2D^2}}$
Amplitude at frequency, f_r	$Y_{mnr} = \dfrac{(F_o/k)}{2D\sqrt{1-D^2}}$	$Y_{mnr} = \dfrac{(m_1 e/m)}{2D\sqrt{1-D^2}}$
Transmissibility factor	$T = \dfrac{\sqrt{1+(2Dr)^2}}{\sqrt{(1-r^2)^2 + (2Dr)^2}}$	$T = \dfrac{r^2\sqrt{1+(2Dr)^2}}{\sqrt{(1-r^2)^2 + (2Dr)^2}}$

Note: Where $r = \omega/\omega_n$, ω_n is the undamped natural circular frequency (in $k/m)^{0.5}$; D is the damping ratio (C/C_c); C_c is the critical damping $[2(km)^{0.5}]$; T_r is the force transmitted/ F_o, and T_r is the force transmitted/$m_1 e\omega_n^2$.

in the absence of a detailed dynamic analysis, the foundation weight shall be designed to be at least 5 times the total machinery weight, unless otherwise specified by the manufacturer.

4.3.3 Soil Parameter

The allowable soil bearing or pile capacity for foundations designed for dynamic load equipment shall be a maximum of half what is normally allowable for static loads. The allowable soil bearing will be provided by the geotechnical consultant's office, which will perform soil investigation tests as illustrated in Chapter 13. Table 4.2, from *Installation of Waukesha Engines and Generator Systems Manual* (Dresser Waukesha, 2004), provides us with a guideline for the allowable soil-bearing capacity.

The maximum eccentricity between the center of gravity of the combined weight of the foundation and machinery and the bearing surface shall be 5% in each direction.

Structures and foundations that support vibrating equipment shall have a natural frequency that is outside the range of 0.80 to 1.20 times the exciting frequency. For this type of foundation, there are some required soil parameters for vibrating equipment:

- The allowable bearing capacity
- The soil density, ρ
- The shear modulus, G
- The shear wave velocity shall be calculated from the formula:

TABLE 4.2

Expected Approximate Values for Soil Bearing Capacity Based on Soil Type

Nature of Soil	Safe Bearing Capacity (ton/m²)
Hard rock—granite, etc.	240–980
Medium rock, shale, etc.	100–150
Hard pan	80–100
Soft rock	50–100
Compacted sand and gravel	50–60
Hard clay	40–50
Gravel and coarse sand	40–50
Loose, medium, and coarse sand, compacted fine sand	30–40
Medium clay	20–40
Loose fine sand	10–20
Soft clay	15

Source: Dresser Waukesha, *Installation of Waukesha Engines and Generator Systems,* Dresser Waukesha, Waukesha, WI, 2004.

$$V_s = \sqrt{\frac{G}{\rho}}$$

- The dynamic coefficient of subgrade reaction for vertical vibration, K_z

For a rigid rectangular footing of plan area, A, on a semi-infinite elastic half-space (Barkan, 1962; Richard et al., 1970)

$$K_z = \frac{B_z}{\sqrt{A}}\left(\frac{G}{(1-v)}\right)$$

where
  V = Poisson's ratio and is equal to 0.3 as a typical value for the sand
  G = Shear modulus
  B_z = Ratio of the length to breadth of footing of about 1.40

Barkan suggests that, for an allowable static bearing capacity of about 0.15 N/mm², K_z should be not less than 0.03 N/mm³:

- Coefficient of subgrade reaction for horizontal vibrations, K_x

$$K_x = 0.5K_z$$

- Coefficient of subgrade reaction for rocking vibrations, K_ϕ

$$K_\phi = 2K_z$$

- Coefficient of subgrade reaction for torsion vibrations, K_θ

$$K_\theta = 0.75K_z$$

Table 4.3 provides the equations to calculate the effective radius of footing in cases of horizontal and vertical vibration, rocking, and torsional vibration.

TABLE 4.3

Effective Radius of Footing, r_0

Vertical and horizontal vibration	$\sqrt{BL/\pi}$
Rocking	$\sqrt[4]{(BL^3)/3\pi}$
Torsional	$\sqrt[4]{\dfrac{BL(B^2+L^2)}{6\pi}}$

Note: B, width of foundation; L, length of foundation.

TABLE 4.4
Main Parameters

Vibration Mode	Mass Ratio (B)	Damping Parameters	Damping Ratio	Spring Constant, K	Natural Frequency, f_n
Vertical	$B_z = \dfrac{(1-v)m}{4\rho r_o^3}$	$c_z = \dfrac{3.4 r_o^2 \sqrt{\rho G}}{1-v}$	$D_z = \dfrac{0.425}{\sqrt{B_z}}$	$k_z = \dfrac{4Gr_o}{1-v}$	$f_{nz} = \dfrac{1}{2\pi}\sqrt{\dfrac{k_z}{m}}$
Horizontal	$B_x = \dfrac{(7-8v)m}{32(1-v)\rho r_o^3}$	$c_x = \dfrac{18.4(1-v)\sqrt{\rho G}}{7-8v}$	$D_x = \dfrac{0.288}{\sqrt{B_x}}$	$k_x = \dfrac{32(1-v)Gr_o}{7-8v}$	$f_{nx} = \dfrac{1}{2\pi}\sqrt{\dfrac{k_x}{m}}$
Rocking	$B_\psi = \dfrac{3(1-v)m}{8\rho r_o^3}$	$c_\psi = \dfrac{0.8 r_o^4 \sqrt{\rho G}}{(1-v)(1+B_\psi)}$	$D_\psi = \dfrac{0.15}{(1+B_\psi)\sqrt{B_\psi}}$	$k_\psi = \dfrac{8Gr_o^3}{3(1-v)}$	$f_{n\psi} = \dfrac{1}{2\pi}\sqrt{\dfrac{k_\psi}{I_\psi}}$
Torsional	$B_\theta = \dfrac{I_\theta}{\rho r_o^5}$	$c_\theta = \dfrac{4\sqrt{B_\theta \rho G}}{1+2B_\theta}$	$D_\theta = \dfrac{0.50}{1+2B_\theta}$	$k_\theta = \dfrac{16}{3}Gr_o^3$	$f_{n\theta} = \dfrac{1}{2\pi}\sqrt{\dfrac{k_\theta}{I_\theta}}$

Table 4.4 provides a summary by which to calculate the mass ratio, damping ratio, spring constant, and natural frequency for different modes of vibration based on the foundation dimensions and the geotechnical data from the soil investigation tests.

Example 1: Centrifugal Machine

The machine data are given below.

Machine Data
Compressor weight, W_c = 35,270 lb = 16 tons
Rotor weight, W_r = 2100 lb = 0.952 tons
Operating speed, ω = 6949 rpm; (f) = 727.7 rad/s
Critical speed, ω_c = 3400 and 9000 rpm
Eccentricity of unbalanced mass, e' = 0.0015 in (0.0381 mm) (this value is provided by the manufacturer for the static condition)

It is worth noting that the value of eccentricity due to the unbalance of mass is usually not available from the manufacturer, but there were some studies performed in 1962 that presented some values that can assist us in predicting this value, as shown in Table 4.5.

The equation delivered by the American Petroleum Institute for centrifugal compressors is

$$e \text{ (mil)} = \alpha(12,000/\omega)0.5 < 1.0 \text{ mil}$$

where
α = 0.5 at installation time
α = 1.0 after several years of operation
1 mil = 0.001 in. = 0.0254 mm

The dynamic eccentricity $(e) = e'/(1 - (\omega/\omega_c)^2)$ = 0.000472 in = 0.01199 mm
Centrifugal force $F_o = (W_r/g)ef^2$ = 0.616 tons

Turbine Data
Weight, W_t = 7.26 tons
Rotor weight, W_r = 0.247 tons
Operating speed, ω = 6949 rpm

TABLE 4.5

Design Eccentricity for Centrifugal Machines

Operating Speed (rpm)	Eccentricity (mm)
750	0.356–0.812
1500	0.203
3000	0.051

Source: API Standard 617.

Critical speed, ω_c, at 2000 and 9000 rpm
Dynamic eccentricity at operating speed:

$$e = 0.0381/[1 - (6949/2000)^2] = 0.00343 \text{ mm}$$

Centrifugal force, F_o = 0.0458 tons
Total centrifugal force = 0.616 + 0.0458 = 0.662 tons
Skid weight, W_b = 2.27 tons
Total machine weight = $W_c + W_t + W_b$ = 16 + 7.26 + 2.27 = 25.53 tons

Soil Data
Soil density, g = 20 kN/m³
Shear modulus, G = 6500 psi = 44.82 N/mm²
Poisson's ratio, v = 0.45
Soil internal damping ratio, D = 0.05
Static allowable bearing capacity, q = 72 kN/m²

Selection of Foundation Criteria
The foundation data are shown in Figure 4.4.
Weight of the footing, W_f = 45.58 tons
Total static weight = machine weight + footing weight = 25.53 + 45.58 = 71.11 tons
Actual soil pressure = 71.11/3.8(6.0) = 3.11 ton/m²

$$<0.5 \ q_{all} \quad \text{OK}$$

W_f/W_m = 1.78, which is close to 2   OK

Dynamic Analysis
The axis of rotation of the shaft is located 6 ft (1.8 m) above the foundation bottom. The dynamic force will excite the structure in three different modes: vertical, horizontal, and rocking.

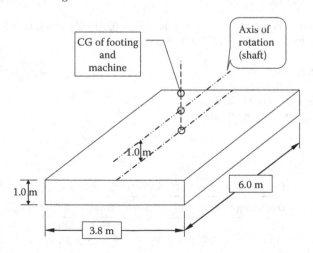

FIGURE 4.4
Sketch for centrifugal machine.

TABLE 4.6

Summary of the Dynamic Analysis for Three Modes of Failure

Item	Parameters	Source	Vertical	Horizontal	Rocking Oscillation
1	r_o	Table 4.3	$r_o = 2.71$ m	$r_o = 2.71$ m	$r_o = 2.44$ m
2	Mass	$M = W/g$	72.5 kN s²/m	72.5 kN s²/m	105,128
3	Mass ratio	Table 4.4	0.24	0.35	0.166
4	Geometric damping ratio	Table 4.4	0.868	0.494	0.361
5	Spring coefficient	Table 4.4	2.15	0.95	0.46
6	Equivalent spring constant	Table 4.4	57.85×10^4 lb/ft	40.77×10^6 lb/ft	2446.36×10^6 lb ft/rad
7	ω_n	Table 4.4	1040.9 rpm	873.9 rpm	1449.8 rpm
8	Resonance frequency, f_{mr}	Table 4.1	Not possible	1221.4 rpm	1686.1 rpm
9	Magnification factor, M	Table 4.1	0.022	0.016	0.045
10	Dynamic force		6.5 kN	6.5 kN	13 m kN
11	Vibration amplitude	Table 4.1	1.42×10^{-5} mm	1.45×10^{-5} mm	0.1611×10^{-6} rad
12	Component of rocking oscillation		2.558×10^{-5} mm	2.47×10^{-5} mm	
13	Resultant vibration amplitude		0.000019 in. (0.000483 mm)	0.000018 in. (0.000457 mm)	
14	Transmissibility factor, T_r		0.256	0.127	0.162

V—static condition check

1. Static bearing capacity is 50% of the allowable bearing capacity.
2. Static settlement must be uniform, as the *CG* of footing and machine coincides.
3. Bearing capacity for static plus dynamic loads will be less than 0.75 of the allowable bearing capacity.
4. The magnification factor should be less than 1.5.

Limiting Dynamic Condition

After calculating the dynamic analysis parameters, the level of vibration should be within the following limits. If it is beyond the limits, another trial of foundation dimensions should be performed.

1. By knowing the vibration amplitude and operating frequency, Figure 4.5 presents the allowable and safe limits for the amplitude and operating frequency (rpm). In our case, the vertical vibrating amplitude is 0.000019 in. at 6949 rpm. From the figure, we can determine that it is within safe limits.

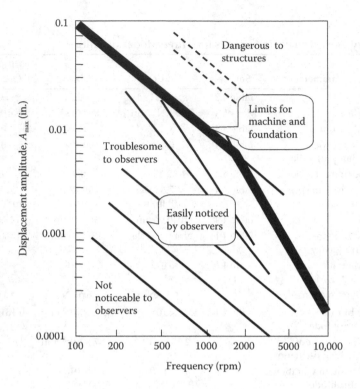

FIGURE 4.5
Frequency limits.

2. The velocity = $2\pi f$ (cps) × displacement amplitude as calculated above. Compare with the limiting value in Table 4.7 to determine that the velocity is equal to 0.0138 in./s and that this velocity falls within smooth operation parameters.
3. The magnification factor should be less than 1.5, which is applicable in this example.
4. The machine frequency should have at least a difference of ±20% from the resonance frequency.

$$0.8\,\omega_{mr} \geq \omega \geq 1.2\,\omega_{mr}$$

$$\omega/\omega_{mr} = 6949/1221.4 = 5.69 > 1.2 \quad \text{OK}$$

Example 2: Reciprocating Machine

The following data are for the machine. The data for the auxiliary equipment should be supplied by the machine vendor.

TABLE 4.7

General Machinery Vibration Severity Data

Horizontal Peak Velocity, in./s (mm/s)	Machine Operation
<0.005 (<0.127)	Extremely smooth
0.005–0.010 (0.127–0.245)	Very smooth
0.010–0.020 (0.254–0.508)	Smooth
0.020–0.040 (0.508–1.016)	Very good
0.04–0.080 (1.016–2.032)	Good
0.080–0.160 (2.032–4.064)	Fair
0.160–0.315 (4.046–8.00)	Slightly rough
0.315–0.630 (8.00–16.00)	Rough
>0.630 (>16.00)	Very rough

Source: Baxter, R. L., and Bernhard, D. L., *Vibration Tolerances for Industry*, ASME Paper 67-PEM-14, Plant Engineering and Maintenance Conference, Detroit, MI, April, 1967.

Machine Data
Compressor = 12.75 tons
Gas coolers = 1.972 tons
Scrubbers = 3.179 tons
Motor = 8.164 tons
Rotor weight = 2.721 tons (included in compressor weight)
Total machine load = 26.07 tons

Dynamic Forces

1. Compressor speed, primary (operating) = 585 rpm, secondary = 1170 rpm
2. Vertical, primary force, F_z = 0.603 tons
3. Vertical, secondary force F_z = 0.251 tons
4. Horizontal, primary force, F_x = 0.329 tons
5. Rocking, primary moment, T_ψ = 1.563 mt
6. Pitching, primary moment, T_ϕ = 4.70 mt
7. Pitching, secondary moment, T_ϕ = 1.71 mt

Soil and Foundation Parameters
Soil density = 18.7 kN/m³
Shear modulus, G = 96.53 N/mm²
Poisson's ratio, v = 0.35
Soil internal damping ratio, D = 0.05
Static allowable bearing capacity, q_{all} = 120 kN/m²
Permanent settlement of soil = 5 mm at 120 kN/m²

Trial Sizing for the Foundation
The trial-size footing, using the previous trial described in Section 4.3.1 as a guide, will be as shown in Figure 4.6.

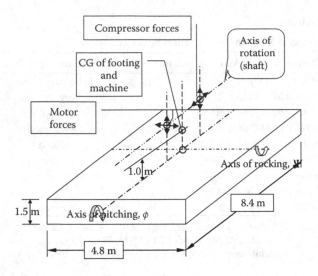

FIGURE 4.6
Sketch for reciprocating machine.

Weight of the footing, W_F = 147.35 tons
Total static load = machine weight + footing weight = 173.41 tons
Footing weight/machine weight = 5.65 > 5   OK
Actual soil pressure = 173.41/8.4(4.8) = 4.3 ton/m²

$$\leq q_{all} \text{ OK}$$

This trial dimension is OK.

1. Mass and mass moment of inertia

$$M = W/g$$

$$I_{\psi+\phi} = \sum_{1}^{n}\left[\frac{m_t}{12}(a_i^2 + b_i^2) + m_1 k_1^2\right]$$

where
m_1 = machine mass
k_1 = distance from machine CG and bottom of the foundation
m_t = foundation mass

Vertical Excitation (z Direction)	Horizontal Direction (x Direction)	Rocking Excitation (ψ Direction)	Pitching Excitation (φ Direction)
m = 173.41 kN s²/m	m = 173.41 kN s²/m	I_ψ (machine) = 162.9 I_ψ (footing) = 976.93 Total = 1139.83 kN s²/m	I_ϕ (machine) = 162.9 I_ϕ (footing) = 393.4 Total = 556.3 kN s²/m

2. Spring constant

   This data will be calculated from the following table:
   In rock direction, $L = 8.4$ m and $B = 4.8$ m
   In pitching direction, $L = 4.8$ m and $B = 8.4$ m

Parameter	Vertical Excitation (z Direction)	Horizontal Direction (x Direction)	Rocking Excitation (ψ Direction)	Pitching Excitation (ϕ Direction)
Equivalent radius, r_o	3.58 m	3.58 m	4.16 m	3.15 m
Embedment factor, η	1.1	1.232	1.175	1.235
Spring constant coefficient, B	2.15	0.95	0.58	0.45
Equivalent spring constant, k	227,167 t/m'	197,302 t/m'	3481×10^4 m.t/rad	1626×10^4 m.t/rad

3. Damping ratio

Parameter	Vertical Excitation (z Direction)	Horizontal Direction (x Direction)	Rocking Excitation (ψ Direction)	Pitching Excitation (ϕ Direction)
Embedment factor, α	1.254	1.623	1.024	1.041
Mass ratio, B	0.328	0.408	0.115	0.222
Effective damping coefficient			1.6	1.58
Geometrical damping ratio, D	0.931	0.732	0.302	0.195
Internal damping, D	0.05	0.05	0.05	0.05
Total damping	0.981	0.782	0.352	0.245

4. Natural frequency

$$f_n = \frac{60}{2\pi}\sqrt{\frac{k}{m}}$$

5. Machine parameter
   $F_{zo}(P) = 0.603$ tons
   $F_{zo}(S) = 0.25$ tons
   $F_{xo}(p) = 0.329$ tons
   $T_{\psi o}(P) = 1.563 + 0.329 \times 2.44 = 2.366$ mt
   $T_{\phi o}(P) = 4.7$ mt
   $T_{\phi o}(S) = 1.71$ mt

Displacement Response
The displacement response will be calculated from the following equations:

In vertical direction, z

$$Z = \sum M_z \cdot F_{zo}/K_z$$

In horizontal direction, x

$$X = \sum M_x \cdot F_{xo}/K_x$$

In rocking excitation

$$\Psi = \sum M_\psi \cdot F_{\psi o}/K_\psi$$

In pitching excitation

$$\phi = \sum M_\phi \cdot F_{\phi o}/K_\phi$$

Total displacement in z direction = $Z + \psi \cdot L/2 + \phi \cdot B/2$

Parameter	Vertical Excitation (z Direction)	Horizontal Direction (x Direction)	Rocking Excitation (ψ Direction)	Pitching Excitation (φ Direction)
Natural frequency	1082.8	1009.1	1677.6	1611.2
Resonance frequency	$2D^2 > 1$	$2D^2 > 1$	$f_{m\psi} = 1934.3$	$f_{m\phi} = 1717.6$
Vibrating force	$F_{zo}(P) = 0.603$ tons $F_{zo}(S) = 0.25$ tons	$F_{xo}(p) = 0.329$ tons	$T_{\psi o}(P) =$ 2.366 mt	$T_{\phi o}(P) = 4.7$ mt $T_{\phi o}(S) = 1.71$ mt
Magnification factor	M_z, max not possible	M_x, max not possible	$M_\psi(P) = 1.111$ at 585 rpm $M_\psi(S) = 1.454$ at 1170 rpm	$M_\phi(P) = 1.128$ at 585 rpm $M_\phi(S) = 1.690$ at 1170 rpm
Displacement response	0.0026 mm × 10^{-2} in.	0.0015 mm × 10^{-2} in.	0.8128 × 10^{-6} rad	5.0369 × 10^{-6} rad
Total displacement	0.018 mm × 10^{-3} in.	0.011 mm × 10^{-3} in.		

Limiting the Dynamic Condition

1. The maximum vibrations for the horizontal and vertical dimensions are 0.000717 in. (0.0182 mm) and 0.000443 in. (0.01125 mm), respectively. If you check Figure 4.5, you will find that these numbers fall within the safe zone at the operating frequency of 585 rpm.
2. The velocity is equal to $2\pi f/(cps) \times$ displacement as calculated above and compared to the limiting values in Table 4.7.

3. Acceleration is calculated as $4\pi^2f^2/(\text{cps}) \times$ displacement. This check should be considered if the above checks are not satisfactory and the acceleration should be compared with the same figure.
4. The magnification factor should be less than 1.5, which was used in this example.
5. The acting frequency of the machine should have a difference of at least ±20 from the resonance frequency. For horizontal and vertical directions, there is no resonance.

For rocking mode:

$$0.8\, f_m = 0.8 \times 1934.4 = 1547.4 \text{ rpm}$$

$$1.2\, f_m = 1.2 \times 1934.4 = 2321.2 \text{ rpm}$$

The primary and secondary frequencies are 585 and 1170 rpm, respectively, and are considered safe and without resonance.

For pitching mode:

$$0.8\, f_m = 0.8 \times 1717.6 = 1374.1 \text{ rpm}$$

$$1.2\, f_m = 1.2 \times 1717.6 = 2061.1 \text{ rpm}$$

The primary and secondary frequencies are 585 and 1170 rpm, respectively, and are considered safe and without resonance.

6. The transmissibility factor is usually considered for high-frequency machines mounted on a spring. The transmissibility factor should be less than 1.

Example 3: Design of Steel Structure Support for Machine Skid

Check the steel frame support gas generator package with the following data:

Dry weight = 12.38 tons
Wet weight = 13.02 tons
Skid dimension = 5842 × 1067 mm
Machine has 900 rpm
Beam section = W10 × 22 with steel ASTM A36 with $F_y = 240$ N/mm²

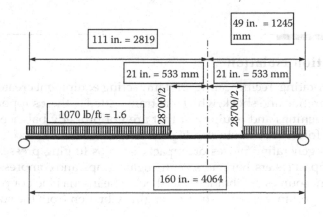

Maximum beam design moment = 4.07 mt
Maximum beam stress = $4.07/3.8 \times 10^{-4}$ = 10,705.5 t/m^2 = 105 N/mm^2
Allowable beam stress = 0.66×240 = 158 N/mm^2
Unity check ratio, UC = 105/158 = 0.66 < 1.0   OK

A local check has been performed to ensure that no resonance will occur when sitting the new gas generator package down over the deck.

$$\Delta = \frac{Pa^2 b^2}{3EIL}$$

Beam stiffness:

$$(K) = P/\Delta = 3EIL \big/ a^2 b^2$$

$$= 3 \times 29 \times 106 \times 118 \times 160 / (111)^2 (49)^2$$

$$= 55,525 \text{ lb/in.}$$

$$T_n = 2\pi \sqrt{\frac{M}{K}} = 0.103 \text{ s}$$

$$\text{Mass } (M) = W/g = 13.02 \text{ kN s}^2/\text{m}$$

$$f_n = 1/T_n = 4.35 \text{ Hz}$$

$$\omega_n = 2\pi f_n = 27.32 \text{ rad/s} = 1640 \text{ rpm}$$

The new generator frequency (ω) = 900 rpm
The frequencies ratio is:

$$\omega/\omega_n = 900/1640 = 0.55 < 0.8 \quad \text{OK}$$

4.4 Vibration Isolation

Vibrating, rotating, reciprocating, and impacting equipment create machine-induced vibration and shock, which are transmitted to their support systems. Rotating machines and equipment that are not properly balanced produce centrifugal forces that create steady-state and random vibration.

Machines generating pulses or impacts, such as forging presses, injection molding, impact testers, hammers, centrifugal pumps, and compressors are the predominant sources of vibration and shock. The inertia block or pad of isolation is an important factor in isolating engine vibration from the surrounding structure. However, the proper design of the foundation is not enough. There are several additional techniques that can be used to isolate the vibration.

4.4.1 Isolating Liners

A liner can be fabricated and used to line the pit into which the concrete inertia block is poured, as shown in Figure 4.7. A number of suitable liners are available commercially.

It is important to consult the liner manufacturer for specific information. The principle for all liners is the same: Line the bottom and sides of the pit and pour the concrete inertia block inside the isolator lining. The engine or common mounting skid will still vibrate, but the vibration is dampened and largely confined within the liner.

Be sure to construct the liner so that no liquid concrete can flow into the gaps between the liner slabs. If concrete seeps between the inertia block and the pit, the vibration absorption value of the liner will be greatly reduced.

Other materials such as sand or gravel may be used as isolating mediums. One foot of well-tamped, settled gravel under the inertia block will be satisfactory. Do not bridge the gap between the inertia block and the surrounding floor with concrete or similar solid material. If, for reasons of neatness or appearance, it is necessary to close this gap, use an expansion joint or a similarly resilient material.

4.4.2 Spring and Rubber Mounts

Based on Fabreeca (2009), spring and rubber mounts of various sizes and resiliencies are available for installation purposes. These mounts can be positioned

FIGURE 4.7
Liner isolation before concrete pouring.

between the common skid and the inertia block or pad or between the inertia block and bottom of the pit. As with the isolating liners, we recommend contacting the manufacturer of the mounts for specific instructions.

Isolation systems have natural frequencies as low as 6.0 Hz and can be designed to exhibit low or high internal damping. These solutions are used under large concrete foundations supporting heavy machinery, buildings, measuring machines, roller mills, and similar equipment. Isolators are available in two types to suit a wide range of design options when considering the formwork required to provide proper support during the foundation construction process. Figure 4.8 presents a type of isolation made of a molded elastomer, which has been compounded and designed to offer low-frequency isolation and high load capacity.

For units installed in basements or ground floors (no other floors beneath), neoprene waffle-type pads, which provide 50% vibration reduction, or the sandwich-type pad of rubber and cork, which provides 75% vibration reduction, can be used.

Where engine-generator sets are to be installed above the ground floor, the more critical type of isolators should be used. For units up to 200 kW, the type of isolator made of rubber bonded to metal can be used and will provide about 90% isolation. Larger units should use spring-type vibration isolators that provide about 95% isolation. All percentages are approximate and exact information for your particular application should be discussed with the machine manufacturer to ensure that the right type of isolator is selected.

4.4.3 Inertia Block Bolt or Pad Mounting Bolt Installation

The inertia block or pad mounting bolts should be of SAE grade 5 bolt materials.

FIGURE 4.8
Isolations from mold elastomer under foundation.

The bolt diameter will be determined by the hole diameter in the engine mounting base or common skid frame. The bolts should be long enough to provide a minimum embedded length of 30 times the bolt diameter, plus 3–4 in. (76–102 mm) for a hook. (The bolt should have a "J" or "L" shaped hook on the nonthreaded end to increase its holding power.) Approximately 7 in. (178 mm) more are needed to protrude above the top surface of the inertia block or pad. These 7 in. (178 mm) will provide the length needed for:

- The grout (if used), 2 in. (51 mm)
- Sole plate (if used), 3/4 in. (19 mm)
- Chock, 1/2 in. (13 mm)
- Shims and engine base, 1-3/4 in. (44.5 mm)
- Washer, nut, and small variations in levelness, 7/8 in. (22 mm)

For a common skid-mounted engine, only 5-1/2 in. (140 mm) of bolt should protrude above the inertia block or pad surface.

Bolt placement in the inertia block or pad can be determined by making a template from 1 × 6 in. (25 × 1 52 mm) boards. Suspend the template over the inertia block or pad and hang bolts and sleeves through the template holes. The bolt must extend 7 in. (178 mm) from the top surface of the inertia block or pad. A sleeve of convoluted plastic tubing 2–3 in. (51–76 mm) in diameter should be placed around the bolts before they are embedded in the concrete. This will allow the bolts to bend and conform to the dimensions of the sole plate (if used) if the template was not exact. The sleeve may be 10–12 in. (254–305 mm) long. The top end of the sleeve should be slightly above the top level of the inertia block or pad so that the concrete will not spill into the sleeve and interfere with bolt adjustments.

4.4.4 Grouting

Grouting can be done only after the installation of the inertia block or pad has fully cured and the sole plates have been positioned and leveled (see Figure 1.12). On sole-plate installations, grouting is important because it anchors the sole plates in place. Since the sole plates support the engine, it is important that the grout be installed properly to hold the plates level.

Engines and common skids can be mounted directly to the grout without the use of sole plates. When this is done, the engine must be mounted and leveled before the grout is poured, then Shim and level the engine, as described in Chapter 2. Pour the grout under the engine base or common skid. After all grout has cured, back out the jacking screws and fill with grout. The grouting procedure is very important and requires good supervision and a competent applicator.

The first step is to make a form around the inertia block or pad. If possible, pour the grout from one point on the inertia block or pad only and allow the grout to flow under the common skid or engine base rails. This pouring procedure will help lessen the chances of air pockets being trapped between the engine and the inertia block or pad. Air pockets will lessen the contact area between the grouting and the engine base or common skid, reducing support for the engine. Also, a metallic-based grout will expand into these spaces and force the engine out of alignment. If the pour point on the engine or common skid is slightly higher than the rest of the inertia block or pad, the grout will flow more easily under the engine or common skid.

The best way to install a concrete, metallic-based grout is to form wedge-shaped grout pads. These pads should run the length of the engine or common skid. Slope the grout outward in a wedge shape toward the inertia block or pad to provide better support. Sole plates can be embedded in this run of grout, or the engine base can be installed directly on it.

The advantage of this grouting technique is that it will keep grout out from under the engine. The grout will not be able to expand up into the hollow area under the engine base and force the engine out of alignment.

Grouting should be worked into place using rods or chain lengths. Work the material gently to avoid air entrapment. When using sole plates, pour

TABLE 4.8

Checklist for Machine Foundation Design

Item	Design Condition	Procedure
1.	Static-bearing capacity	Fifty percent of the allowable soil pressure
2.	Settlement	The eccentricity of the machine center of gravity and the foundation center of gravity is within 5% of any linear dimension
3.	Static plus dynamic loads	Seventy-five percent of the allowable soil pressure
4.	Settlement of static plus dynamic load	Satisfy item 2 and, in the case of a rocking motion, the axis of rocking should coincide with the principal axis of the footing
5.	Magnification factor (applied for machines generating unbalanced forces)	Calculated values of M should be less than 1.5 at resonance frequency
6.	Resonance	The acting frequencies of the machine should not be within $\pm20\%$ of the resonance frequency
7.	Transmissibility factor	Transmissibility factor should be less than 3%
8.	Resonance of individual structure component	The resonance condition with the lowest natural frequency shall be avoided by maintaining the frequency ratio <0.5 and >1.5

enough grout to embed the plates 1/2 in. (13 mm) into the grout. When sole plates are not used, never allow the grout to come up over the engine base or common skid to allow for future adjustments.

Follow the grout manufacturer's instructions for applying the grout and their recommendations for curing times. Concrete grouts must be sealed after curing. All metallic-based grouts should be sealed to prevent rust from destroying the grout. If the grout is allowed to settle at a slight outward slope, oil and water will be able to run off the inertia block or pad.

After the grout has cured, remove the leveling screws and remove any accumulation from the common skid or engine base. Save enough grout to pour into the inertia block bolt sleeves after the engine has been aligned. Many epoxy grouts are also available that provide superior performance for these applications.

4.5 Design Checklist

This checklist is very important to follow if you review the calculation for the foundation design for vibrating equipment or if you perform the design by yourself. It will be the final step of the checking process (see Table 4.8).

References

API Standard 617 (R2009). 2009. *Axial and Centrifugal Compressors and Expander-Compressors for Petroleum, Chemical, and Gas Industry Services*, 7th Ed., Section 2.18.4, API, Washington, D.C.

Arya, S. C., M. W. O'Neil, and G. Pincus. 1984. *Design of Structures and Foundations for Vibrating Machines*. Houston, TX: Gulf Publishing Company.

Barkan, D. D. 1962. *Dynamics of Bases and Foundations*. Tschebotarioff (translator) Mc-Grawhill.

Baxter, R. L., and D. L. Bernhard. 1967. *Vibration Tolerances for Industry*. ASME Paper 67-PEM-14. Plant Engineering and Maintenance Conference, Detroit, MI, April 1967.

Dresser Waukesha. *Installation of Waukesha Engines and Generator Systems Manual*. Waukesha, WI: Dresser Waukesha, 2004.

FABREEKA. 2009. *Vibration and Shock Control Manual*. Boston: FABREEKA International.

Major, A. *Vibration Analysis and Design of Foundations*. Budapest: Akademiai Kiado, 1962. Budapest: Akademiai Kiado, 1982.

Richard Jr., F. E., J. R. Hall Jr., and R. D. Woods. 1970. *Vibrations of Soils and Foundations*, Englewood Cliffs, New Jersey: Prentice-Hall Inc. 414 pp.

5

Storage Tank Design

5.1 Introduction

Tanks are well-known structures in industrial projects. There are different types of storage tanks used for different purposes in industrial projects. The dimensions of the tank and its material depend on the fluid that will be stored in it. Tanks for oil storage usually have a steel structure for their shell and reinforced concrete ring beams that transfer the load to the soil. Some applications use a reinforced concrete tank for storage of salt water, which will be treated and used in the water flood system traditionally used in some process plants.

The process of treating the oil can be performed using reinforced concrete weir tanks to separate oil and water in the treatment process. Liquified natural gas (LNG) projects use prestress tanks with steel shells inside to cool the gas to liquid form for transportation and export. In this chapter, the types of retaining walls and their design methodology will be presented, as will the main features of design and construction of reinforced concrete and steel storage tanks.

5.2 Concrete Storage Tanks

There are two classifications of tanks based on their load effects: elevated tanks under hydrostatic pressure and underground tanks that will be affected by both hydrostatic loads and earth loads. This section will discuss the categories used for the classification of tank shapes (Figure 5.1).

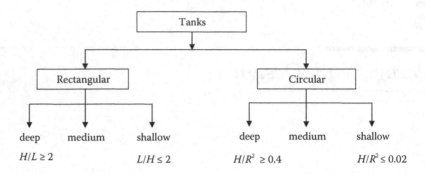

FIGURE 5.1
Types of reinforced concrete tanks.

(a) A cantilever wall (up to 3 m)

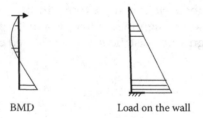

BMD   Load on the wall

(b) This type of wall is simply supported at the top and fixed at the bottom, $H \geq 3$ m

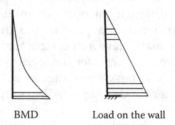

BMD   Load on the wall

(c) A wall fixed at the top and bottom

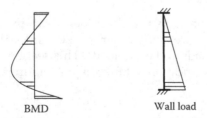

BMD   Wall load

FIGURE 5.2
(a) Cantilever wall. (b) Wall with fixed support at bottom and simple support at top. (c) Different types of tank joints.

5.2.1 Rectangular Wall—Concrete

Rectangular tanks are usually used for special applications in industrial projects. The walls of rectangular tanks have a different static system. Types of rectangular walls for tanks are:

1. Cantilever wall (height of up to 3 m) (Figure 5.2a)
2. Wall with fixed support at bottom and simple support at top (height greater than or equal to 3 m) (Figure 5.2b)
3. Wall with fixed support at top and bottom

Example: Pump Room

This is a pump room embedded in the earth with a volume of 5 × 5 × 5 m and a control room above ground with dimensions of 5 × 5 × 3 m, as shown in Figure 5.3.

Soil Parameters
Soil density, γ_s = 1.8 ton/m³
Water density, γ_w = 1.0 ton/m³
K_a = 0.333
Φ = 30°

The earth pressure diagram will be as follows:

r = 5/5 = 1.0 $\alpha = \beta = 0.5$ according to Grashoff tables (see Table 5.1)
e_1 = $\gamma_s h_1 K_a$ = 1.8 × 3.75 × 0.3 = 2.25 ton/m² (see Figure 5.4)
e_2 = $(\gamma_s - \gamma_w)h^2 k_a$ = (1.8 – 1) × 1.25 × 0.3 = 0.33 ton/m², underground water level (GWT)
e_3 = $\gamma_w h_2$ = 1 × 1.25 = 1.25 ton/m²
e_4 = αe_1 = 0.5 × 2.25 = 1.125 ton/m²
e_4/e_5 = 3/4
e_5 = 1.5 ton/m²
e_6 = (2.25 + 0.33 + 1.25) – 1.5 = 2.33 ton/m²
e_7 = βe_1 = 0.5 × 2.25 = 1.125 ton/m²

Loads on the Roof
Assume that the thickness of the roof and the pump room are 100 and 120 mm, respectively.

Weight of slab on roof = 250 kg/m² self-weight

150 kg/m² plastering

200 kg/m² live load

Total 600 kg/m²

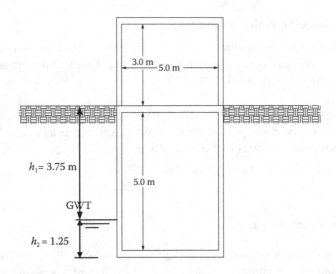

FIGURE 5.3
Sketch for pump room.

$$\text{Weight on pump room slab} = 300 \text{ kg/m}^2 \quad \text{self-weight}$$

$$150 \text{ kg/m}^2 \quad \text{plastering}$$

$$200 \text{ kg/m}^2 \quad \text{live load}$$

$$\text{Total} \quad 650 \text{ kg/m}^2$$

Total load from slab = $(600 + 650) \times 5 \times 5 = 31{,}250$ kg = 31.25 tons
Weight of brick walls = $(4 \times 3 \times 5 \times 0.25) \times 1.7 = 25.5$ tons
Weight of reinforced concrete (RC) walls = $(4 \times 5 \times 5 \times 0.2) \times 2.5 = 50$ tons
Total load = 106.75 tons
Load on floor/m² = 106.75/25 = 4.27 ton/m²

$$r = 55 = 1.0 \quad \alpha = \beta = 0.5$$

TABLE 5.1

Grashoff Values for the Distributed Load in Both Directions, α, β

r	1.0	1.1	1.2	1.3	1.4	1.5	1.6	1.7	1.8	1.9	2.0
α	0.5	0.595	0.672	0.742	0.797	0.834	0.867	0.893	0.914	0.928	0.914
β	0.5	0.405	0.328	0.258	0.203	0.166	0.107	0.107	0.086	0.072	0.059

Note: r, is the ratio between the slab or wall dimensions, length, and width.

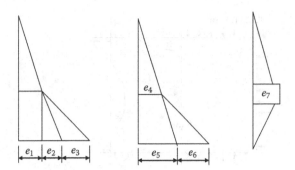

FIGURE 5.4
Distributed load on walls.

Load in each direction = $0.5 \times 4.27 = 2.135$ ton/m²
Straining Action
The first step in design is to calculate the straining action in vertical and horizontal direction. Figure 5.5 presents a longitudinal section with a load and bending moment diagram.

In a horizontal direction, consider a 1.0-m strip at $h/4$ from the room floor. The structure analysis can provide us with the bending moment diagram as shown in Figures 5.5 and 5.6.
Design of Sections
Assume a thickness of the floor (t_f) equal to the wall thickness (t_w) and equal to 200 mm for the bottom corner and, due to a maximum moment, haunches of 350 mm with a 3:1 slope. Sections 2, 3, and 6 will be designed as watertight because the tension zones are adjacent to water (Figure 5.7).

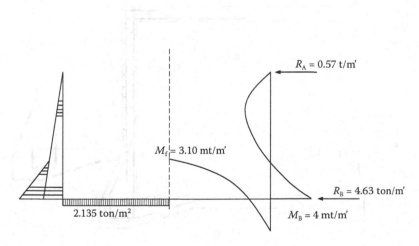

FIGURE 5.5
Longitudinal section showing bending moment and loads.

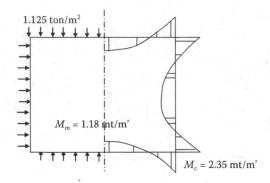

FIGURE 5.6
Shape of bending moment and load in cross section.

5.2.2 Circular Tank

Based on Hilal (1988), the American Portland Cement Association has published the following series of tables, giving the ring tension and cantilever moments in rectangular walls of cylindrical tanks that are free at top and fixed or hinged at bottom for different cases of loading. The tables include very useful data for the design of circular tanks.

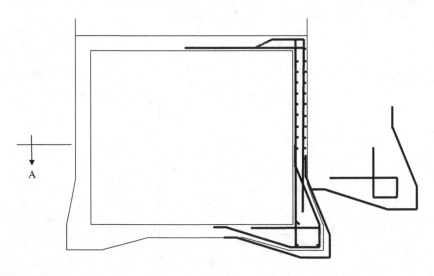

FIGURE 5.7
Reinforcement detail for the basement of a pump room.

TABLE 5.2

Coefficient, χ, Tension in Circular Rings with Triangular Load and Fixed Base and Free Top

H^2/Dt	0.0H	0.1H	0.2H	0.3H	0.4H	0.5H	0.6H	0.7H	0.8H	0.9H
0.4	+0.149	+0.134	+0.120	+0.101	+0.082	+0.066	+0.049	+0.029	+0.014	+0.004
0.8	+0.263	0.239	0.215	+0.190	0.160	+0.130	+0.096	+0.063	+0.034	+0.010
1.2	+0.283	+0.271	+0.254	+0.234	+0.209	+0.180	+0.142	+0.099	+0.054	+0.016
1.6	+0.265	+0.268	+0.268	+0.266	+0.250	+0.226	+0.185	+0.134	+0.075	+0.023
2.0	+0.234	+0.251	+0.273	+0.285	+0.285	+0.274	0.232	0.172	+0.104	+0.031
3.0	+0.134	+0.203	+0.267	+0.322	+0.357	+0.362	+0.330	+0.262	0.157	+0.052
4.0	+0.067	+0.164	+0.256	+0.339	+0.403	+0.429	+0.409	+0.334	+0.210	+0.073
5.0	+0.025	+0.137	+0.245	+0.346	+0.428	+0.477	+0.469	+0.398	+0.259	+0.092
6.0	+0.018	+0.119	+0.234	+0.344	+0.441	+0.504	+0.514	+0.447	+0.301	+0.112
8.0	−0.011	+0.104	+0.218	+0.335	+0.443	+0.534	+0.575	+0.530	+0.381	+0.151
10.0	−0.011	+0.098	+0.208	+0.323	+0.437	+0.542	+0.608	+0.589	+0.440	+0.179
12.0	−0.005	+0.097	+0.202	+0.312	+0.429	+0.543	+0.628	+0.633	+0.494	+0.211
14.0	−0.002	+0.098	+0.200	+0.306	+0.420	+0.539	+0.639	+0.666	+0.541	+0.241
16.0	0.000	+0.099	+0.199	+0.304	+0.412	+0.531	+0.641	+0.687	+0.582	+0.265

Tables 5.2 and 5.3 present the tension force and bending moments, respectively, of the wall for a tank fixed at bottom, and free at top, where $T = \chi \times WH \times (D/2)$. A positive sign indicates tension.

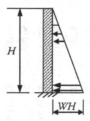

where

H = Tank height
D = Tank diameter
T = Wall thickness

To calculate moment at the fixed end:

$$M = \mu WH^3$$

TABLE 5.3

Coefficient, μ, Moment in Cylindrical Wall with Triangular Load, Fixed Base, and Free Top

H²/Dt	0.1H	0.2H	0.3H	0.4H	0.5H	0.6H	0.7H	0.8H	0.9H	1.0H
0.4	+0.0005	+0.0014	+0.0021	+0.0007	-0.0042	-0.0150	-0.0302	-0.0529	-0.0816	-0.1205
0.8	+0.0011	+0.0037	+0.0063	+0.0080	+0.0070	+0.0023	-0.0068	-0.0224	-0.0465	-0.0795
1.2	+0.0012	+0.0042	+0.0077	+0.0103	+0.0112	+0.0090	+0.0022	-0.0108	-0.0311	-0.0602
1.6	+0.0011	+0.0041	+0.0075	+0.0107	+0.0121	+0.0111	+0.0058	-0.0051	-0.0232	-0.0505
2.0	+0.0010	+0.0035	+0.0068	+0.0099	+0.0120	+0.0115	+0.0075	-0.0021	-0.0185	-0.0436
3.0	+0.0006	+0.0024	+0.0047	+0.0071	+0.0090	+0.0097	+0.0077	+0.0012	-0.0189	-0.0333
4.0	+0.0003	+0.0015	+0.0028	+0.0047	+0.0066	+0.0077	+0.0069	+0.0023	-0.0080	-0.0268
5.0	+0.0002	+0.0008	+0.0016	+0.0029	+0.0046	+0.0059	+0.0059	+0.0028	-0.0058	-0.0222
6.0	+0.0001	+0.0003	+0.0008	+0.0019	+0.0032	+0.0046	+0.0051	+0.0029	-0.0041	-0.0187
8.0	0.0000	+0.0001	+0.0002	+0.0008	+0.0016	+0.0028	+0.0038	+0.0029	-0.0022	-0.0146
10.0	0.0000	0.0000	+0.0001	+0.0004	+0.0007	+0.0019	+0.0029	+0.0028	-0.0012	-0.0122
12.0	0.0000	-0.0001	+0.0001	+0.0002	+0.0003	+0.0013	+0.0023	+0.0026	-0.0005	-0.0104
14.0	0.0000	0.0000	0.0000	0.0000	+0.0001	+0.0008	+0.0019	+0.0023	-0.0001	-0.0090
16.0	0.0000	0.0000	-0.0001	-0.0002	+0.0001	+0.0004	+0.0013	+0.0019	-0.0001	-0.0079

5.3 Retaining Wall

A retaining wall is used to confine water or soil. In industrial projects, it is usually used around the tanks to confine any oil seepage in case of tank failure or leak.

The following are types of retaining walls:

1. Gravity retaining walls
2. Cantilever walls
3. Counterfort walls
4. Buttress walls
5. Bridge abutments
6. Basement walls
7. Semigravity walls
8. Walls with relieving slabs
9. Precast and prestressed walls

In processing plants, it is traditional to use cantilever walls. The forces affecting the retaining walls after the definition of the preliminary dimensions of the walls are as follows:

1. Wall height
2. Dead and live loads on the wall
3. Lateral pressure affecting the wall caused by the weight of soil behind the wall
4. Lateral pressure caused by live loads or moving loads on the soil behind the wall

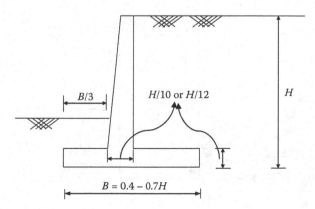

FIGURE 5.8
Preliminary dimensions of a retaining wall.

5. Wave effect in cases of walls adjacent to the sea
6. Earthquake loads
7. Lateral pressure of oil or water in cases of bundle walls
8. Any other load that affects the structure during execution or operation phase

5.3.1 Preliminary Retaining Wall Dimensions

After defining the preliminary dimensions as shown in Figure 5.8, the design of the retaining wall will begin by checking the stability of the retaining wall against the following three criteria:

1. Stability against overturning
2. Stability against sliding
3. Stability against bearing capacity

After checking the stability of the retaining wall, calculate the stress on the wall in order to design the reinforced concrete section of the wall. If unsafe, try another size and check the stability again until the dimensions of the reinforced concrete design sections become compatible with the stability.

5.3.1.1 Check Stability against Overturning

Checking of the stability of the retaining wall against overturning will be done by applying Equation (5.1). The sum of moments attributable to forces tending to resist overturning should be higher than the sum of the moments attributable to forces tending to overturn the walls.

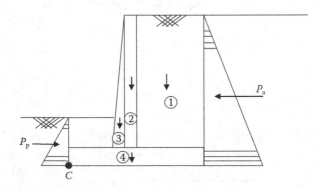

FIGURE 5.9
Forces tending to overturn and resisting overturning the retaining wall, preliminary dimensions.

TABLE 5.4

Resisting Overturning Moment

Section	Area	Weight	Moment Arm Measure from C	Moment about C
1	A_1	$W_1 = A_1 \times \gamma_s$	X_1	M_1
2	A_2	$W_2 = A_2 \times \gamma_c$	X_2	M_2
3	A_3	$W_3 = A_3 \times \gamma_c$	X_3	M_3
4	A_4	$W_4 = A_4 \times \gamma_c$	X_4	M_4
		ΣV		ΣM_R

Notes: Where
γ_c = Concrete density
γ_s = Soil density

$$F_S = \frac{\sum M_R}{\sum M_O} \tag{5.1}$$

where
M_R = Sum of the moments attributable to forces tending to resist overturning the wall at point C, as shown in Figure 5.9
M_O = Sum of the moments attributable to forces tending to overturn the wall at point C, as shown in Figure 5.9
F_S = Factor of safety, which is about 1.5–2

Figure 5.9 presents the moments affecting the wall's tendency to overturn and resistance to overturning.

$$P_a = K_a \gamma H \tag{5.2}$$

where
P_a = Active force
$K_a = 0.3$

$$\sum M_O = P_a(H/3) \tag{5.3}$$

where M_R can be calculated based on the data shown in Table 5.4.

Calculate F_S. If it is within 1.5–2.0, it is acceptable. If it is less, change the retaining wall dimensions to achieve the required overturning stability factor of safety.

5.3.1.2 Check Stability against Sliding

$$F_S = \frac{\sum F_R}{\sum F_d} \tag{5.4}$$

where
$\sum F_R$ = Sum of forces that resist sliding
$\sum F_d$ = Sum of horizontal forces driving the wall to slide
$\sum F_d = P_a$

$$\sum F_R = \sum V \tan \phi_2 + P_P$$

$$= \sum V \tan(k_1 \phi_2) + k_2 C_2 \tag{5.5}$$

where $k_1 = k_2 = 2/3$.

P_P is the passive force, but from a practical point of view it should not be seen to increase the factor of safety. During construction, filling in front of the wall is usually the last stage. If the safety factor is below the limit, you can increase the stability against sliding by using a key lock, as shown in Figure 5.10.

5.3.1.3 Check Stability against Bearing Capacity

The soil-bearing capacity is obtained from the soil report. The retaining wall should be checked to ensure that the load transferred from the wall to the soil is less than the soil-bearing capacity.

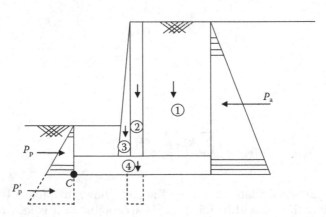

FIGURE 5.10
Forces that drive and resist sliding.

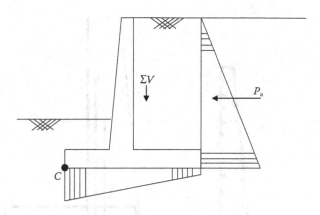

FIGURE 5.11
Checking stress against bearing capacity.

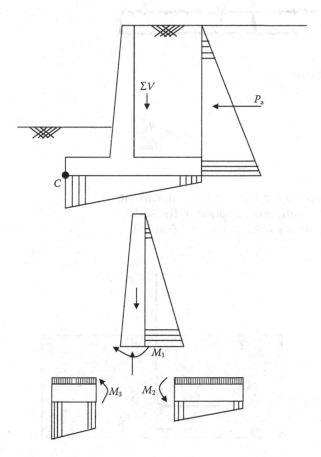

FIGURE 5.12
Checking stress against bearing capacity.

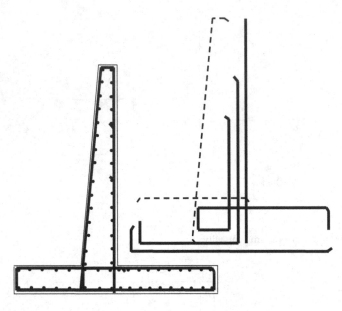

FIGURE 5.13
Retaining wall steel details.

$$F_S = \frac{q_u}{q_{max}}$$ (5.6)

where

q_{all} = Maximum allowable bearing capacity
q_{max} = Maximum load applied to the soil
F_S = Factor of safety, between 2.5 and 3

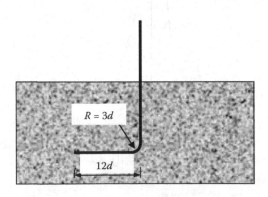

FIGURE 5.14
Steel bending details.

$$M_n = \sum M_R - \sum M_O$$

$$X = M_n \Big/ \sum V$$

(5.7)

$$e = B/2 - X$$

$$q_{min} = \frac{\sum V}{B}\left(1 - \frac{6e}{B}\right)$$

(5.8)

$$q_{max} = \frac{\sum V}{B}\left(1 + \frac{6e}{B}\right)$$

(5.9)

The second step is to begin the design of the concrete sections. In Figures 5.11 and 5.12, the main forces that affect each element are presented.

$$M_1 = M_2 + M_3$$

(5.10)

The shape of the steel reinforcement is shown in detail in Figure 5.13. The steel bending for the steel bars embedded in the foundation and extended into the stem is presented in detail in Figure 5.14.

5.4 Steel Storage Tank

Petrochemical plants usually use steel storage tanks. Figure 5.15 presents a storage tank in an oil plant and Figure 5.16 presents condensate storage tanks with a concrete dike around them in a gas plant.

The design, construction, and manufacture of steel storage tanks is usually performed based on American Petroleum Institute codes. The following sections present the guidelines for best practices in interaction with international codes for the design of steel tanks.

5.4.1 Tank Capacity

The client should specify the maximum capacity and the overfill protection level or volume requirement based on API RP 2350. Maximum capacity is defined as the volume of product in a tank when the tank is filled to its

FIGURE 5.15
Oil storage tank.

designed liquid level, as shown in Figure 5.17. The volume of available product under normal operating conditions should also be defined. From the designed level, a normal level can be defined in empty, operating, and test levels, as the load depends on the liquid level.

5.4.2 Bottom Plates

All bottom plates shall have a minimum nominal thickness of 6 mm (1/4 in.) and its weight shall be about 49.8 kg/m². You should add any corrosion allowance specified by the owner for the bottom plates based on his statement of

FIGURE 5.16
Condensate tanks inside a dike from retaining the wall.

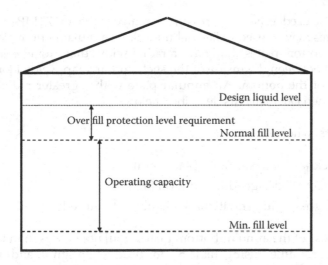

FIGURE 5.17
Steel storage tank levels.

requirement (SOR). Unless otherwise agreed to by the owner, all rectangular sketch plates shall have a minimum nominal width of 1800 mm (72 in.).

Bottom plates of sufficient size shall be ordered so that, when trimmed, at least a 50-mm (2-in.) width will project outside the shell. The bottom plates shall be welded based on the welding procedure.

According to API STD 650, the tank bottoms shall have a minimum slope of 1:120 upward and toward the center of the tank. Unless otherwise specified on the data sheet, the material shall be carbon steel with a minimum thickness of 3 mm (1/8 in.).

5.4.3 Annular Bottom Plates

Based on the API STD 650, if the bottom shell course is designed using the allowable stress for materials shown in Table 5.8, butt-welded annular bottom

TABLE 5.5

Plate Thickness for Annular Bottom, t_{br}

Plate Thickness[a] of Shell Course (mm)	Stress in First Shell Course (MPa)			
	≤190	≤210	≤220	≤250
$t \leq 19$	6	6	7	9
$19 < t \leq 25$	6	7	10	11
$25 < t \leq 32$	6	9	12	14
$32 < t \leq 38$	8	11	14	17
$38 < t \leq 45$	9	13	16	19

[a] Plate thickness refers to shell plate thickness exclusive of corrosion allowance for product design and thickness as constructed for hydrostatic test design.

plates shall be used. When the stress is less than or equal to 172 MPa, lap-welded bottom plates may be used in lieu of butt-welded annular bottom plates.

Annular bottom plates shall have a radial width that provides at least 600 mm (24 in.) between the inside of the shell and any lap-welded joints in the remainder of the bottom. An annular plate with a greater radial width is required when calculated as described below.

In SI units, where:

t_b = Thickness of the annular plate (mm)

H = Maximum designed liquid level (m)

G = Design-specific gravity of the liquid to be stored

The thickness of the annular bottom plates shall not be less than the greater thickness determined using Table 5.5 for product design in addition to any specified corrosion allowance or for hydrostatic test design.

Table 5.5 is applicable for an effective product height of $H \times G \leq 23$ m. Beyond this height, an elastic analysis must be made to determine the annular plate thickness.

The thicknesses specified in Table 5.6 are based on the foundation providing uniform support under the full width of the annular plate. It is worth mentioning—and we will discuss this further later—that unless the foundation is properly compacted inside a concrete ring wall, settlement will produce additional stresses in the annular plate.

The ring of annular plates shall have a circular outside circumference but may have a regular polygonal shape inside the tank shell, with the number of sides equal to the number of annular plates to which these pieces shall be welded. In lieu of annular plates, the entire bottom may be butt-welded, provided that the requirements for annular plate thickness, welding, materials, and inspection are met for the annular requirement.

5.4.4 Shell Design

The required shell thickness should be the design shell thickness with a greater value, including any corrosion allowance, and the hydrostatic test

TABLE 5.6

Relation between Tank Diameter and Minimum Plate Thickness

Nominal Tank Diameter, m (ft)	Nominal Plate Thickness, mm (in.)
<15 (50)	5 (3/16)
15–36 (50–120)	6 (1/4)
36–60 (120–200)	8 (5/16)
>60 (200)	10 (3/8)

shell thickness, but in any case, the shell thickness shall not be less than the data given in Table 5.6.

Unless otherwise specified by the owner, the nominal tank diameter shall be the centerline diameter of the bottom shell-course plate. Nominal plate thickness refers to the tank shell as constructed. The thickness will be specified based on erection requirements. In some cases, the client specified a plate with minimum nominal thickness of 6 mm. If the tank diameter is less than 15 m (50 ft), but greater than 3.2 m (10.5 ft), for the lowest shell course only, the minimum thickness is 6 mm.

The shell plates shall have a minimum nominal width of 1800 mm (72 in.) unless specified by another value from the owner or the available factories. Plates that are to be butt-welded shall be properly squared. The calculated stress for each shell course shall not be greater than the stress permitted for the particular material used for the course. For any case, and under any circumstances, no shell course shall be thinner than the course above it.

The tank shell shall be checked against the wind speed load for stability due to buckling. If required for stability, intermediate girders, increased shell-plate thicknesses, or both shall be used.

It is essential to isolate the radial loads on the tank shell, such as those caused by heavy loads on platforms and elevated walkways between tanks, and ensure that they shall be distributed by rolled structural sections, plate ribs, or built-up members or any other structure configuration to transfer load to ground.

5.4.4.1 Allowable Stress

The maximum allowable product design stress, S_d, shall be as shown in Table 5.8. The net plate thicknesses—the actual thicknesses minus any corrosion allowance—shall be used in the calculation.

According to API STD 650, the design stress basis, S_d, shall be either two-thirds the yield strength ($2/3F_y$) or two-fifths the tensile strength ($2/5F_t$), whichever is less.

The maximum allowable hydrostatic test stress, S_t, shall be as shown in Table 5.2. The gross plate thicknesses, including any corrosion allowance, shall be used in the calculation. The hydrostatic test basis shall be either three-fourths the yield strength ($3/4F_y$) or three-sevenths the tensile strength ($3/7F_y$), whichever is less. Shell thickness less than or equal to 1/2 in. permits an alternative shell design with a fixed allowable stress of 145 MPa (21,000 psi^2) and a joint efficiency factor of 0.85 or 0.70.

5.4.4.2 Calculation of Thickness by the 1-Foot Method

The 1-ft method is required to calculate the thicknesses at design points 0.3 m (1 ft) above the bottom of each shell course. It is important to know that this method shall not be used for tanks larger than 60 m (200 ft) in diameter.

The required minimum thickness of shell plates shall be the values computed by the following formulas:

$$t_d = \frac{4.9D(H-0.3)G}{S_d} + C_A \tag{5.11}$$

$$t_t = \frac{4.9D(h-0.3)}{S_t} \tag{5.12}$$

where
t_d = Design shell thickness (mm)
t_t = Hydrostatic test shell thickness (mm)
D = Nominal tank diameter (m)
H = Design liquid level (m)
h = Height from the bottom of the course under consideration to the top of the shell, including the top angle, if any, to the bottom of any overflow that limits the tank-filling height, to a point restricted by an internal floating roof, or to a point controlled to allow for seismic wave action
G = Design-specific gravity of the liquid to be stored, as defined by the owner
C_A = Corrosion allowance (in mm) as specified by the owner
S_d = Allowable stress for the design condition (MPa), as in Table 5.7
S_t = Allowable stress for the hydrostatic test condition (MPa), as in Table 5.7

5.4.4.3 Calculation of Thickness by the Variable-Design-Point Method

Before using this method, we must note that this procedure normally provides a reduction in shell-course thicknesses and total material weight, but more important is its potential to permit construction of larger diameter tanks within the maximum plate thickness limitation.

This calculation is based on the study performed by Zick and McGrath (1968) for *Design of Large Diameter Cylindrical Shells*. Design by the variable design point method gives shell thicknesses at design points that result in the calculated stresses being relatively close to the actual circumferential shell stresses. This method may only be used when the client has not specified that the 1-ft method be used and when the following equation is verified. In SI units:

$$\frac{L}{H} \leq \frac{1000}{6} \tag{5.13}$$

TABLE 5.7

Plate Materials and Allowable Stresses

Plate Specification	Grade	Minimum Yield Strength (MPa)	Minimum Tensile Strength (MPa)	Product Design Stress, S_d (MPa)	Hydrostatic Test Stress, S_t (MPa)
A 283M (A 283)	C	205	380	137	154
A 285M (A 285)	C	205	380	137	154
A 131M (A 131)	A, B, CS	235	400	157	171
A 36M (A 36)		250	400	160	171
A 131M (A 131)	EH36	360	490a	196	210
A 573M (A 573)	400	220	400	147	165
A 573M (A 573)	450	240	450	160	180
A 573M (A 573)	485	290	485a	193	208
A 516M (A 516)	380	205	380	137	154
A 516M (A 516)	415	220	415	147	165
A 516M (A 516)	450	240	450	160	180
A 516M (A 516)	485	260	485	173	195
A 662M (A 662)	B	275	450	180	193
A 662M (A 662)	C	295	485a	194	208
A 537M (A 537)	1	345	485a	194	208
A 537M (A 537)	2	415	550a	220	236
A 633M (A 633)	C, D	345	485a	194	208
A 678M (A 678)	A	345	485a	220	208
A 678M (A 678)	B	415	550a	194	236
A 737M (A 737)	B	345	485a	194	208
A 841M (A 841)	Class 1	345	485a	194	208
A 841M (A 841)	Class 2	415	550a	220	236

where

$L = (500D_t)^{0.5}$

D = Tank diameter (m)

t = Bottom course shell thickness, excluding any corrosion allowance (mm)

H = Maximum design liquid level (m)

The minimum plate thicknesses for both the design condition and the hydrostatic test condition shall be determined as outlined. The complete calculations shall be made for all the courses of design condition, exclusive of any corrosion allowance, and hydrostatic test condition. The required shell thickness for each course shall be the higher design shell thickness, plus any corrosion allowance, and the hydrostatic test shell thickness.

To calculate the bottom course thicknesses, preliminary values t_{pd} and t_{pt} for the design and hydrostatic test conditions shall first be calculated. The bottom course thicknesses, t_{1d}, and t_{1t} for the design and hydrostatic test conditions shall be calculated using Equations (5.11) and (5.12), respectively.

$$t_{d1} = \left(1.06 - \frac{0.0696D}{H}\sqrt{\frac{HG}{S_d}}\right)\left(\frac{4.9HGD}{S_d}\right) + C_A \qquad (5.14)$$

$$t_{1t} = \left(1.06 - \frac{0.0696D}{H}\sqrt{\frac{H}{S_t}}\right)\left(\frac{4.9HD}{S_t}\right) \qquad (5.15)$$

The design condition, t_{d1}, need not be greater than t_{pd}, and, for the hydrostatic test condition, t_{1t} need not be greater than t_{pt}.

To calculate the second-course thicknesses for both the design condition and the hydrostatic test condition, the value of the following ratio shall be calculated for the bottom course:

$$R = \frac{h_1}{(rt_1)^{0.5}} \qquad (5.16)$$

where
h_1 = Height of the bottom shell course, in mm (in.).
r = Nominal tank radius, in mm (in.).
t_1 = Calculated thickness of the bottom shell course, less any thickness added for the corrosion allowance, in mm (in.) and is used to calculate t_2 (design). The calculated hydrostatic thickness of the bottom shell course shall be used to calculate t_2 (hydrostatic test).

$$\text{If } R \leq 1.375 \qquad t_2 = t_1$$

$$R \geq 2.625 \qquad t_2 = t_{2a}$$

$$1.375 \leq R < 2.625$$

$$t_2 = t_{2a} + (t_1 - t_{2a})\left[2.1 - \frac{h_1}{1.25(rt_1)^{0.5}}\right] \qquad (5.17)$$

where
t_2 = Minimum design thickness of the second shell course, excluding any corrosion allowance, in mm (in.).
t_{2a} = Thickness of the second shell course, in mm (in.), as calculated for an upper shell course, as described in Equations (5.18) and (5.19), exclusive of any corrosion allowance. Use dn to calculate the second shell course thickness, t_2, for the design case, and, for the hydrostatic test case, applicable values of t_{2a} and t_1 shall be used.

The preceding formula for t_2 is based on the same allowable stress being used for the design of the bottom and second courses. For tanks where the value of the ratio is greater than or to 2.625, the allowable stress for the second course may be lower than the allowable stress for the bottom course.

To calculate the upper course thicknesses for both the operating condition and the hydrostatic test condition, a preliminary value, t_u, for the upper course thickness shall be calculated using Equations (5.11) and (5.12), excluding any corrosion allowance, and then the distance, x, of the variable design point from the bottom of the course shall be calculated using the lowest value obtained from the following:

$$X_1 = 0.61(rt_u)^{0.5} + 320CH$$

$$X_2 = 1000CH$$

$$X_3 = 1.22(rt_u)^{0.5}$$

where
t_u = Thickness of the upper course at the girth joint calculation from Equations (5.11) and (5.12), exclusive of any corrosion allowance (in mm)
$C = [K^{0.5}(K-1)]/(1+K^{1.5})$
$K = t_L/t_u$
t_L = Thickness of the lower course at the girth joint (in mm), exclusive of any corrosion allowance
H = Design liquid level (m)

The minimum thickness, t_x, for the upper shell courses shall be calculated for both the design condition (t_{dx}) and the hydrostatic test condition (t_{tx}) using the minimum value of x.

$$t_{d_x} = \frac{4.9D\left(H - \dfrac{x}{1000}\right)G}{S_d} + C_A \tag{5.18}$$

$$t_{t_x} = \frac{4.9D\left(H - \dfrac{x}{1000}\right)}{S_t} \tag{5.19}$$

5.4.5 Roof System

The designer will select the type of the roof system based on the tank diameter and the intended use of the tank. In some tanks, there will be a floating roof, but this section will focus on the structure system for the fixed roof. The fixed roof types are as follows:

(1) A supported cone roof is formed to approximate the surface of a right cone that is supported principally either by rafters on girders and columns or by rafters on trusses with or without columns.

(2) A self-supporting cone roof is formed to approximate the surface of a right cone that is supported only at its periphery.

(3) A self-supporting dome roof is formed to approximate a spherical surface that is supported only at its periphery.

(4) A self-supporting umbrella roof is a modified dome roof formed so that any horizontal section is a regular polygon with as many sides as there are roof plates. It is supported only at its periphery.

All roofs and supporting structures shall be designed for load combinations (1), (2), (3), and (4). Roof plates shall have a minimum nominal thickness of 5 mm (3/16 in.) or a 7-gauge sheet. Thicker roof plates may be required for self-supporting roofs. Corrosion allowance should be incorporated in the design of roof thickness.

In general, the roof plates of supported cone roofs shall not be attached to the supporting members unless otherwise approved by the client. Continuously attaching the roof to cone supporting members may be beneficial when interior coating systems are required; however, the tank roof cannot be considered frangible.

All internal and external structural members shall have a minimum nominal thickness of 4 mm in any component. The method of providing a corrosion allowance, if any, for the structural members shall be a matter of agreement between the client and the contractor.

Roof plates shall be attached to the top angle of the tank with a continuous fillet weld on the top side only.

A roof is considered frangible for an emergency venting requirement if the roof-to-shell joint will fail before the shell-to-bottom joint in the event of excessive internal pressures. When a client specifies a tank with a frangible roof, the tank design should comply with all of the following:

1. The tank should be 15 m in diameter or greater.

2. The slope of the roof at the top-angle attachment does not exceed 1:6.

3. The roof is attached to the top angle with a single continuous fillet weld that does not exceed 5 mm.

4. The roof support members shall not be attached to the roof plate.

5. All members in the region of the roof-to-shell junction, including insulation rings, are considered as contributing to the cross-sectional area (A).

6. The cross-sectional area (A) of the roof-to-shell junction is less than the limit shown below (in SI units):

$$A = \frac{D_{ls}}{1390 \tan \theta} \qquad (5.20)$$

For all types of roofs, the plates may be stiffened by sections welded to the plates but may not be stiffened by sections welded to the supporting rafters or girders. The contractor shall provide a roof designed and constructed to be as safe as that of the standard. In the roof design, particular attention should be given to preventing failure caused by instability.

5.4.5.1 Allowable Stresses

All parts of the roof structure shall be proportioned so that the sum of the maximum static and dynamic stresses shall not exceed the limitations specified in the American Institute of Steel Construction Specification for Structural Steel Buildings or with the agreement of the client, an equivalent structural design code recognized by the government of the country where the tank is located. The portion of the specification *Allowable Stress Design* shall be used in determining allowable unit stresses. Based on API STD 650, a plastic design is not allowed to be used in tank design.

The minimum thickness of any structural member, including any corrosion allowance on the exposed side or sides, shall not be less than 6 mm for columns, knee braces and beams, or stiffeners, which, by design, normally resist axial compressive forces of 4 mm for any other structural member.

For columns, the value L/r shall not exceed 180. For other compression members, the value L/r shall not exceed 200. For all other members, except tie rods whose design is based on tensile force, the value L/r shall not exceed 300, where L is the unbraced length (mm) and r is the governing radius of gyration (mm).

5.4.5.2 Supported Cone Roofs

The slope of the roof shall be 1:16, or greater if specified by the client. If the rafters are set directly on chord girders, producing slightly varying rafter slopes, the slope of the flattest rafter shall conform to the specified or ordered roof slope. Main supporting members, including those supporting the rafters, may be rolled or fabricated sections or trusses.

Although these members may be in contact with the roof plates, the compression flange of a member or the top chord of a truss shall be designed with no lateral support from the roof plates. It shall be laterally braced, if necessary, by other acceptable methods. Structural members serving as rafters may be rolled or built up. Rafters shall be designed for the dead load of the rafters and roof plates where the compression flange of the rafter has no lateral support from the roof plates and shall be laterally braced.

When considering additional dead loads or live loads, the rafters in direct contact with the roof plates applying the loading to the rafters may be considered as receiving adequate lateral support from the friction between the roof plates and the compression flanges of the rafters, with the following exceptions:

1. Trusses and open-web joints used as rafters
2. Rafters with a nominal depth greater than 375 mm (15 in.)
3. Rafters with a slope greater than 1:6

Rafters shall be spaced so that in the outer ring their centers are not more than 0.6π m (6.28 ft) apart measured along the circumference of the tank. Spacing on inner rings shall not be greater than 1.7 m. Roof columns shall be pipe or structural shape sections. Pipe columns shall either be sealed or have openings on both the top and bottom of the column.

Rafter clips for the outer row of rafters shall be welded to the tank shell. Roof support columns shall be provided at their bases with details that provide for the following:

1. Column loads shall be distributed over a bearing area based on the specified soil-bearing capacity or foundation design. Where an unstiffened horizontal plate is designed to distribute the load, it shall have a minimum thickness of 12 mm (1/2 in.). Alternatively, the column load may be distributed by an assembly of structural beams. The plate or members shall be designed to distribute the load without exceeding the allowable stresses.

2. At each column, a wear plate with a minimum 6-mm thickness shall be welded to the tank bottom with a 6-mm (1/4-in.) minimum fillet weld. A single plate of adequate thickness may be designed for the dual functions of load distribution and corrosion and abrasion protection.

3. The design shall allow the columns to move vertically relative to the tank bottom without restraint in the event of tank overpressure or bottom settlement.

4. The columns shall be effectively guided at their bases to prevent lateral movement.

5. The guides shall remain effective in the event of vertical movement of the columns, relative to the tank bottom, of up to 75 mm. The guides shall be appropriately located so that they are not welded directly to the tank bottom plates.

In cases where supporting members are attached to the roof plate, consideration shall be given to the design of the supporting members and their attachment details when considering internal pressure.

5.4.5.3 Self-Supporting Cone Roofs

Self-supporting roofs whose roof plates are stiffened by sections welded to the plates need not conform to the minimum thickness requirements. However, the thickness of the roof plates should not be less than 5 mm when designed by the contractor, subject to the approval of the engineer.

Self-supporting cone roofs should conform to the following requirements:

$\theta \le 37°$ (slope = 3:4)
$\theta \ge 9.5°$ (slope = 1:6)

In SI units,

$$t_{min} = \frac{D}{4.8\sin\theta}\sqrt{\frac{T}{2.2}} \ge 5 \text{ mm}$$

t_{max} = 12.5 mm, exclusive of corrosion allowance.

where
D = Nominal diameter of the tank shell (m)
T = Greater of load combinations in Section 5.4.7 for gravity loads (kPa)
θ = Angle of cone elements to the horizontal (°)

5.4.5.4 Self-Supporting Dome and Umbrella Roofs

Self-supporting dome and umbrella roofs should conform to the following requirements:

Minimum radius = 0.8D
Maximum radius = 1.2D

$$t_{min} = \frac{r}{2.4}\sqrt{\frac{T}{2.2}} + C_A \ge 5 \text{ mm}$$

t_{max} = 12.5 mm, exclusive of corrosion allowance.
The following are certain restrictions on types of top-angle joints:

TABLE 5.8

Tank Diameter and Top-Angle Size

Tank Diameter, D	Minimum Top Angle Size[a] (mm)
≤11 m (≤35 ft)	51 × 51 × 4.8
11 m < D ≤ 18 m (35 ft < D ≤ 60 ft)	51 × 51 × 6.4
>18 m (>60 ft)	76 × 76 × 76

[a] You can choose an equivalent section size to accommodate the availability of local materials.

1. Roof plates shall, at a minimum, be welded on the top side with a continuous full-fillet weld on all seams. Butt-welds are also permitted.

2. For frangible roofs, roof plates should be attached to the top angle of a tank with a continuous fillet weld on the top side only. For non-frangible roofs, alternate details are permitted.

3. The top-angle sections, tension rings, and compression rings should be joined by butt-welds having complete penetration and fusion.

4. For self-supporting roofs of the cone, dome, or umbrella type, the edges of the roof plates may be flanged horizontally to rest flat against the top angle to improve welding conditions.

5. Except as specified for open-top tanks, for tanks with frangible joints for self-supporting roofs, and for tanks with the flanged roof-to-shell detail, tank shells should be supplied with top angles of not less than the sizes listed in Table 5.8.

5.4.6 Tank Design Loads

It is important that the tank design loads be defined by the client in his SOR. For example, in some countries, there is snow or earthquakes, while other countries do not experience these conditions.

Loads are defined as follows:

1. *Dead load* (DL)—the weight of the tank or tank component, including any corrosion allowance unless otherwise noted.

2. *Design external pressure* (P_e)—should not be less than 0.25 kPa (1 in. of water). This standard does not contain provisions for external pressures exceeding 0.25 kPa (1 in. of water). Design requirements for vacuums exceeding this value and to resist flotation and external fluid pressure should be agreed on by the client and the manufacturer.

3. *Design internal pressure* (P_i)—shall not exceed 18 kPa (2.5 lbf/in.²).

4. *Hydrostatic test* (H_t)—the load caused by filling the tank with water to the design liquid level.

5. *Minimum roof live load* (L_r)—1.0 kPa (20 lb/ft^2) on the horizontal projected area of the roof.

6. *Seismic load* (*E*)—seismic loads are determined based on the seismic zone of the tank location.

7. *Snow* (*S*)—the ground snow load should be determined from ASCE 7-02, Figure 7.1, or Table 7.1 unless a ground snow load that equals or exceeds the value based on a 2% annual probability of being exceeded (50-year mean recurrence interval) is specified by the client. The design snow load shall be 0.84 times the ground snow load. The design snow load shall be reported to the client.

8. *Stored liquid* (*F*)—the load caused by filling the tank to the design liquid level with liquid of the design-specific gravity, which should be specified by the client.

9. *Test pressure* (P_t)—the pressure caused by filling the tank to the design liquid level with water for hydrostatic test purposes.

10. *Wind* (*W*)—The design wind speed (*V*) should be 190 km/h (120 mph), the 3-s gust design wind speed determined from ASCE 7-02, Figure 6.1, or the 3-s gust design wind speed specified by the client this specified wind speed shall be for a 3-s gust based on a 2% annual probability of being exceeded; 50-year mean recurrence interval. The design wind pressure should be 0.86 kPa ($V/190$)^2 ([18 lbf/ft^2] [$V/120$]^2) on vertical projected areas of cylindrical surfaces, and 1.44 kPa ($V/190$)^2, ([30 lbf/ft^2] [$V/120$]^2) uplift (see item (b) below) on horizontal projected areas of conical or doubly curved surfaces, where *V* is the 3-s gust wind speed. The 3-s gust wind speed used shall be reported to the client.

   (a) These design wind pressures are in accordance with ASCE 7-02 for wind exposure category C. As an alternative, pressures may be determined in accordance with ASCE 7-02 (exposure category and importance factor provided by the client) or a national standard for the specific conditions for which the tank is being designed.

   (b) The design uplift pressure on the roof (wind plus internal pressure) need not exceed 1.6 times the design pressure.

   (c) Windward and leeward horizontal wind loads on the roof will not be included in the above pressures, as they are conservatively equal and opposite.

   (d) Fastest mile wind speed times 1.2 is approximately equal to 3-s gust wind speed.

5.4.7 Load Combination

The purpose of load combination is to consider all the possibilities along the tank lifetime.

1. Fluid and internal pressure
   $DL + F + P_i$
2. Hydrostatic test
   $DL + (H_t + P_t)$
3. Wind and internal pressure
   $DL + W + 0.4P_i$
4. Wind and external pressure
   $DL + W + 0.4P_e$
5. Gravity loads
   (a) $DL + (L_r \text{ or } S) + 0.4P_e$
   (b) $DL + P_e + 0.4(L_r \text{ or } S)$
6. Seismic
   $DL + F + E + 0.1S + 0.4P_i$

If the ratio of operating pressure to design pressure exceeds 0.4, the client should consider specifying a higher factor on design pressure in (3), (4), (5a), and (6).

5.4.8 Design Basis for Small Tanks

The definition of a small tank is a tank in which the stressed components have a maximum nominal thickness of 12 mm, including any corrosion allowance specified by the client. The stressed components include the shell and reinforcing plates, the shell-reinforcing plates for flush-type cleanout fittings and flush-type shell connections, and the bottom plates that are welded to the shell.

For small tanks, API STD 650 provides Tables 5.5 and 5.6 to present the tank volume capacity based on the height and the diameter of the tank and the corresponding tank shell thickness, respectively.

Tables 5.9 and 5.10 are applicable only when specified by the client and are limited to design metal temperatures above −30°C (−20°F) or above −40°C (−40°F) when a killed, fine-grain material is used. The maximum tensile stress before the joint efficiency factor is applied should be 145 MPa. The stresses shall be computed on the assumption that the tank is filled with water (specific gravity = 1.0) or with the liquid to be stored that has a higher specific gravity than water. The tension in each ring shall be computed at 300 mm (12 in.) above the centerline of the lower horizontal joint of the course in question. When these stresses are computed, the tank diameter shall be taken as the nominal diameter of the bottom course.

TABLE 5.9

Typical Sizes and Corresponding Nominal Capacities (m³) for Tanks with 1800-mm Courses

Tank Diameter (m)	Capacity per m of Height (m³)	Tank Height (m)/Number of Courses in Completed Tank								
		3.6/2	5.4/3	7.24	9/5	10.8/6	12.6/7	14.4/8	16.2/9	18/10
3	7.07	25	38	51	64	76				
4.5	15.9	57	86	115	143	172				
6	28.3	102	153	204	254	305	356	407		
7.5	44.2	159	239	318	398	477	557	636	716	795
9	63.6	229	344	458	573	687	802	916	1,031	1,145
10.5	86.6	312	468	623	779	935	1,091	1,247	1,403	1,559
12	113	407	611	814	1,018	1,221	1,425	1,629	1,832	2,036
13.5	143	515	773	1031	1,288	1,546	1,804	2,061	2,319	2,576
15	177	636	954	1272	1,590	1,909	2,227	2,545	2,863	3,818
18	254	916	1,374	11,832	2,290	2,748	3,206	3,664	4,122	4,580
21	346	1,247	1,870	2,494	3,117	3,741	4,364	4,988	5,089	D = 18
24	452	1,629	2,443	3,257	4,072	4,886	5,700	5,474	D = 20	
27	573	2,061	3,092	4,122	5,153	6,184	6,690	D = 22		
30	707	2,545	3,817	5,089	6,362	7,634	D = 26			
36	1,018	3,664	5,497	7,329	9,161	D = 30				
42	1,385	4,988	7,481	9,975	D = 36					
48	1,810	6,514	9,772	11,966						
54	2,290	8,245	12,367	D = 46						
60	2,827	10,179	15,268							
66	3,421	12,316	16,303							
		D = 62								

TABLE 5.10

Typical Sizes and Corresponding Nominal Capacities (Barrels) for Tanks with 72-in. Courses

Tank Diameter (m)	Capacity per Meter of Height, Barrels (m³)	Tank Height (ft)/Number of Courses in Completed Tank								
		12/2	18/3	24/4	30/5	36/6	42/7	48/8	54/9	60/10
10	14.0	170	250	335	420	505				
15	31.5	380	565	755	945	1,130				
20	56.0	670	1,010	1,340	1,680	2,010	2,350	2,690		
25	87.4	1,050	1,570	2,100	2,620	3,150	3,670	4,200	4,720	5,250
30	126	1,510	2,270	3,020	3,780	4,530	5,290	6,040	6,800	7,550
35	171	2,060	3,080	4,110	5,140	6,170	7,200	8,230	9,250	10,280
40	224	2,690	4,030	5,370	6,710	8,060	9,400	10,740	12,100	13,430
45	283	3,400	5,100	6,800	8,500	10,200	11,900	13,600	15,300	17,000
50	350	4,200	6,300	8,400	10,500	12,600	14,700	16,800	18,900	21,000
60	504	6,040	9,060	12,100	15,110	18,130	21,150	24,190	27,220	D = 58
70	685	8,230	12,340	16,450	20,580	24,700	28,800	32,930	D = 64	
80	895	10,740	16,120	21,500	26,880	32,260	37,600	D = 73		
90	1,133	13,600	20,400	27,220	34,030	40,820	D = 83			
100	1,399	16,800	25,200	33,600	42,000	D = 98				
120	2,014	24,190	36,290	48,380	D = 118					
140	2,742	32,930	49,350	65,860						
160	3,581	43,000	64,510	D = 149						
180	4,532	54,430	81,650							
200	5,595	67,200	100,800							
220	6,770	81,310	D = 202,335							

The joint efficiency factor should be 0.85 with the spot radiography. By agreement between the client and the contractor, the spot radiography may be omitted and a joint efficiency factor of 0.70 shall be used.

5.4.9 Piping Flexibility

Piping systems connected to tanks should consider the potential movement of the connection points during earthquakes and provide sufficient flexibility to avoid release of the product by failure of the piping system. The piping system and supports should be designed so as not to impart significant mechanical loading on the attachment to the tank shell. Local loads at piping connections should be considered when designing the tank shell. Mechanical devices that add flexibility such as bellows, expansion joints, and other flexible apparatus may be used when they are designed for seismic loads and displacements.

Unless otherwise calculated, piping systems shall provide for the minimum displacements in Table 5.11, at working stress levels (with a 33% increase for seismic loads) in the piping, supports, and tank connection. For attachment points located above the support or foundation elevation, the displacements in Table 5.11 shall be increased, to account for drift of the tank or vessel.

The values given in Table 5.11 do not include the influence of relative movements of the foundation and piping anchorage points due to foundation movements (such as settlement or seismic displacements). The effects of foundation movements shall be included in the design of the piping system design, including the determination of the mechanical loading on the tank or vessel and consideration of the total displacement capacity of the mechanical devices intended to add flexibility.

TABLE 5.11

Design Displacement for Piping Connecting to the Tank

Condition	ASD[a] Design Displacement (mm)
Mechanically anchored tanks	
Upward vertical displacement relative to support or foundation	25
Downward vertical displacement relative to support or foundation	13
Range of horizontal displacement (radial and tangential) relative to support or foundation	13
Self-anchored tanks	
Upward vertical displacement relative to support or foundation	
Anchorage ratio less than or equal to 0.785	25
Anchorage ratio greater than 0.785	100
Downward vertical displacement relative to support or foundation	
For tanks with a ring wall/mat foundation:	13
For tanks with a berm foundation	25
Range of horizontal displacement	50

[a] ASD is the allowable stress design.

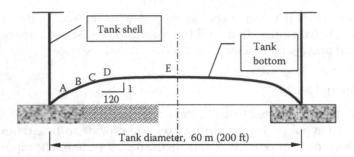

FIGURE 5.18
Tank bottom fabrication.

5.4.10 Differential Settlement Tank Bottom Designs

Based on the degree and type of settlement expected, as determined from similar installations in the area or from soil surveys, there are different methods that provide increasing effectiveness but also increasing costs, shown in the order of increasing cost (low to high):

1. Standard lap-welded bottom
2. Annular plates with lap-welded bottom
3. Butt-welded bottoms

Unless needed for large differential settlement, the butt-welded tank bottom design is usually rejected on a cost/benefit basis. Additional design features such as the following can provide further effectiveness:

4. Deeper levels of soil compaction
5. Crushed stone ring walls
6. Reinforced concrete ring walls
7. Slabs on ring wall foundations

TABLE 5.12

Tank Bottom Coordinates Based on EMUEA

Point	Distance from Tank Shell, m (ft)	Height above Elevation (0,0), m (ft)
A	3 (10)	0.22 (0.71)
B	6 (20)	0.39 (1.28)
C	9 (30)	0.53 (1.73)
D	12 (40)	0.61 (2.0)
E	30.5 (100)	0.76 (2.5)

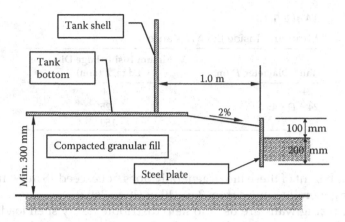

FIGURE 5.19
Compacted granular fill foundation.

Tank Size

Based on EEMUA, the bottom of the tank should be campered by elevating its level during construction as shown in Figure 5.18 and by the values in Table 5.12.

- For large tanks (15 m or more), a concrete ring beam (preferred) or a crushed stone ring wall should be used as shown in Figure 5.19.
- For small tanks (6 m or less), a concrete slab foundation (preferred) or compacted granular fill foundation should be used.
- For medium tanks (6–15 m), the type of foundation should be at the discretion of the foundation design engineer.

5.5 Ring Beam Design Consideration

The load that will be used to design the tank shell will be the same as that for the ring beam, except the static equipment expert or the manufacturer that is designing and building the steel tank will deliver the wind load and seismic load that will affect the ring beam (Table 5.13). The forces on the ring beam will be as shown in Figure 5.20.

The design and specifications for construction of storage tank ring beam foundations should be adequate for the structure's intended use, in accordance with API STD 650.

The types of storage tank normally encountered in gas and oil plants have cylindrical shells, essentially flat bottoms, and either cone roofs or floating

TABLE 5.13

Minimum Inside Edge Distance

Tank Diameter, D (m)	Minimum Inside Edge Distance Edge, L (mm)
≤24	100
24 < D ≤ 45	125
≥45	150

roofs. The height of these tanks generally does not exceed 18 m (60 ft) and the diameter generally ranges from 3 to 110 m (10 to 360 ft).

Concrete ring walls are used to help distribute heavy shell loads such as in large dome roof tanks where the roof is supported entirely by the shell. Concrete ring walls are also used when the tank must be anchored for internal pressure, earthquake, or wind loads. Ring wall foundation design shall conform to this engineering guideline and to applicable requirements of the soil report.

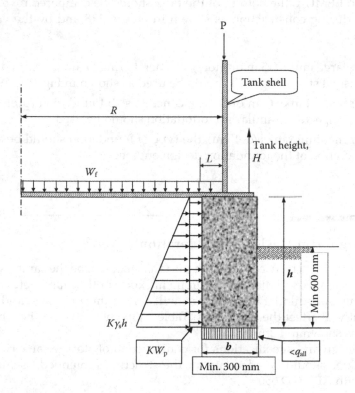

FIGURE 5.20
Loads affecting the ring beam.

A geotechnical investigation is required before constructing the tanks. Soil investigation will be conducted as described in Chapter 13. The allowable soil-bearing pressure shall be based on the results of the geotechnical investigation and consideration of permissible total and differential settlements. Soil pressures shall be calculated under the action of vertical and lateral loads using load combinations that result in maximum soil pressures.

Foundations shall be founded on either undisturbed soil or compacted fill. In the case of foundations supported on compacted fill, the geotechnical investigation shall govern the type of fill material and the degree of compaction required. In cases of sites that have firm soils, the recommended steps for site preparation are as follows:

1. Excavate the site of all topsoil and organic material.
2. Build the ring wall at a minimum of 15 m above finished grade elevation to ensure that adequate drainage is about 150 mm.
3. Slope interior tank pad surface to match the tank bottom. A minimum of 1 in. to 10 ft is recommended, taking into account settlement and differential settlement between the tank center and the ring beam to ensure adequate tank drainage.
4. Provide a concrete ring wall to reduce differential settlement along the circumference of the tank shell as shown in Figures 5.21 and 5.22.

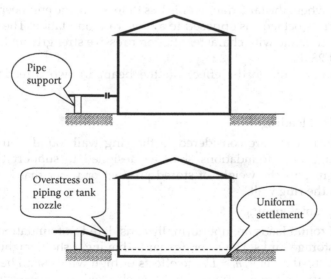

FIGURE 5.21
Uniform settlement.

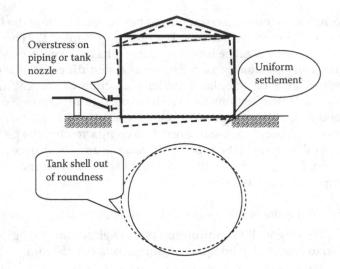

FIGURE 5.22
Effect of differential settlement in tank.

In areas of weak compressible soils, special foundations and design procedures or soil improvement may be required to prevent failures or excessive settlement, as recommended in the project soil investigation report. Ring beam depth below grade should be determined from allowable soil-bearing pressures. For tanks utilizing interior roof supports, the tank bottom shall be reinforced as required to distribute column loads and reduce local settlement.

In cases where the tank diameter is less than 4.5 m, the engineer may use a solid concrete octagon as opposed to a ring wall foundation. The ring beam is usually concrete with characteristic compressive strength not less than 25 N/mm² at 28 days.

In summary, the loads' effect on the beam in two directions are as follows:

- Vertical loading

  Vertical loads are considered in the ring wall foundation design. Storage tank foundations shall be designed to support the tank weight plus the weight of stored product or water that is directly over the ring wall.

- Horizontal loading

  Horizontal loads are not normally considered in foundation design for storage tanks. However, for tanks having a shell height greater than the diameter, $h_t > D$, the effects of high winds shall be considered and the foundations should be checked for necessary anchor bolts.

5.5.1 Wind and Earthquake Stability and Pressures

In general, wind stability is not a problem with storage tanks; however, if $h_t/D > 1$, the stability should be checked. For an unanchored tank, the overturning moment from wind pressure should not exceed two-thirds of the dead load resisting moment, excluding any tank content, and should be calculated as follows:

$$M_{OT} \leq 2/3(WD/2)$$

where

M_{OT} = Overturning moment from wind pressure (mt)

W = Shell weight available to resist uplift, in tons, less any corrosion allowance, plus dead weight of any portion of roof supported by the shell

D = Tank diameter (m)

For empty, small, high tanks, the resistance to sliding and overturning under the force of wind should be checked. If $M_{OT} > 2/3 \ (WD/2)$, anchor bolts are required.

If anchors are required, the design tension load per anchor shall be calculated as follows:

$$t_B = (4M_{OT})/dN - W/N$$

where

t_B = Design tension load per anchor (tons)

d = Diameter of anchor circle (m)

N = Number of anchors

Noting that, the anchors should be spaced at a maximum of 3 m (10 ft) apart.

5.5.2 Earthquake Stability

The seismic overturning moment of the tank and shear forces shall be provided by the tank's vendor. The minimum safety factor against overturning and sliding should be 1.5. For seismic forces computations, refer to ASCE 7-02 and API STD 650.

5.5.3 Soil Bearing

In selecting the proper type of foundation, the bearing capacity of the soil is the primary factor. A thorough knowledge of soil properties is necessary to avoid excessive differential settlement and possible failure.

Foundations on sandy soil shall be checked using the weight of water for the tank hydrostatic test since settlement under this condition occurs very rapidly and should be essentially complete after the hydrotest weight (which is the first loading). Foundations on clay soils should be checked using the weight of the commodity (i.e., operation load) because clay settles more evenly and slowly due to consolidation. In some instances, it is advisable to use a water test to consolidate clay layers by slowly filling the tank and waiting for settlement to occur.

When considering soil-bearing pressure due to empty plus wind conditions, there are two possible cases. In the first case, the foundation has no uplift, so all of the ring beam is subject to compression. If there is no uplift, then maximum soil pressure for empty plus wind will be less than the maximum soil pressure for operating plus wind. Thus, when there is no uplift, soil pressure for empty plus wind does not need to be checked. The e/D_o ratio can be used to determine whether uplift is present. If uplift is *not* present, then:

$$q_{min} = P/A - M_{OT}/S_{xx}$$

or

$$e/D_o \leq S_{xx}/AD_o$$

where

$$S_{xx} = \pi\left(D_o^4 - D_i^4\right)/32D_o$$

$$A = \pi\left(D_o^2 - D_i^2\right)/4$$

Thus,

$$e/D_o \leq 0.125\left\{1+\left(D_i/D_o\right)^2\right\},$$

or

$$e/D_o \leq 0.20 \text{ for } D_i/D_o \geq 0.8,$$

where
D_o = Ring beam outside diameter (m)
D_i = Ring beam inside diameter (m)
e = Eccentricity (m)

5.5.4 Soil Pressure (Uplift Is Present)

If uplift is present ($e/D_o > 0.20$), soil pressure should be calculated because the maximum pressure for empty plus wind could conceivably be higher than operating plus wind. In computing maximum pressure for a particular ratio (e/D_o), two transcendental simultaneous equilibrium equations must be solved. Since uplift will seldom be of major consequence with storage tanks, no design curves are developed to solve the two equilibrium equations. If $e/D_o > 0.33$, the engineer may either increase h_b or use another foundation type. However, t_b should not be increased to meet wind requirements.

5.5.5 Concrete Ring Beam Design

The ring beam depth shall be determined from the allowable soil-bearing pressure requirements. The thickness of the ring wall, t_b, is determined from consideration of bearing pressure. However, to minimize differential settlements, the bearing pressure under the ring wall must equal the bearing pressure beneath the tank, at the same elevation as the ring wall bottom. The minimum ring wall width shall be 300 mm (12 in.).

Soil pressure under the main portion of the tank (ton/m²)

$$P_o = \gamma_e h_L + \gamma_s h \tag{5.21}$$

Soil pressure beneath the ring beam (ton/m²)

$$P_o = P/b + \gamma_o(Lh_L/b) + h\gamma_C \tag{5.22}$$

Equating Equations (5.21) and (5.22) and solving for b yield:

$$b = (P + Lh_L\gamma_o)/[\gamma_o h_L - h(\gamma_C - \gamma_S)] \tag{5.23}$$

where
  γ_o = Unit weight of tank product (ton/m³)
  γ_S = Unit weight of soil (ton/m³)
  γ_C = Unit weight of concrete (ton/m³) = 2.5 ton/m³
  h_L = Height of tank product (m)
  h = Height of ring beam (m)
  P = Weight of tank wall and roof weight per meter of circumference (ton/m')
  b = Width of ring wall (m)

In any way, the beam width should satisfy the following equation:

$$q = P/b + \gamma_o(L \cdot h_L/b) + h \cdot \gamma_C \le q_{all}$$

where
  q_{all} = Allowable bearing pressure

5.5.6 Ring Wall Reinforcement

Horizontal reinforcing of the ring beam foundations should be designed to resist horizontal active earth pressure in hoop tension with horizontal reinforcing sized to take all of the tension. Passive soil pressure is neglected and should not be used to reduce the horizontal design loads. The normal horizontal (radial) forces acting at every section along the ring wall are shown below.

F_o = horizontal force due to products in tanks = $K_a\gamma_o(h)h_L$
F_s = horizontal active soil = $K_a\gamma_s(h)^2/2$
Total force $F = F_o + F_s$

Thus,

$$F = K_a h(\gamma_o h_L + \gamma_s h/2) \tag{5.24}$$

where
F = Radial outward force on ring beam (ton/m)
K_a = Active earth pressure coefficient = $\tan^2(45° - \Phi/2)$
Φ = Angle of internal soil friction

If Φ is not known, K_a should be taken as 0.3 for sand and 0.7 for clays.
   The radial force, F, causes a hoop tension in the ring, which is presented in Figure 5.23.
   For working stress design:

$$T = \tfrac{1}{2}FD \tag{5.25}$$

For limit state design based on ACI:

$$T_U = 1.7T \tag{5.26}$$

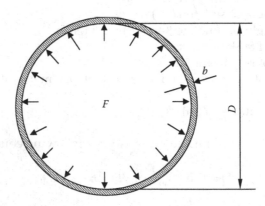

FIGURE 5.23
Hoop stresses.

where
   T = Hoop tension in ring beam
   T_U = Ultimate hoop tension in ring beam
   D = Tank diameter

The horizontal reinforcing should be designed to carry all of the tension. By using ultimate stress design:

$$A_s = T_U/\Phi f_v$$

The minimum horizontal reinforcing should be equal to 0.0025 times the ring cross-section area.

$$A_s \text{ (min)} = 0.0025 \times h \times b$$

All circumferential tension steels should be continuous with splices staggered. All splices should be class B type (1.3 ld) as given in ACI 318-99. For vertical reinforcing, the minimum shall equal 0.0015 times the ring cross-section area, but not less than 5 bars diameter, 8 mm every meter ($5\phi8/\text{m}'$).

$$A_s \text{ (min)} = 0.0015 \times h \times b$$

In special cases for areas of weak, compressible soils, special foundations and design procedures or soil improvement may be required to prevent excessive settlement. As a general guideline, if the differential settlement

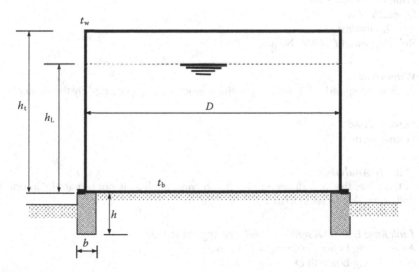

FIGURE 5.24
Example—sketch ring beam design.

approaches 12 to 20 mm at a distance of 9 to 12 m of the shell circumference of an ordinary flat bottom tank, a detailed evaluation of the soil pressure should be made. Under no circumstances should the soil pressure exceed the allowable rate.

The location of drain-out nozzles at the tank bottom should be considered and appropriate block out details should be provided in the concrete ring wall. This will prevent any interference with the drain out nozzle flanges.

Example: Floating Roof Tank

Tank Data (from Certified Vendor Drawing)
Floating roof tank as shown in Figure 5.24
Diameter, $D = 110$ m
Tank height, $h_T = 18$ m
Average wall plate thickness, $t_w = 22$ mm
Bottom plate thickness = 10 mm
Height of liquid, $h_L = 17.7$ m
Unit weight of product, $\gamma_o = 9$ ton/m^3
Unit weight of water to hydrostatic test, $\gamma_o = 1.0$ ton/m^3
Assume weight of stairs and platform to 5% of the weight of tank shell.

Soil Data (from Geotechnical Soil Investigation Report)
Unit weight of soil, $\gamma_S = 2$ ton/m^3
$K_a = 0.30$
Allowable bearing pressure = 0.19 N/mm^2

Foundation Material
$f_c' = 27.5$ N/mm^2
$\gamma_C = 2.5$ ton/m^3
Reinforcement $f_y = 360$ N/mm^2

Wind Load
Basic wind speed = 93 mph as per the Metocean data provided by the owner

Seismic Load
Seismic zone = 0

Stability Analysis
In this case, the tank diameter $D > h_L$, so wind analysis is not required. No uplift will occur due to wind load.

Tank Ring Beam Design and Soil-Bearing Pressure
Assume ring beam depth, $h_b = 1250$ mm
Assume ring beam thickness, $t_b = 450$ mm
Weight of tank wall per meter of circumference

$$W_W = 0.022 \times 7.85 \times 18 \times 1.05 = 3.3 \text{ ton/m}'$$

Forces Acting on Ring Beam
Soil–Bearing Pressure under Ring Beam
$W_R = 0$ for floating root
$L = b/2$
$P_o = (W_W + W_R)/b + \gamma_o h_L/2 + h\gamma_C$
$P_o = (3.3 + 0)/0.45 + (0.9 \times 17.7)/2 + 1.25 \times 2.5$
$P_o = 7.33 + 7.97 + 3.125 = 18.43 \text{ ton/m}^2 < 19.2 \text{ ton/m}^2$ OK
Soil–Bearing Pressure under Tank
$P_o = \gamma_o h_L + \gamma_s h$
$P_o = (0.9 \times 17.7) + (1.8 \times 1.25) = 18.18 \text{ ton/m}^2 < 19.2 \text{ ton/m}^2$ OK
Soil-bearing pressure under ring wall and tank bottom is almost equal; therefore, no settlement problem will occur.

Ringwall Thickness
$b = (W_W + W_R)/\{\gamma_p h_L - 2h(\gamma_C - \gamma_s)\}$
b (product) $= (3.3 + 0)/\{0.9 \times 17.7 - 2 \times 1.25 (2.5 - 1.8)\} = 0.23$ m
b (hydrotest) $= 3.3/\{1.0 \times 17.7 - 2 \times 1.25 (2.5 - 1.81)\} = 0.21 < 0.45$ OK

Ringwall Reinforcement
Horizontal Reinforcement
Radial outward force, F, on ring beam (ton/m)

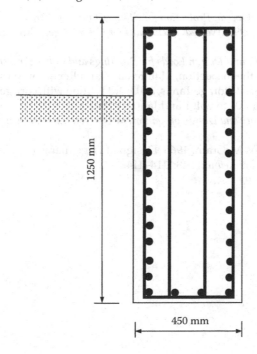

FIGURE 5.25
Reinforced detail for ring beam.

For hydrotest, as a maximum loading case
$F_H = 0.3 \times 1.25 (1.0 \times 17.7 + 2 \times 1.25/2) = 7.11$ ton/m'
Hoop tension, T, in ring beam $= \frac{1}{2}FD$
$T_H = \frac{1}{2} \times 7.11 \times 109.78 = 390$ tons
Using ultimate stress design

$$T_U = 1.7 \times 390 = 663 \text{ tons}$$

A_s (required) $= T_U/\phi f_y$
A_s (required) $= 663 (9800)/0.9 \times 360 = 20{,}054$ mm²
A_s (provided) $= 22 \; \phi \; 25$ mm (two faces) $= 21{,}598$ mm² $> 20{,}054$ mm²   OK
A_s (minimum) $= 0.0025 \times 1250 \times 450 = 1406$ mm² $< 20{,}054$ mm² provided   OK
Vertical Reinforcement
A_s (minimum) $= 0.0015 \times h_R \times t_R$
A_s (minimum) $= 0.0015 \times 1.25 \times 0.45 \times 1{,}000{,}000 = 843.75$ mm²
Choose 6 ϕ 10 mm/m'
$A_s = 2 \times 78 \times 6 = 936$ mm² $> A_s$ min   OK
The detailing of the steel bars in the ring beam is illustrated in Figure 5.25.

References

API RP 2350. 2005. *Overfill Protection for Storage Tanks in Petroleum Facilities.*
API Standard 650. 1998: *Welded Steel Tanks for Oil Storage*, 10th Edition. Washington, DC: API.
ASCE 7. 2002. *Minimum Design Loads for Buildings and Other Structures.*
EEMUA, Guide to the Inspection, Maintenance and Repair of Above Ground Vertical Cylindrical Steel Storage Tanks, 2003. 3rd Edition with corrigenda and ammendment February 04 to vol. 1 and January 2005 to vol. 2.
Hilal, M. 1988. *Theory and Design of Reinforced Concrete Tanks.* Cairo, Egypt: J & Marcou Co.
Zick, L. P., and R. V. McGrath. 1968. Design of large diameter cylindrical shells. *Proc Am Pet Inst, Div Refining.* 48:1114–1140.

6

Static Equipment Foundation Design

6.1 Introduction

The equipment used by oil and gas plants generally falls into two categories: static equipment and rotating equipment. In this chapter, we will focus on static equipment such as vertical and horizontal separators, knockout drums, vertical desalination towers, and similar pieces of equipment, which are called pressure vessels because most contain a fluid under pressure.

This chapter establishes guidelines and recommends procedures for use by engineers in analyzing and designing foundations for static equipment using the heat exchanger and horizontal vessel as examples. These guidelines and recommendations should be used where applicable, unless otherwise specified.

6.2 Design Procedure

The engineer should review project design criteria to determine wind and earthquake loads, corrosion allowances for anchor bolts, anchor bolt types, and any special requirements dictated by the owner. The engineer should verify that the design is based on applicable codes when the foundation drawings are issued.

6.2.1 Dead Loads

If you apply working strength design or limit state design, the dead load will consist of the following items (these loads represent all the types that will affect the pressure vessel during its lifetime):

1. Structure dead load (D_s): the weight of the foundation and soil above the part of the foundation that resists uplift.

2. Erection dead load (D_f): the fabricated weight of the exchanger or vessel, generally taken from certified exchanger or vessel drawings.

3. Empty dead load (D_e): the empty weight of the exchanger or vessel, including all attachments, trays, internals, bundle, insulation, fireproofing, agitators, piping, ladders, platforms, and others. The eccentric load should also be added to the empty dead load weight.

4. Operating dead load (D_o): the load affect during operation. This load will be the empty dead load of the exchanger or vessel, in addition to the maximum weight of contents during normal operation. The eccentric load should also be added to the operating dead load weight.

5. Test dead load (D_t): the load during testing of the vessel. This load is the dead load of an empty vessel plus the weight of the test medium contained in the system. The fluid that will be used in the test should be as specified in the contract documents or by the owner. Unless otherwise specified, a minimum specific gravity of 1.0 should be used for the test medium.

The load during cleaning should be used instead of the test dead load if the cleaning fluid has a specific gravity higher than 1. It should be determined whether testing or cleaning will actually be conducted in the field.

It is generally desirable to design for the test dead load because unforeseen circumstances may occur. The eccentric load should also be added to the test dead load weight. The following guidelines are recommended for use if more exact information about piping supported on the exchanger or horizontal vessel is not available. These guidelines should be verified after obtaining the actual data.

1. Heat exchangers
   (a) A load of an additional 20% of the applicable weight (empty or operating) for heat exchangers with diameters less than 600 mm (24 in.)
   (b) A load of an additional 10% of the applicable weight (empty or operating) for heat exchangers with diameters equal to or greater than 600 mm (24 in.)

2. Vessels

   A load of an additional 10% of the applicable weight (empty, operating, or test) for horizontal vessels. This additional load, which we discussed earlier, should be applied at a perpendicular horizontal distance of $D/2 + 450$ mm (18 in.) from the longitudinal centerline of the vessel. The basic diameter is calculated using the following equation: basic diameter (D) = vessel's internal diameter (ID) + 2(wall thickness) + 2(insulation thickness).

This additional eccentric load (vertical load and moment caused by eccentricity) should be distributed to each pedestal in proportion to the distribution of the operating load to each pedestal. For stacked exchangers, the weight of only the largest exchanger should be used to estimate the eccentric load. For most common shell and tube heat exchangers, vertical dead loads should normally be distributed with 60% to the channel end support and 40% to the shell end support. The actual exchanger shape and support configuration should be reviewed when determining weight distribution because, in many cases, load distribution may vary.

6.2.2 Live Loads

Live load (*L*) data are described in Chapter 3, although the live load data discussed in that chapter concern maintenance of valves, measurement, and other activities. There will be less load effect in these cases than in the case of pressure vessels.

6.2.3 Wind Loads

The engineering office is responsible for determining the wind loads (W) used for the foundation design. Wind loads from vendors or other engineering disciplines should not be accepted without verification. Transverse wind—the wind pressure on the projected area of the side of the exchanger or vessel—should be applied as a horizontal shear at the center of the exchanger or vessel. Including the wind load on projections such as piping, manways, insulation, and platforms during wind analysis is important. The saddle-to-pier connection should be considered fixed for transverse loads.

Longitudinal wind—the wind pressure on the end of the exchanger or vessel—should be applied as a horizontal shear at the center of the exchanger or vessel. The flat surface wind pressure on the exposed area of both piers and both columns should also be included, applied as a horizontal shear at the centroid of the exposed area. The saddle-to-pier connection will be considered pinned for longitudinal loads unless more than one row of anchor bolts exists.

No allowance should be made for shielding from wind by nearby equipment or structures, except under unusual conditions because, in industrial facilities, equipment is regularly moved or changed.

6.2.4 Earthquake Loads

Earthquake loads (*E*) should be calculated in accordance with the relevant code of the country where the plant is located, but in general, UBC is the most popular code for earthquake loads for buildings, whereas SEI/ASCE 7-05 is preferred for nonbuilding structures, as this code focuses specifically on them.

The engineer is responsible for determining the earthquake loads used for the foundation design. Earthquake loads from vendors or other engineering disciplines should not be accepted without verification. For low-friction slide plates ($\mu \leq 0.2$), all longitudinal earthquake loads should be applied at the fixed pier. In higher friction slide plates ($\mu > 0.2$), 70% of the earthquake loads should be applied at the fixed pier. Earthquake loads in transverse and vertical directions should be distributed in proportion to the vertical load applied to both piers.

The piers are normally designed for the fixed end, then the pier for the sliding end is made identical, to avoid potential errors in construction and reduce engineering time. If this proves to be uneconomical, the sliding end should be designed for 30% of the longitudinal earthquake load if using low-friction slide plates and for 50% of the longitudinal earthquake load if using high-friction slide plates. Horizontal loads should be divided equally between piers unless otherwise required by the engineering judgment.

6.2.5 Bundle Pull Load (Exchangers)

Bundle pull loads only exist in the case of heat exchangers, as during operation, it is necessary to remove the bundle tubes inside the heat exchangers for cleaning. This should be defined by the operation group and the data sheet, as there are some types of cleaning that do not require removing the bundle.

In this calculation, reducing the empty weight of the exchanger because of the removal of the exchanger head (channel) to pull the bundle should be considered. The weight of the exchanger head (channel) is typically within the range of 8% to 15% of the empty weight of the exchanger.

6.2.6 Thermal Forces

By knowing the maximum design temperature, you can calculate the thermal growth using the thermal coefficients, as shown in Table 6.1. The thermal force used for design should be the smaller value resulting from the following two calculations:

1. The force required to overcome static friction between the exchanger or vessel support and the slide plate:

$$F_f = \mu(P_o) \tag{6.1}$$

where
F_f = Static friction force
μ = Coefficient of friction; refer to the values given in Section 6.6
Po = Nominal operating compression dead load on slide plate

TABLE 6.1

Total Linear Expansion between 70°F and Indicated Temperature (mm/m) for Different Materials

Temperature (°F)	Carbon Steel Carbon-Moly Low-Chrome (through 3 Cr Mo)	5 Cr Mo through 9 Cr Mo	Austenitic Stainless Steels 18Cr 8 Ni	12 Cr 17 Cr 27 Cr	12 Cr 20 Ni	Monel 67 Ni 30 Cr	3-1/2 Nikel	Ni–Fe–Cr
70	0	0	0	0	0	0	0	0
100	0.18	0.28	0.17	0.27	0.23	0.19	0.23	0.18
125	0.33	0.52	0.30	0.48	0.43	0.35	0.43	0.33
150	0.48	0.75	0.44	0.70	0.63	0.51	0.63	0.48
175	0.63	0.98	0.58	0.92	0.83	0.68	0.83	0.63
200	0.78	1.22	0.72	1.14	1.02	0.84	1.03	0.78
225	0.94	1.46	0.86	1.37	1.22	1.01	1.24	0.94
250	1.11	1.69	1.01	1.59	1.43	1.18	1.47	1.11
275	1.27	1.93	1.15	1.82	1.63	1.36	1.69	1.27
300	1.43	2.18	1.30	2.04	1.84	1.53	1.92	1.43
325	1.58	2.42	1.45	2.27	2.03	1.71	2.16	1.58
350	1.75	2.67	1.61	2.49	2.23	1.88	2.40	1.75
375	1.92	2.92	1.76	2.72	2.43	2.06	2.65	1.92
400	2.08	3.17	1.92	2.94	2.71	2.24	2.90	2.08
425	2.27	3.42	2.08	3.17	2.93	2.43	3.13	2.27
450	2.44	3.68	2.24	3.39	3.16	2.61	3.37	2.44
475	2.62	3.93	2.41	3.62	3.38	2.79	3.59	2.62
500	2.79	4.18	2.57	3.84	3.61	2.98	3.83	2.79
525	2.98	4.43	2.73	4.07	3.84	3.18	4.06	2.98
550	3.17	4.68	2.91	4.29	4.08	3.37	4.30	3.17

(continued)

TABLE 6.1 Continued

Total Linear Expansion between 70°F and Indicated Temperature (mm/m) for Different Materials

Temperature (°F)	Carbon Steel Carbon-Moly Low-Chrome (through 3 Cr Mo)	5 Cr Mo through 9 Cr Mo	Austenitic Stainless Steels 18Cr 8 Ni	12 Cr 17 Cr 27 Cr	12 Cr 20 Ni	Monel 67 Ni 30 Cr	3-1/2 Nikel	Ni-Fe–Cr
				Shell Material				
575	3.35	4.94	3.08	4.52	4.32	3.56	4.53	3.35
600	3.53	5.20	3.25	4.74	4.55	3.75	4.77	3.53
625	3.73	5.46	3.42	4.97	4.79	3.95	5.01	3.73
650	3.91	5.73	3.59	5.19	5.04	4.15	5.25	3.91
675	4.10	5.98	3.77	5.42	5.28	4.35	5.48	4.10
700	4.28	6.25	3.94	5.64	5.53	4.55	5.73	4.28
725	4.48	6.52	4.12	5.87	5.78	4.75	5.98	4.48
750	4.68	6.79	4.30	6.09	6.04	4.95	6.23	4.68
775	4.88	7.06	4.48	6.32	6.29	5.15	6.47	4.88
800	5.08	7.33	4.67	6.54	6.54	5.36	6.72	5.08
825	5.28	7.61	4.85	6.79	6.80	5.57	6.96	5.28
850	5.49	7.88	5.04	7.04	7.07	5.78	7.22	5.49
875	5.69	8.16	5.23	7.29	7.33	5.98	7.46	5.69
900	5.89	8.43	5.41	7.54	7.60	6.19	7.72	5.89
925	6.09	8.72	5.59	7.79	7.87	6.40	7.97	6.09
950	6.30	9.00	5.78	8.04	8.14	6.61	8.23	6.30
975	6.51	9.28	5.98	8.29	8.41	6.81	8.48	6.51
1000	6.72	9.57	6.17	8.54	8.68	7.01	8.74	6.72
1025	6.92	9.85	6.35	8.79	8.96	0.00	9.00	6.92

1050	7.13	10.13	6.63	9.04	9.24	0.00	9.26	7.13
1075	7.33	10.42	6.82	9.29	9.53	0.00	9.52	7.33
1100	7.54	10.70	6.93	9.54	9.81	0.00	9.78	7.54
1125	7.73	10.98	7.11	9.82	10.09	0.00	10.04	7.73
1150	7.93	11.27	7.30	10.09	10.39	0.00	10.32	7.93
1175	8.13	11.55	7.48	10.37	10.68	0.00	10.58	8.13
1200	8.33	11.83	7.67	10.64	10.96	0.00	10.85	8.33
1225	8.55	12.12	7.85	10.92	11.25	0.00	11.13	8.55
1250	8.78	12.40	8.04	11.19	11.55	0.00	11.43	8.78
1275	8.99	12.68	8.23	11.47	11.85	0.00	11.70	8.99
1300	9.22	12.97	8.43	11.74	12.15	0.00	11.99	9.22
1325	9.42	13.25	8.61	11.99	12.45	0.00	12.28	9.42
1350	9.63	13.53	8.80	12.24	12.75	0.00	12.58	9.63
1375	9.83	13.82	8.98	12.49	13.05	0.00	12.87	9.83
1400	10.04	14.10	9.18	12.74	13.35	0.00	13.17	10.04
1425	0.00	14.42	0.00	0.00	0.00	0.00	13.47	0.00
1450	0.00	14.74	0.00	0.00	0.00	0.00	13.78	0.00
1475	0.00	15.07	0.00	0.00	0.00	0.00	14.07	0.00
1500	0.00	15.39	0.00	0.00	0.00	0.00	14.38	0.00

2. The force required to deflect the pier or column is equal to half of the thermal expansion between exchanger or vessel saddles:

$$T = \frac{3\Delta EI}{2H^3}$$ (6.2)

where

T = Force from thermal expansion required to deflect pier or column
Δ = Total growth between exchanger or vessel saddles = εL
ε = Thermal expansion coefficient in accordance with Table 6.1
L = Length of exchanger or vessel between saddles
E = Modulus of elasticity of concrete pier
I = Pier moment of inertia
H = Pier height

The thermal force should be applied at the top of the piers. Horizontal loads should be divided equally between piers, unless otherwise required.

6.2.7 Load Combinations

Heat exchangers and horizontal vessel foundations should be designed using the load combinations presented in Chapter 3. The piping thermal loads should be included in combinations when deemed advisable and should be considered dead loads when applying load factors.

6.3 Anchor Bolts

Friction force at the bottom of the saddle should be overcome before a lateral load is assumed to produce shear in the anchor bolts. For earthquake loads, horizontal shear forces should be applied to the anchor bolts, assuming no frictional resistance.

6.4 Slide Plates

A steel slide plate or low-friction slide plate assembly should typically be provided at the sliding end of every exchanger or vessel, regardless of the flexibility inherent in the structural support. Small, lightly loaded exchangers or vessels may not require slide plates. Low-friction manufactured slide plate assemblies should be used to reduce high-frictional resistance, especially in cases of heavy exchangers or for exchangers with significant thermal growth. For exchangers with bundle pull, steel slide plates instead of low-friction slide plate assemblies may be more cost efficient.

Typically, a low-friction slide plate assembly consists of multiple individual slide plate components spaced out along the length of the saddle. Each slide plate component consists of an upper element and a lower element, and the sliding surface is at the interface of the upper and lower elements. The elements should be fabricated with a carbon steel backer plate attached to the elements to facilitate welding of the upper elements to the saddles and the lower elements to the steel bearing plate.

Typical coefficients of friction are as follows. For low-friction slide plate assemblies, the manufacturer's data sheet should be read carefully, as the coefficients of friction vary with slide plate material, temperature, and pressure. Use the following figures for coefficients of friction:

1. No slide plate (steel support on concrete): 0.60
2. Steel slide plate: 0.40
3. Low-friction slide plate assemblies: 0.05 to 0.20

It is suggested that the criteria for sizing low-friction slide plate elements should be as follows. The manufacturer's data sheet and specifications should be consulted for temperature restrictions, pressure limitations, and other requirements that may affect the size and types of materials used for the slide plate elements. These data should be available to the mechanical department, which is always responsible for delivering these data. By considering Δ as the total thermal growth between exchanger or vessel saddles, the element widths will be defined by the following limits:

1. Upper element = saddle width + 25 mm minimum to allow for downhand welding on the element-to-saddle weld (a larger upper element width may be required for exchangers or vessels with large Δ values).
2. Lower element = upper element width – 2(Δ) – 25 mm (minimum of 25 mm narrower than upper element).

Element lengths should be defined as follows (use a 450-mm maximum clear distance between lower elements):

1. The lower element is based on the allowable contact pressure, in accordance with the manufacturer's literature and the lower element width.
2. Upper element = lower element length + 25 mm.

Plates should be aligned with saddle stiffeners where practical. A continuous steel bearing plate should be provided under the lower elements so that the lower elements can be welded to the bearing plate. The minimum width of the bearing plate should be 25 mm larger than the width of the lower elements. The minimum length of the bearing plate should be 25 mm larger than the saddle length. Bearing stress on concrete should be checked in accordance with ACI 318-05/318R-05.

In summary, the suggested criteria for sizing steel slide plates are as follows:

1. Minimum width = saddle width + 2Δ + 25 mm
2. Minimum length = saddle length + 25 mm

Bearing stress on concrete should be checked in accordance with ACI 318-05/318R-05.

6.5 Pier Design

The pier dimensions should be sized on the basis of standard available forms for the project. When information is not available, pier dimensions should be sized in 2-in. (50-mm) increments to allow use of standard contractor wood forms. Minimum pier dimensions should equal the maximum of the saddle, the bearing plate, or the steel slide plate dimensions, plus 100 mm, and should be sized to provide an adequate anchor bolt edge distance based on the anchor bolt design section. Minimum pier width should be no less than 250 mm or 10% of the pier height.

6.5.1 Anchorage Considerations

It is normally desirable to make the pier high enough to contain the anchor bolts and to keep them out of the footing. Consideration must be given to anchor bolt development and foundation depth requirements.

6.5.2 Reinforcement for Piers

In general, the piers should normally be designed as cantilever beams with two layers of reinforcement. If the pier is a compression-controlled member, the pier should be designed as a column. In any case, the size and reinforcement for each pier should normally be the same. Dowel splices are not required if the vertical pier reinforcing projection is less than 1.8 m or the rebar size (in feet) above the top of the footing. For example, the #8 rebar (25 mm) can extend up to 8 ft (2.5 m) above the mat without dowel splices.

For cases that exceed this limit, dowels with minimum projections required for tension splices should be used in accordance with ACI 318-05/318R-05 or BS8110.

Vertical reinforcement in piers may need to be increased to account for shear friction. The following formula should be used to calculate the area of reinforcement required for shear friction:

$$A_{vf} = [V_u/(\mu\varphi) - P_u \text{ pier}]/f_v \tag{6.3}$$

where V_u is the strength design factored shear force at the bottom of the pier.

6.6 Foundation Design

Soil-bearing pressure should be computed for the footing design and checked against the allowable bearing capacity, which is available from the geotechnical report based on the following equation (the total footing area will be under compression in the case of $e \leq (b/6)$):

$$e = M/P$$

$$q = \frac{P}{A}\left(1 \pm \left(\frac{6e}{b}\right)\right) \tag{6.4}$$

In the case of $e > (b/6)$, the total footing area is not under compression and some of the area is under tension (Figure 6.1), where

e = Eccentricity of the vertical service load caused by the horizontal service load
a = Size of footing perpendicular to the direction of the horizontal load
b = Size of footing parallel to the direction of the horizontal load
P = Total vertical service load (exchanger or vessel, pier, footing, and soil)
A = Area of footing

In some cases, eccentricity exists in both directions (Figure 6.2), and under such a situation, the above equations cannot apply.

The numerical solutions can be found in many soil mechanics textbooks, and there are many commercial software available for such calculations. The following principal equation will be applied as follows:

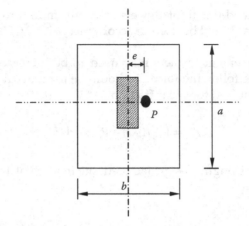

FIGURE 6.1
Sketch of load eccentricity.

$$e_1 = \frac{M_1}{p} \text{ and } e_2 \frac{m_2}{p} \quad q_{max} = \frac{p}{a} + \frac{6pe_1}{ba^2} + \frac{6pe_2}{ab^2} \quad q_{min} = \frac{q}{a} + \frac{6pe_1}{ba^2} + \frac{6pe_2}{ab^2}$$

6.6.1 Foundation Reinforcement

6.6.1.1 Bottom Reinforcement

The strength design factored moment and shear should be figured on a unit width strip, assuming a simple cantilever. The critical section for moment and diagonal tension shear should be taken at the pier or column face. If

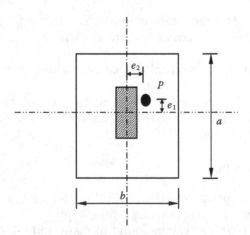

FIGURE 6.2
Sketch of load eccentricity in two directions.

shear is excessive, the strength design factored shear should be rechecked using the critical section for shear specified in ACI 318-05/318R-05.

The resulting reinforcing steel should be placed continuously across the entire footing. The minimum amount of bottom reinforcement is 16 mm at every 300 mm based on ACI 318-05/318R-05. In a normal design for a concrete section procedure, it will be according to the code or standard that you are using in the design. In this book, we will apply ACI 318-05/318R-05.

6.6.1.2 Top Reinforcement

Except where seismic effects create tensile stresses, top reinforcement in the footing is not necessary if the factored tensile stress at the upper face of the footing does not exceed the flexural strength of structural plain concrete, as follows:

$$f_t' = 0.4167\varphi(f_c') \tag{6.5}$$

where

f_t' = Flexural strength of structural plain concrete (MPa)
f_c' = Compressive strength of concrete (MPa)
φ = Strength reduction factor for structural plain concrete = 0.55

The actual thickness, t, specified in the drawing should be higher than the calculated thickness of the footing for tensile stress calculations by about 50 mm to allow for the unevenness of excavation and contamination of the concrete adjacent to the soil. For footings cast against plain concrete, the actual thickness of the footing may be used for the calculated depth. If the factored tensile stress exceeds the flexural strength of structural plain concrete, top reinforcement should be used if an increase in footing thickness is not feasible.

The following formulas are used for calculating the required footing thicknesses with no top reinforcing steel.

For footings cast against soil:

$$t = t_c + 50 \text{ mm} \tag{6.6}$$

With t_c calculated as follows:

$$t_c = (M_u/f_t')^{1/2} \tag{6.7}$$

where

t = Required footing thickness with no top reinforcing steel (mm)
t_c = Calculated footing thickness (mm)
M_u = Strength design factored moment caused by the weight of soil and concrete acting on a 1-m strip in the footing at the face of the pier (mt/m'), using a load factor of 1.4
f_t = Flexural strength of structural plain concrete (MPa)

TABLE 6.2

Approximate Heat Exchanger Weight

Heat Exchanger Diameter (in.)	Heat Exchanger Pressure Class		
	150	300	450
15	2.2	2.5	2.9
18	2.8	3.2	3.8
20	3.4	3.9	4.6
22	4.0	4.4	5.4
24	4.8	5.4	6.2
26	5.6	6.4	7.2
28	6.4	7.2	8.2
30	7.2	8.3	8.4
32	8.2	9.4	10.8
34	9.2	10.6	12.3
36	10.4	12.0	13.9
38	11.6	13.6	15.4
40	12.8	14.6	17.0
42	14.2	16.2	18.8
44	15.4	17.8	
46	16.8	19.5	
48	18.5	21	

If tensile stress in the upper face of the footing exceeds the ACI plain concrete design requirements, top steel should be used if increasing the footing thickness is unfeasible. If top reinforcement is required, minimum reinforcement is 12-mm diameter at every 300 mm.

Table 6.2 gives the approximate weight of standard heat exchangers, all in tons. The curves are for a 4.8-m (192-in.) type ET exchanger with two passes in the tubes. The tubes are 3/4 in. on a 90° layout. The tube material is 14-gauge steel. For the weights of heat exchangers with other tube lengths, multiply by the factors given in Table 6.3.

TABLE 6.3

Modification Factor for Different Tube Lengths

Tube Length (m)	Modification Factor
2.4	0.80
3.0	0.85
3.6	0.90
4.2	0.95
4.8	1.00
6.0	1.10

TABLE 6.4

Approximate Tube Bundle Weight

Heat Exchanger Diameter (in.)	Heat Exchanger Pressure Class			
	150	300	450	600
15	0.9	0.9	0.9	1.0
18	1.2	1.3	1.3	1.4
20	1.5	1.6	1.6	1.7
22	1.8	2.0	2.0	2.1
24	2.3	2.4	2.5	2.7
26	2.7	2.8	2.9	3.0
28	3.2	3.3	3.4	3.6
30	3.7	3.9	4.0	4.2
32	4.2	4.45	4.55	4.8
34	4.8	5.0	5.25	5.45
36	5.4	5.7	6.0	6.15
38	6.1	6.4	6.75	6.9
40	6.7	7.0	7.3	7.6
42	7.4	7.7	8.0	8.4
44	8.1	8.4	8.95	9.3
46	8.9	9.2	9.75	10.2
48	9.6	10.0	10.5	11.0

Table 6.4 gives the approximate weight of standard tube bundles, all in tons. The tubes are 3/4 in., 14 gauge, and 4.8 m (192 in.) long. The tubes are two pass on a square pitch. Baffle spacing ranges from 200 mm on the 375-mm (15 in.) exchanger to 400 mm on the 1200-mm (48 in.) exchanger. For the weight of bundle with other lengths, multiply by the factors given in Table 6.5.

TABLE 6.5

Modification Factor for Different Tube Lengths

Tube Length (m)	Modification Factor
3.0	0.70
3.6	0.80
4.2	0.90
4.8	1.00
6.0	1.20

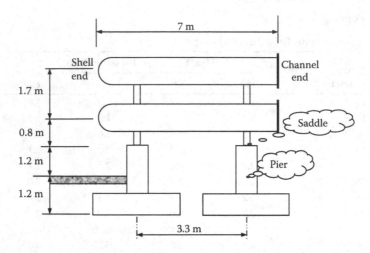

FIGURE 6.3
Example—sketch of heat exchanger.

6.7 Example: Heat Exchanger Data

Heat exchanger dimensions as shown in Figure 6.3.

6.7.1 Design Data

Empty weight = 14.5 tons
Operating weight = 20 tons
Bundle weight = 8.6 tons
Channel weight = 1.6 tons
Basic diameter = 1070 mm
Maximum design temperature = 550°F
Exchanger material: carbon steel
Bolts: Two 1-1/4-in. diameter, ASTM F1554, Grade 36 (galvanized) per pier
Bolt spacing = 80 mm
Saddle = 0.9 × 0.25 m
Load distribution = 60% at channel end, 40% at shell end

6.7.2 Design Criteria

Concrete, $f_c' = 27.5$ MPa
Reinforcing, $f_y = 248$ MPa
Soil unit weight, $\gamma = 2.0$ ton/m³
Allowable net soil-bearing pressure, $q_{all} = 27$ ton/m² at 1.2-m depth
Wind load: SEI/ASCE 7-05
Earthquake load: SEI/ASCE 7-05

6.7.3 Loads Calculation

- Empty and operating loads

  The exchanger weight supplied by outside manufacturers does not include the weight of attached accessories such as pipes and insulation. Hence, it is recommended that you increase the exchanger weight by 10% of the larger exchanger to account for these attached items, as the basic diameter is larger than 600 mm.

  Empty dead load, $D_e = 14.5 + 14.5(1.1) = 30.45$ tons

  Operating dead load, $D_o = 20 + 20(1.1) = 42$ tons

	Empty (per Exchanger) (ton)	Operating (per Exchanger) (ton)	Pier (W)
Transverse	2.5	3.4	0.154
Longitudinal	4	5.5	0.250

- Transverse moment from pipe eccentricity

  Eccentricity = (basic diameter)/2 + (0.45 m) = (1.05)/2 + 0.45 = 0.975 m

  Empty M_{Te} (channel end) = 14.5(0.1)(0.6 channel end)(0.975) = 1.41 mt

  Empty M_{Te} (shell end) = (14.5)(0.1)(0.4 shell end)(0.975) = 0.566 mt

  Operating M_{TO} (channel end) = (20)(0.1)(0.6 channel end)(0.975) = 1.17 mt

  Operating M_{TO} (shell end) = (20)(0.1)(0.4 shell end)(0.975) = 0.78 mt

- Wind loads

  Wind load will be calculated based on the load calculation discussed in Chapter 3 and the data will be obtained from the vendor data. The lateral force due to wind load will affect each exchanger at the center.

  1. Transverse wind, $H_w = 0.58$ tons (per exchanger)
  2. Longitudinal wind, $H_w = 0.11$ tons (per exchanger)
  3. Transverse or longitudinal wind on each pier, $H_w = 1.87$ kN/m^2

- Earthquake loads

  The lateral force due to earthquakes will also be calculated based on the methodology described in Chapter 3. The strength design load will be as described in the following table:

	Empty (per Exchanger) (ton)	Operating (per Exchanger) (ton)	Pier (W)
Transverse	1.72	2.37	0.108
Longitudinal	2.8	3.84	0.175

For calculations based on allowable stress design (service loads), the strength design loads shown in the preceding table should be converted to service loads by multiplying by 0.7.

	Empty (per Exchanger) (ton)	Operating (per Exchanger) (ton)	Pier (W)
Transverse	1.72 tons	2.37 tons	0.108
Longitudinal	2.8 tons	3.84 tons	0.175

- Bundle pull

  The bundle weight, which is applied at the centerline of the top exchanger, will be as follows:

  $$V_B = 8.62 \text{ tons}$$

  Note that a reduction in the empty load of the exchanger owing to the removal of the exchanger head (channel) to pull the bundle is not included in this foundation calculation because the reduction in empty load is not considered to have a significant effect on the design.

- Thermal force

  1. Compute the sliding force (assume that a steel slide plate is used). Then, the coefficient of friction, $\mu = 0.40$.

     The operating load is calculated as follows:

     $$F_f = \mu(P_o) = 0.4(25.2) = 10.08 \text{ tons (at channel end)}$$

  2. Compute the force required to deflect the pier.

     Assume that the pier is 1.05 m (length) by 0.4 m (width) by 1.95 m (height).

     Moment of inertia is as follows:

     $$I = b(h)^3/12 = (1.05)(0.4)^3/12 = 0.0056 \text{ m}^4$$

     Based on ACI 318-05/318R-05, Section 8.5.1, the modulus of elasticity will be calculated as:

     $$E_c = 4700\sqrt{f_c'} \text{ MPa}$$

     $$E_C = 4700(f_c')^{0.5}$$

     $$= 4700(27.5)^{0.5}$$

     $$= 24{,}781 \text{ MPa}$$

Use Table 6.1 as a guideline to obtain the thermal expansion coefficient from the vessel data sheet. The thermal expansion coefficient for carbon steel at 550°F is $\varepsilon = 3.4$ mm/m.

The thermal growth between saddles is as follows:

$$\Delta = (\varepsilon)(L) = (3.4 \text{ mm/m})(3.35) = 11.48 \text{ mm}$$

$$T = 3\Delta EI/2H^3$$

$$= 3(11.48/1000)(24 \times 10^6)(0.0056)/2(1.95)^3$$

$$= 322,182 \text{ N} = 31 \text{ tons}$$

Because $F_f < T$ and because a lower friction factor will not help the distribution of earthquake and bundle pull loads, it is advisable to use a steel slide plate.

6.7.4 Design Elements

6.7.4.1 Size Steel Slide Plate

Width = (saddle width) + 2(Δ) + 25 = 225 + 2(11.48) + 25 = 273 mm (make it 280 mm)

Length = (saddle length) + 25 mm = 900 + 25 = 925 mm

Check bearing stress (operating and longitudinal earthquake):

$$P_{Eo} = 5.5(0.85 + 2.5)/3.4 = 5.5 \text{ tons}$$

Based on Table 3.26 in Chapter 3 for load combination 3:

$$P_u = 1.2(P_o) + 1.0(P_{Eo})$$

$$P_u = 1.2(25.2) + 1.0(5.5) = 35.74 \text{ tons}$$

Based on ACI 318-08, Section 10.17:

$$P_n = \varphi 0.85 f'_c A1 = (0.65)(0.85)(27.5 \text{ m})(275 \text{ m})(925 \text{ m})$$

$$= 3864.910 \text{ kN} = 3864.9 \text{ kN} > P_u \quad \text{OK}$$

Choose a steel slide plate of thickness 925 × 280 × 10 mm.

6.7.4.2 Pier Size

- Pier length
  - c/c bolts + [2 mm(minimum anchor bolt edge distance)] = 800 + [2(125 mm)] = 1050 mm
  - (steel slide plate length) + (100 mm) = (925) + (100) = 1025 mm

  Therefore, choose a pier length that is not less than 1050 mm.
- Pier width

  Choose the greater of:
  - 250 mm
  - 10% of the pier height, in mm = (0.10)(1.95) = 0.195 mm (pier height is assumed)
  - 2 × (minimum anchor bolt edge distance, in mm) = 2(125) = 25 mm
  - (steel slide plate width, in mm) + (100) = (280) + (100) = 380 mm ← controls

The wooden form should be in increments of 50 mm; therefore, the pier size will be

$$110 \text{ mm (length)} \times 400 \text{ mm (width)}$$

The anchor bolt design is illustrated in Chapter 7. In this example, assume the use of two 25-mm, ASTM F1554, Grade 36 anchor bolts per pier.

6.7.4.3 Pier Design

Assume that footing thickness is equal to 0.5 m.

$$\text{Pier height} = 2.45 - 0.5 = 1.95 \text{ m}$$

The pier height information is constrained by the piping attached to the vessel.

$$\text{Pier weight} = 2.5(0.4)(1.1)(1.95) = 2.15 \text{ tons}$$

Use load combinations and strength design load factors as shown in Table 3.26 in Chapter 3.

1. Operating and longitudinal earthquake at fixed end

   Apply 70% of exchanger earthquake loads at fixed end.

   Horizontal load at fixed end (based on Table 3.26 in Chapter 3, load combination 3),

$$V_{FX} = 1.0[(0.7)(5.5)(2 \text{ exchangers}) + (0.25)(2.15)] = 7.7 + 0.54 = 8.25 \text{ tons}$$

Therefore, shear and moment at the bottom of pier will be

$$V_{FX} = 8.24 \text{ tons}$$

$$M_{FX} = 7.7(1.95) + 1.14(1.95/2) = 16.13 \text{ mt}$$

2. Empty and bundle pull at fixed end

Bundle pull force (using Table 3.26 in Chapter 3, load combination 9 or 10)

$$V_B = 1.6(8.6) = 13.76 \text{ tons}$$

Vertical load at the top of pier due to bundle pull on top exchanger

$$P_B = 13.76(0.85 + 1.68)/3.4 = 10.24 \text{ tons}$$

Net vertical load on sliding pier pushing top bundle in (use for load factor 0.9D from Table 3.26 in Chapter 3, load combination 10)

$$P_{SL} = (0.9)(18.27) - (10.24) = 6.2 \text{ tons}$$

Horizontal load at sliding end

$$V_{SL} = \mu(P_{SL}) = 0.40 \ (6.2) = 2.48 \text{ tons} < 1/2 \text{ bundle pull force } (V_B)$$

Horizontal load at fixed end

$$V_{FX} = (V_B) - (V_{SL}) = (13.76) - (2.48) = 11.28 \text{ tons}$$

Shear and moment at bottom of pier,

$$V_{FX} = 11.28 \text{ tons}$$

$$M_{FX} = (11.28)(1.95 \text{ m}) = 22 \text{ mt}$$

3. Operating and thermal at fixed end

Thermal force

Horizontal force due to thermal effect (V_T) = 1.4(10.08 kips) = 14.1 tons

Horizontal load at fixed end

$$V_{FX} = V_T = 14.1 \text{ tons}$$

Shear and moment at bottom of pier

$$V_{FX} = 14.1 \text{ tons}$$

$$M_{FX} = (14.1)(1.95) = 27.5 \text{ mt}$$

Therefore, the pier will be designed on

Shear force, $V = 14.1$ tons

Bending moment, $M = 27.5$ mt

Proceed with the design based on your applicable code in the project.

The following is a design guide by ACI.
Check diagonal tension shear

$$d = 400 - (50 \text{ clear}) - (12.5 \text{ ties}) - (12.5 \text{ bar thickness}) = 325 \text{ mm}$$

Based on ACI 318-08M,

$$\varphi V_c = \varphi (f_c')^{0.5} b_w d / 6$$

$$= 0.75(27.5)^{0.5}(325)(1100)/6/10,000$$

$$= 23.43 \text{ tons} > V = 14.1 \text{ tons} \quad \text{OK}$$

$$0.5 \ \varphi V_c = (0.5)(23.43) = 11.7 \text{ tons} < V = 14.1 \text{ tons}$$

The minimum tie requirements from a practical point of view are 12-mm ties at 300-mm spacing; however, because $V > 0.5 \ \varphi V_c$, spacing requirement should be checked for $\phi 12$-mm ties to meet the minimum shear reinforcement requirements of ACI 318-08M, Section 11.5.5:

$$A_v = (1/16)(f_c')^{0.5} b_w s / f_v, \text{ but not less than } 0.33 b_w s / f_v$$

where b_w and s are expressed in mm.

$$s = A_v f_v / (1/16)(f_c')^{0.5}, b_w = (113)(2)(248)/(1/16)(27.5)^{0.5}(1100)$$

$$= 155 \text{ mm, but not more than } A_v f_v / 0.33 b_w$$

$$= 133(2)(248)/(0.33)(1100) = 154 \text{ mm}$$

Choose ϕ12-mm ties at 15-mm spacing.

Design for Moment

After designing the section by ACI, choose 6 ϕ25 mm at each face.

6.7.4.4 Footing Size

Determine the minimum footing thickness to develop a standard hook for ϕ25 mm and, by applying the development length equation, the footing thickness will not be less than 45-mm pier reinforcing:

$$q_{all} = (25 \text{ ton/m}^2 \text{ net})$$

Try a 2.45 × 1.7 (m) footing, 0.45-m thickness as shown in Figure 6.4

$$\text{Total Pier Weights, } P_s = 2.5\ (1.1)(0.4)(1.95) = 2.15 \text{ tons}$$

Soil-bearing and stability ratio checks
Use load combinations for allowable stress design (service loads) from Table 3.5 in Chapter 3.

1. Check operating, thermal, and eccentric forces (channel/sliding end)
   Thermal force at top of pier, $V_T = 10.08$ tons
   Maximum axial load at bottom of footing,

$$P_{max} = P_s + P_o = (2.15) + (25.2) = 27.3 \text{ tons}$$

   Moments at bottom of footing (from Table 3.25 in Chapter 3, load combination 1),

$$M_L = (10.08)(2.4) = 24.2 \text{ mt}$$

$$M_{TO} \text{ (from pipe eccentricity)} = 1.17 \text{ mt}$$

   Check soil-bearing using maximum axial load,

$$P = 27.3 \text{ tons}$$

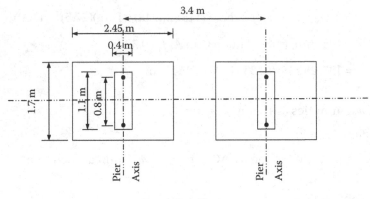

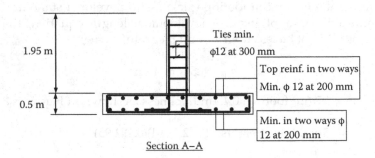

FIGURE 6.4
Example—sketch for foundation plan and sections.

$$M_L = 24.2 \text{ mt}$$

$$M_{TO} = 1.18 \text{ mt}$$

$$q = \frac{27.3}{(2.45)(1.7)} + \frac{6(24.2)}{1.7(2.45)^2} + \frac{6(1.18)}{2.45(1.7)^2} = 22 \text{ ton/m}^2$$

$$q < q_{all} \quad \text{OK safe}$$

2. Check operating and longitudinal earthquake and eccentric loads (shell/fixed end):

   Longitudinal operating earthquake load on exchangers

   $0.7V_{LEo} = 3.84$ tons applied at the center of each exchanger

Vertical load at the top of piers from longitudinal operating earth-quake loads on exchangers due to overturning moment

$$0.7P_{Eo} = (3.84)(0.85 + 2.5)/(3.4) = \pm 3.78 \text{ tons}$$

Axial loads at the bottom of footing

$$P_{max} = P_s + P_o + 0.7P_{Eo} = (2.15) + (16.8) + (3.78) = 22.73 \text{ tons}$$

From Table 3.25, load combination 3,

$$P_{min} = 0.9 (P_s + P_o) - 0.7P_{Eo} = (0.9)(2.15 + 16.8) - (3.78) = 13.28 \text{ tons}$$

Moments at bottom of footing (Table 3.25, load combination 5a)

$$M_L = (0.7 \text{ at fixed end})(3.78)(2 \text{ exchangers})(2.45) + (0.175)(2.15)(1.95/2 + 0.5)$$

$$= 13.52 \text{ mt}$$

M_{To} (from pipe eccentricity) = 0.78 mt

Check the soil-bearing capacity

$$q = \frac{22.73}{(2.45)(1.7)} + \frac{6(13.52)}{1.7(2.45)^2} + \frac{6(0.78)}{2.45(1.7)^2} = 14.1 \text{ ton/m}^2$$

$$q < q_{all} \quad \text{OK}$$

Stability ratio check using minimum axial load:
Overturning moment

$$M_{OT} = \left[M_L + (0.7P_{Eo})(a/2) \right]$$

$$= \left[(13.52) + (3.84)(2.45)/2 \right] = 18.22 \text{ mt}$$

Resisting moment

$$M_R = 0.9 \left(P_s + P_o \right)(a/2)$$

$$= (0.9)(2.15 + 16.8)(2.45)/2 = 20.9 \text{ mt}$$

Stability ratio

$$M_R/M_{OT} = 20.9/(18.22)$$

$$= 1.15 > 1.0 \quad \text{OK}$$

3. Check operating and longitudinal earthquake and eccentric loads (channel/sliding end)

Longitudinal operating earthquake load on exchangers

$0.7V_{LEo} = 3.84$ applied at the center of each exchanger

Vertical load at the top of piers from the longitudinal operating earthquake load on exchangers due to overturning moment

$$0.7P_{Eo} = (3.84)(0.85 + 2.5)/(3.4) = \pm 3.78 \text{ tons}$$

Axial loads at the bottom of footing

$$P_{max} = P_s + P_o + 0.7P_{Eo} = (2.15) + (25.2) + (3.78) = 31.13 \text{ tons}$$

From Table 3.25 in Chapter 3, load combination 3

$$P_{min} = 0.9(P_s + P_o) - 0.7P_{Eo} = (0.9)(2.15 + 25.2) - (3.78) = 20.84 \text{ tons}$$

From Table 3.25 in Chapter 3, load combination 5a, moments at the bottom of footing

$$M_L = (0.3 \text{ at sliding end})(3.84)(2 \text{ exchangers})$$

$$(2.45) + (0.175)(2.15)(1.95/2 + 0.5) = 3.4 \text{ mt}$$

$$M_{TO} \text{ (from pipe eccentricity)} = 1.18 \text{ mt}$$

$$q = \frac{20.84}{(2.45)(1.7)} + \frac{6(3.4)}{1.7(2.45)^2} + \frac{6(1.18)}{2.45(1.7)^2} = 8.0 \text{ ton/m}^2$$

$$q < q_{all} \quad \text{OK safe}$$

Stability ratio check using minimum axial load:
  Overturning moment

$$(M_{TO}) = [M_L + (0.7P_{Eo})(a/2)] = [(3.4) + (3.84)(2.45)/2] = 8.1 \text{ mt}$$

Resisting moment

$$(M_R) = 0.9(P_s + P_o)(a/2) = (0.9)(2.15 + 25.2)(2.45)/2 = 30.15 \text{ mt}$$

Stability ratio

$$M_R/M_{TO} = (30.15)/(8.1) = 3.72 > 1.0$$

4. Check stability for empty and longitudinal earthquake and eccentric loads (channel/sliding end)

Longitudinal empty earthquake load on exchangers

$0.7V_{LEe} = 2.8$ tons applied at the center of each exchanger

Vertical load at the top of piers from longitudinal empty earthquake load on exchangers due to overturning moment

$$0.7P_{Ee} = (2.8)(0.85 + 2.5)/(3.4) = \pm 2.8 \text{ tons}$$

Minimum axial load at the bottom of footing

$$P_{min} = 0.9(P_s + P_e) - 0.7P_{Ee} = (0.9)(2.15 + 18.27) - (2.8) = 15.6 \text{ tons}$$

Longitudinal moment at bottom of footing, from Table 3.25, load combination 5b

$$M_L = (0.3 \text{ at sliding end})(2.8)(2 \text{ exchangers})(2.45)$$

$$+ (0.175)(2.15 \text{ wt})(1.95/2 + 0.45) = 4.65 \text{ mt}$$

Stability ratio check using minimum axial load:
  Overturning moment

$$(M_{TO}) = [M_L + (0.7P_{Ee})(a/2)] = [(4.65) + (2.8)(2.45)/2] = 8.08 \text{ mt}$$

Resisting moment

$$(M_R) = 0.9(P_s + P_e)(a/2) = (0.9)(2.15 + 18.27)(2.45)/2 = 22.5 \text{ mt}$$

Stability ratio

$$M_R/M_{TO} = (22.5)/(8.08) = 2.78 > 1.0 \quad \text{OK}$$

5. Check stability for empty and longitudinal earthquake and eccentric loads (shell/fixed end):

Longitudinal operating earthquake load on exchangers

$0.7V_{LEe} = 2.8$ tons applied at the center of each exchanger

Vertical load at the top of piers from longitudinal empty earthquake load on exchangers due to overturning moment

$$0.7P_{Ee} = (2.8)(0.85 + 2.5)/(3.4) = \pm 2.8 \text{ tons}$$

Minimum axial load at the bottom of footing

$$P_{min} = 0.9(P_s + P_e) - 0.7P_{Ee} = (0.9)(2.15 + 12.18) - (2.8) = 10.1 \text{ tons}$$

From Table 3.25 in Chapter 3, load combination 5b, longitudinal moment at bottom of footing

$$M_L = (0.7 \text{ at fixed end})(2.8)(2 \text{ exchangers})(2.45)$$

$$+ (0.175)(2.15)(1.95/2 + 0.45 \text{ ft}) = 10.14 \text{ mt}$$

Stability ratio check using minimum axial load:

Overturning moment

$$M_{TO} = [M_L + (0.7P_{Ee})(a/2)] = [(10.14) + (2.8)(2.45)/2] = 13.57 \text{ mt}$$

Resisting moment

$$(M_R) = 0.9(P_s + P_e)(a/2) = (0.9)(2.15 + 12.18)(2.45)/2 = 15.8 \text{ tons}$$

Stability ratio

$$M_R/M_{TO} = (15.8)/(13.57) = 1.16 > 1.0$$

6. Check operating and transverse earthquake and eccentric loads (channel/sliding end)

Transverse operating earthquake load on exchangers,

$0.7V_{TEo} = 2.37$ tons, kips applied at the center of each exchanger

Axial loads at bottom of footing

$$P_{max} = P_s + P_o = (2.15) + (25.2) = 27.35 \text{ tons}$$

From Table 3.25 in Chapter 3, load combination 3

$$P_{min} = 0.9(P_s + P_o) = (0.9)(2.15 + 25.2) = 24.62 \text{ tons}$$

From Table 3.25, load combination 5a, transverse moment at the bottom of footing

$$M_T = (2.37)[(0.85 + 2.45) + (1.75 + 0.85 + 2.45)](0.6 \text{ channel end})$$

$$+ (0.108)(2.15)(1.95/0.45) + (1.17 \text{ mt pipe eccentricity})$$

$$= 13.23 \text{ mt}$$

Soil-bearing check using maximum axial load

$$e = M_T/P_{max} = (13.23)/(27.3) = 0.48 \text{ m} > b/6 = (5.5)/6 = 0.283 \text{ m}$$

$$q_{max} = 2P_{max}/[3a(b/2 - e)] = 2(27.35)/[3(2.45)(1.7/2 - 0.48)] = 20.11 \text{ ton/m}^2$$

Stability ratio check using minimum axial load

$$M_R = (P_{min})(b/2) = (24.62)(1.7/2) = 20.9 \text{ tons}$$

Stability ratio

$$M_R/M_{TO} = 20.9/13.23 = 1.58 > 1.0 \quad \text{OK}$$

7. Check operating and transverse earthquake and eccentric loads (shell/fixed end)
   Transverse operating earthquake load on exchangers
   $0.7\, V_{TEo} = 2.37$ tons, applied at the center of each exchanger
   Axial loads at bottom of footing

$$P_{max} = P_s + P_o = (2.15) + (16.8) = 18.95 \text{ tons}$$

From Table 3.25, load combination 3

$$P_{min} = 0.9(P_s + P_o) = (0.9)(18.95) = 17.1 \text{ tons}$$

Since the load will be 0.4 at the shell, the overturning moment, M_T, will be less than the above case, thus, this case will be safe.

$$M_R = (P_{min})(b/2) = (17.1)(1.7/2) = 14.54 \text{ mt}$$

8. Check stability for empty and transverse earthquake and eccentric loads (channel/sliding end)

   Transverse empty earthquake load on exchangers

   $0.7V_{TEe} = 1.72$ tons, applied at the center of each exchanger

   Minimum axial load at bottom of footing

$$P_{min} = 0.9(P_s + P_e) = (0.9)(2.15 + 18.27) = 18.4 \text{ tons}$$

From Table 3.25, load combination 5a, transverse moment at the bottom of footing

$$M_T = (3.79)[(0.85 + 2.45) + (1.7 + 0.85 + 2.45)] \text{ (0.6 channel end)}$$

$$+ (0.108)(2.15)(1.95 + 0.45) + (1.41 \text{ pipe eccentricity}) = 10.31 \text{ mt}$$

Stability ratio check using minimum axial load

$$M_R = (P_{min})(b/2) = (18.4)(1.65/2) = 15.2 \text{ tons}$$

Stability ratio

$$M_R/M_{TO} = 15.2/10.31 = 1.47 > 1.0 \quad \text{OK}$$

9. Check stability for empty and transverse earthquake and eccentric loads (shell/fixed end)

   The load will be 0.4 at the shell, so this case is covered by the case presented in item 7.

10. Check empty, bundle pull, and eccentric loads (channel/sliding end, top bundle pulled out)

    Vertical load from bundle pull on top exchanger

$$P_B = (8.62)(0.85 + 1.7)/(3.4) = 6.47 \text{ tons}$$

Vertical load on sliding end at the top of pier

$$P_{SL} = P_e + P_B = (18.27) + (6.47) = 24.74 \text{ tons}$$

Horizontal load at sliding end

$$V_{SL} = \mu(P_{SL}) = (0.4)(24.74) = 9.9 \text{ tons}$$

Note that the horizontal load on the sliding end, computed on the basis of friction, is greater than half of the total bundle pull (8.62 tons). Therefore, because the two pedestals and footings are equal in size, and thus, even in stiffness, the actual horizontal load will be the same on both pedestals.

$$V_{SL} = V_{FX} = 8.62/2 \text{ piers} = 4.31 \text{ tons}$$

Maximum axial load at bottom of footing,

$$P_{max} = P_s + P_{SL} = (2.15) + (24.75) = 26.9 \text{ ton}$$

Load combination 8, as in Table 3.25, moments at the bottom of footing:

$$M_L = (V_{SL})(8.0) = (4.31)(2.45) = 10.6 \text{ mt}$$

$$M_{Te} \text{ (from pipe eccentricity)} = 1.41 \text{ mt}$$

Soil-bearing check using maximum axial load

$$q_{max} = \frac{26.9}{2.45(1.7)} + \frac{6(10.6)}{1.7(2.45)^2} + \frac{6(1.41)}{2.45(1.7)^2} = 19.2 \text{ ton/m}^2$$

$$q_{max} < q_{all} = 25 \text{ ton/m}^2 \quad \text{OK}$$

11. Check stability for empty, bundle pull, and eccentric loads (shell/ fixed end; top bundle pulled out)

Vertical load from bundle pull on the top exchanger

$$P_B = (8.6)(0.85 + 1.65)/(3.4) = 6.32 \text{ tons}$$

Vertical load on fixed end at the top of pier

$$P_{FX} = P_e - P_B = (12.2) - (6.32) = 5.88 \text{ tons}$$

Horizontal load at fixed end

$$V_{FX} = V_{SL} = 8.62/2 \text{ piers} = 4.31 \text{ tons}$$

Minimum axial load at the bottom of footing

$$P_{min} = P_s + P_{FX} = (12.56) + (5.88) = 18.44 \text{ tons}$$

In this case, as a critical factor in overturning, we add the weight of the foundation to the pedestal weight.

From Table 3.25, load combination 8, moments at the bottom of footing

$$M_L = (V_{FX})(2.45) = (4.31)(2.45) = 10.6 \text{ mt}$$

$$M_{Te} \text{ (from pipe eccentricity)} = 0.575 \text{ mt}$$

Stability ratio check using minimum axial load

$$M_R = P_{min}(a/2) = (18.44)(2.45/2) = 22.6 \text{ mt}$$

Stability ratio

$$M_R/M_{TO} = 22.6/11.17 = 2.0 > 1.0 \quad \text{OK}$$

The footing dimensions will be 2.45 × 1.7 m, with a thickness 0.5 m.

6.7.4.5 Footing Design

Use load combinations and strength design load factors from Table 3.26. From the previous analysis for foot sizing, the following are the maximum stresses on the soil, for operating, thermal, and eccentric loads (channel/sliding end) (load factors are from Table 3.26 in Chapter 3, load combination 1):

Thermal force at the top of pier

$$V_{uThermal} = 1.4(V_T) = 1.4(10.1) = 14.14 \text{ tons}$$

Axial load at the bottom of footing

$$P_u = 1.4(P_s + P_o) = 1.4(2.15 + 25.2) = 38.3 \text{ tons}$$

Moments at the bottom of footing

$$M_{uL} = (V_T)(2.45) = (14.14)(2.45) = 34.64 \text{ mt}$$

$$M_{uTO} \text{ (from pipe eccentricity)} = 1.4(M_{TO}) = 1.66 \text{ mt}$$

This value is small with respect to the moment at the bottom.
Maximum factored soil-bearing

$$e_{u1} = M_{uL}/P_u = (34.64)/(38.3) = 0.9 \text{ m} > a/6 = (2.45)/6 = 0.4 \text{ m}$$

Because transverse eccentricity is very small, it can be ignored in calculations of factored soil-bearing for the design of footing reinforcing.

$$S_{max} = 2(P_u)/(3b)(a/2 - e_{u1}) = (2)(38.3)/(3)(1.7)[(2.45)/2 - (0.9)] = 46.2 \text{ ton/m}^2$$

Bearing length (longitudinal direction) = $3 (a/2 - e_{u1})$

$$= 3[(2.45)/2 - (0.9)] = 0.975 \text{ m}$$

Factored soil-bearing at the face of the pier (for checking moment) is equal to zero.
The force effecting moment

$$F = \frac{1}{2}(0.975)(46.2) = 22.5 \text{ tons}$$

$$M = \frac{1}{2}(0.975)(46.2)2/3(0.975) = 14.63 \text{ mt/m}'$$

Check for two-way action (punching shear) according to ACI 318-08M, Section 11.12 (at distance $d/2$ from the face of the pier).
By inspection, the operating weight at the channel end will govern, as shown in Equation (6.10).

$$P_u = 1.4(P_s + P_o) = 1.4(27.3) = 38.22 \text{ tons}$$

$$q = P_u/(ab) = (38.22)/(2.45)(1.7) = 9.18 \text{ ton/m}^2$$

$$V_u = P_u - (q)(1.1 + d)(0.4 + d)$$

$$= 38.22 - (9.18)(1.1 + 0.4)(0.4 + 0.4) = 27.2 \text{ tons}$$

$$\beta = (1.1)/(0.4) = 2.75$$

$$b_o = 2(1.1 + 0.4) + 2(0.4 + 0.4) = 4.6 \text{ m}$$

$$\alpha_s = 40$$

$$V_c = 0.17(1 + (2/\beta))(f_c')^{0.5}b_o d/6$$

$$= 0.17(1 + (2/2.750)) + (27.5)^{0.5}(4600)(400)/10,000$$

$$= 283.3 \text{ tons} \quad (\text{ACI 318M-08, Equation (11.31)})$$

$$V_c = 0.083(\alpha_s d/b_o + 2)(f_c')^{0.5} b_o d$$

$$= 0.083[(40)(400)/(4600) + 2](27.5)^{0.5}(4600)(400)/10,000$$

$$= 440 \text{ tons}$$

ACI 318M-08, Equation (11.32)

$$V_c = (1/3)(f_c')^{0.5} b_o d$$

$$= (1/3)(27.5)^{0.5}(4600)(400) = 321.6 \text{ tons}$$

[ACI 318M-08, Equation (11.33)]
ACI 318-08, Equation (11.31), governs $V_c = 283.3$ tons

$$V_n = \varphi V_c = 0.75(283.3) = 212.5 \text{ tons} > V_u = 27.2 \text{ tons} \quad \text{OK}$$

Design for Moment

The section is subjected to a bending moment, so by applying a procedure for designing sections for single reinforcement,

$$\rho = 0.0030,$$

where $\rho = A_s/bd$. It is higher than the minimum reinforcement ratio, which is 0.002 based on ACI 318M-08, Section 7.12.2.

$$A_s = \rho bd = (0.0030)(1100)(400) - 1320 \text{ mm}^2/\text{m}$$

Use 6 ϕ18 mm/m' ($A_s = 1526$ mm²/m').

Top Steel

From a practical viewpoint, the bottom of the foundation is not in full bearing for some loading combinations (Figure 6.5). Because the footing is designed for earthquake loads, ductility is required; therefore, the top of the foundation needs to be reinforced with steel bars. To conservatively calculate the moment for top steel due to the effect of the weight of the soil and concrete with a 1.4 load factor, assuming no soil-bearing under the portion of the footing extending from the edge of the pier, the process is as follows:

Factored soil and concrete weight

$$W_u = (1.4)[(2.5)(0.4) + (2.0)(0.75)] = 2.1 \text{ ton/m}^2$$

$$d = 0.35 \text{ m}$$

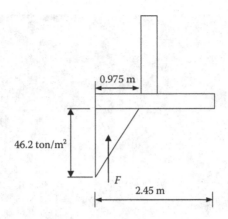

FIGURE 6.5
Sketch of stresses on foundation.

Factored moment at the face of the pier

$$M_u = (2.1)(1.025)^2 \, (1/2) = 1.1 \text{ mt/m}'$$

Design for Moment

From a practical point of view, the steel in the foundation for top or bottom, by any means, will be not less than $5\phi12$ mm/m'.

6.8 Separator Design Example

A photo of the separator and sketch of the foundation dimensions and steel reinforcement is presented in Figures 6.6–6.8.

6.8.1 Design Data

Vessel Data

Empty weight = 44.4 tons
Operating weight = 152 tons
Test weight = 178.7 tons
Basic diameter = 3.66 m
Maximum design temperature = 500°F
Vessel material: carbon steel
Bolts: two 25-mm diameter, ASTM F1554, Grade 36 (galvanized) per pier
Bolt spacing: 800 mm, the saddle dimensions are 3.55 × 0.25 m

FIGURE 6.6
Photo of separator.

Design Criteria

Concrete, (f_c') = 27.5 N/mm²
Reinforcing, f_y = 414 N/mm²
Soil unit weight, γ = 2.0 ton/m³
At a depth of 1.5 m, the allowable net soil-bearing capacity, q_{all} = 0.18 N/ mm²
Wind loads: SEI/ASCE 7–05

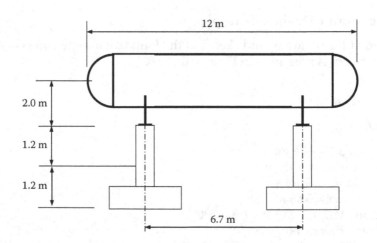

FIGURE 6.7
Sketch of horizontal separator.

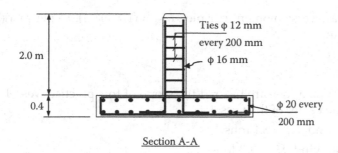

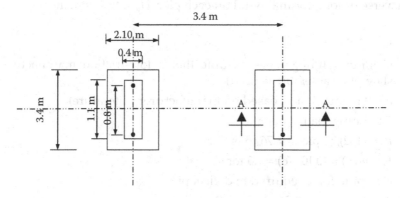

FIGURE 6.8
Foundation sketch.

Earthquake loads: SEI/ASCE 7–05
Use a 20% increase in allowable soil pressure for test load combinations.

6.8.2 Loads Calculation

- Empty, operating, and test loads

  Include an additional 10% of the applicable weight (empty, operating, or test) to account for piping supported on the horizontal vessel (refer to PIP Standard STC01015, Section 4.2, "Vertical Loads"):

  Total empty load, $D_e = (1.10)(44.4) = 48.84$ tons

  Total operating load, $D_o = (1.10)(152) = 167.2$ tons

  Total test load, $D_t = (1.10)(178.7) = 196.6$ tons

- Transverse moment from pipe eccentricity

  Eccentricity = (basic diameter)/2 + (0.45) = (3.6)/2 + (0.45) = 2.25 m

  Empty transverse moment per pier, $M_{Te} = (44.4)(0.1)(0.5 \text{ per pier})(2.25) = 5.0$ mt

  Operating transverse moment per pier, $M_{To} = (152)(0.1)(0.5 \text{ per pier})(2.25) = 17.1$ mt

Test transverse moment per pier, $M_{Tt} = (178.7)(0.1)(0.5$ per pier$)(2.25)$
$= 20.1$ mt

Wind Loads

Wind load calculations are beyond the scope of this practice. Vessel wind is applied at the center of the vessel.

Transverse wind, $H_w = 6.1$ tons

Longitudinal wind, $H_w = 1.3$ tons

Transverse or longitudinal wind on each pier, $H_w = 0.37$ ton/m²

Thermal Force

1. Compute sliding force (assume that a low-friction manufactured slide plate assembly is used)

   $\mu = 0.10$ (maximum based on manufacturer's literature)

   Operating load at one pier:

   $P_o = (152)/(2$ piers$) = 76$ tons

   $F_f = \mu(P_o) = (0.10)(76) = 7.6$ tons

2. Compute force required to deflect pier

   Assume pier is $3.7 \times 0.35 \times 2.0$ m

   Moment of inertia, $I = b(h)3/12 = (3.7)(0.35)^3/12 = 0.01322$ m⁴

   Based on ACI 318-08, Section 8.5.1

   Modulus of elasticity, $E_c = 57,000 f_c' = 24,518.4$ N/mm²

   Thermal expansion coefficient for carbon steel at 500°F, $\varepsilon = 1.96$ mm/m, as obtained from Table 6.1

   Thermal growth between saddles, $\Delta = (\varepsilon)(L) = (2.94$ mm/m$)(6.7) = 19.88$

   $T = 3\Delta EI/2H^3 = 3(0.0202)(2500000)(0.01322)/2(2.0)^3 = 125$ tons

   Because $F_f < T$ and to reduce high-friction forces, it is advisable to use a low-friction manufactured slide plate assembly.

6.8.3 Design Elements

Size Low-Friction Manufactured Slide Plate Elements

Upper element width = (saddle width) + 25 mm = 250 + 25 = 275 mm = 300 mm

Lower element width = upper element width − 2Δ − 25 mm = (280) − (2)(20) − 25 ≈ 220 mm

Maximum load on sliding end (from test weight), $P_t = (196.6)/(2$ piers$) = 98.3$ tons

Operating load on sliding end, P_o = (167.2)/(2 piers) = 83.6 tons

According to the manufacturer's recommendations, seven slide plate components are required for each assembly, with the lower element being 220 × 65 mm and the upper element being 280 × 90 mm.

Total length of lower elements provided = 7(65) = 455 mm

Maximum bearing pressure on elements = $(98.3) \times 10^6/[(7)(200)(65)]$ = 1080.2 ton/m^2

Operating bearing pressure on elements = $(83.6) \times 10^6/[(7)(200)(65)]$ = 918.7 ton/m^2

From the vendor specifications, we obtain a bearing around μ = 0.055, based on the bearing load.

Revised operating frictional force, F_f = (0.055)(83.6) = 4.6 tons

Size Steel Bearing Plate

Steel bearing plate dimensions
    Width = (lower slide plate element width) + (25) = (200) + (25) = 225 mm
    Length = (saddle length) + (25 mm) = (3.55) + (0.025) = 3.6 m

Check bearing stress (test load case)
    P_u = 1.4P_t = 1.4(196.6/2 piers) = 137.6 tons

Based on ACI 318-09 Section 10.17
    $P_n = \varphi 0.85\, f_c' A_1$ = (0.65)(0.85)(27.5)(225)(3600)/1000 = 12,307 kN = 1255 tons
    OK

Choose a bearing plate of dimensions 3.6 × 0.25 × 10 m.

Pier Size

Choose the pier length from:
    (c/c bolts) + (2) (minimum anchor bolt edge distance) = (3350) + 2 (125) = 3600 mm
    (bearing plate length) + (100) = (3600) + (100) = 3700 mm

Choose a pier length of 3.7 m.

Choose the pier width from:

- 250 mm
- 10% of pier height = (0.10)(2.05) = 0.205 m (based on assumed pier height)
- (2)(5-in. minimum anchor bolt edge distance) = 2 (125) = 250 mm
- (bearing plate width) + (100) = (250) + (100) = 350 mm

Choose a pier size with dimensions of 0.35 × 3.7 m.

Anchor Bolt Design, Pier Design, and Footing Design

Anchor bolt design, pier design, and footing design are very similar to the previous example.

6.9 Vertical Vessel Foundation Design

The vertical vessel is usually shown in gas facilities and in refinery plant as well, the shape of this vertical vessel is shown in Figure 6.9.

6.9.1 Dead Loads

The following nominal loads shall be considered dead loads when applying load factors used in strength design.

1. Structural dead load, D_s—a vessel's foundation weight, which is defined as the combined weight of footing, pedestal dead load (D_p), and the overburdened soil.

FIGURE 6.9
Photo of vertical vessel.

2. Erection dead load, D_f—fabricated weight of a vessel, generally taken from the certified vessel drawing.

3. Empty dead load, D_e—empty weight of a vessel, including all attachments, trays, internals, insulation, fireproofing, agitators, piping, ladders, platforms, etc. This is generally taken from the certified vessel drawing.

4. Operating dead load, D_o—empty dead load of a vessel plus the maximum weight of its contents (including packing and catalyst) during normal operation. The operating dead load shall be taken from the certified vessel drawing and data sheet.

5. Test dead load, D_t—empty dead load of a vessel plus the weight of the test medium contained in the system. The test medium shall be as specified in the contract documents.

Unless otherwise specified, a minimum specific gravity of 1.0 shall be used for the test medium. The cleaning load shall be used for the test dead load if the cleaning fluid is heavier than the test medium. Whether to test or clean in the field should be determined. Depending on the design, the test dead load is generally desirable because unforeseen circumstances may occur. Eccentric vessel loads caused by large pipes or reboilers shall be considered for the applicable load cases.

- Live Loads

  Live loads (L) shall be calculated in accordance with Chapter 3. The data in Section 3.2.2 are to be considered a guideline if pertinent data are not available from the client. In general, the live load has a minor effect with respect to the dead load of the vessel, so it will have less influence on the design of the vessel, horizontal and vertical.

- Wind Loads

  Wind loads (W) shall be calculated in accordance with the requirements of the client based on site location, *Meteorological and Seismic Design Data, Structural Design Criteria for Non-Building Structures*, and the guidelines of ASCE presented in *Wind Load and Anchor Bolt Design for Buildings and Other Structures*.

  The engineer is responsible for determining the wind loads to be used for the foundation design. The partial wind load (W_p) shall be based on the requirements of ASCE 37-02 Section 6.2.1, for the specified test or erection duration. The design wind speed shall be 75% of the actual wind speed.

  When calculating or checking wind loads, due consideration shall be given to factors that may significantly affect total wind loads, such as the application of dynamic gust factors or the presence of spoilers, platforms, ladders, piping, etc., on the vessel. If detailed information (number of platforms, platform size, etc.) is unavailable at the time of foundation design, the following simplified method may be used:

1. For the projected width, add 5 ft (1.52 m) to the diameter of the vessel, or add 3 ft (0.91 m), plus the diameter of the largest pipe, to the diameter of the vessel, whichever is greater. This will account for platforms, ladders, nozzles, and piping below the top tangent line.

2. The vessel height should be increased by 1 vessel diameter to account for a large diameter and platform attached above the top tangent, as is the case with most tower arrangements.

3. The increases in vessel height or diameter to account for wind on appurtenances should not be used in calculating the h/D ratio for force coefficients or flexibility.

4. The force coefficient (C_f) should be determined as discussed in Chapter 3.

5. If most design detail items such as platforms, piping, ladders, and other accessories of the vessel are known, the detailed method of the guidelines of the ASCE's *Wind Load and Anchor Bolt Design for Buildings and Other Structures* shall be used.

- Earthquake Loads

  Seismic forces shall be calculated in accordance with ASCE or UBC standards, according to the basis of design. The seismic zone will be defined by the owner based on the site location or from meteorological and seismic design data. The earthquake load (E) is usually calculated by the vessel vendor. This information should be transferred to the structural engineer to be included in the design of the foundation.

  For a skirt-supported, vertical vessel classified as SUG III in accordance with ASCE 7-02, Section 9, the critical earthquake provisions and implied load combinations of ASCE 7-02 should be observed, as illustrated in Chapter 3.

- Other Loading

  Thrust forces caused by thermal expansion of piping shall be included in the operating load combinations if deemed advisable. Dead load factors shall be applied to the resultants of piping thermal loadings. The pipe stress engineer shall be consulted for any thermal loads that are to be considered. Consideration shall be given to upset conditions that could occur and increase loading on the foundation.

- Load Combinations

  The structure, equipment, and foundations shall be designed for the appropriate load combinations from ASCE 7-02 and any other probable realistic combination of loads. The load combination for vertical vessels is discussed in Chapter 3. Table 3.23 is a load combination guide for designing by the allowable stress design and Table 3.24 is a load combination guide to be used in designing by the strength design method or load resistance factor design.

6.9.2 Pedestal Design

The concrete pedestal dimensions shall be sized as we usually size the foundation based on the available form on site. It is usually accepted that octagon pedestal dimensions shall be sized with pedestal faces in increments up to 50 mm. The following criteria shall be used to determine the size and shape for the pedestal. The face-to-face pedestal size shall not be less than the largest of the following:

- If you will not use a sleeve, choose the largest between:

$$D_{BC} + 250 \text{ mm} \tag{6.8}$$

$$D_{BC} + 12(d) \text{ (for high-strength anchor bolts)} \tag{6.9}$$

- If using a sleeve:

$$D_{BC} + ds + 11(d) \text{ (for high-strength anchor bolts)} \tag{6.10}$$

where
D_{BC} = Diameter for the circle of bolts (mm)
d = Bolt diameter (mm)
d_s = Sleeve diameter (mm)

Pedestals 1.8 m and larger shall be octagonal. Dimensions for octagon pedestals are provided in Table 6.4. The octagon faces have 50-mm increments to be matched with the wooden form. If the pedestals are smaller than 1.8 mm, they shall be square or, if forms are available, round.

Anchorage. It is normally desirable to make the pedestal deep enough to contain the anchor bolts and keep them out of the footing. Consideration shall be given to anchor bolt development and foundation depth requirements. Pedestal size may need to be increased to provide adequate projected concrete area for anchor bolts when additional reinforcement is not used.

Pedestal reinforcement shall be tied to the footing with sufficient dowels around the pedestal perimeter to prevent separation of the pedestal and footing. Development of reinforcing steel shall be checked. Dowels shall be sized by computing the maximum tension existing at the pedestal perimeter due to overturning moments. The following formula may be used conservatively. More exact tension loads may be obtained by using ACI 318 strength design or BS 8110 methodology based on project specifications and standards. In this case, we will take ACI 318 equations into consideration.

$$\text{Tension } F_u = 4(M_{up})/[(N_d)(D_C)] - 0.9[D_e \text{ or } D_o) + D_p]/N_d \qquad (6.11)$$

where

D_e or D_o = Nominal empty or operating vessel weight

Use the empty weight for wind loads. Use the empty or operating weight for earthquake loads, depending on which condition is used to calculate M_u.

$$A_s \text{ (required)} = \text{tension/design stress} = F_u/\varphi f_v \qquad (6.12)$$

where

F_u = Maximum ultimate tension in reinforcing bar
M_{up} = Maximum factored overturning moment at base of pedestal, calculated by using load factors in load combinations for uplift cases in Table 3.24
N_d = Number of dowels (assumed); shall be a multiple of 8
D_C = Dowel circle diameter (assume a pedestal size minus 150 mm)
$D_e + D_p$ = Nominal empty weight of vessel and pedestal weight
$D_o + D_p$ = Nominal operating weight of vessel and pedestal weight
Φ = Strength reduction factor = 0.90
f_y = Yield strength of reinforcing steel

The minimum pedestal reinforcement shall be as follows:

1. Octagons from 1.8 to 2.5 m

   A minimum of 16 bars with 12-mm diameter verticals and 10-mm diameter steel ties at 300-mm maximum spacing.

2. Octagons from 2.5 to 3.5 m

   A minimum of 24 bars with 16-mm diameter verticals with 10-mm diameter steel ties at 300-mm maximum spacing.

3. Octagons larger than 3.5 mm

   The minimum bar diameter is 16 mm at 450-mm maximum spacing with 10-mm ties at 300-mm maximum spacing.

Top reinforcement, a mat of reinforcing steel at the top of the pedestal, shall be provided. This entails a minimum of steel bars with 12-mm diameter at 300-mm maximum spacing across the flats in two directions only.

Anchor Bolts

Conservatively, the maximum tension on an anchor bolt may be determined using the following formula. More exact tension loads may be obtained by using ACI 318 strength design methodology.

$$N_u = 4M_u/[(N_b)(B_C)] - 0.9(D_e \text{ or } D_o)/N_b \qquad (6.13)$$

where

N_u = Factored maximum tensile load on an anchor bolt

M_u = Factored moment at the base of vessel, calculated using load factors in load combinations for uplift cases in Table 3.24 (*Loading Combinations and Load Factors—Strength Design*)

N_b = Number of anchor bolts

B_C = Bolt circle diameter

$D_e \text{ or } D_o$ = nominal empty or operating vessel weight (Use the empty weight for wind loads. Use the empty or operating weight for earthquake loads, depending on which condition is used to calculate M_u.)

For most cases, there is no shear on the anchor bolts because the load is resisted by friction caused primarily by the overturning moment. If friction cannot resist the load, the bolts shall be designed to resist the entire shear load, or other methods may be used to resist the shear load. The friction resistance can be calculated using the following formulas:

$$P_u = M_u/L_A + 0.9(D_e \text{ or } D_o) \qquad (6.14)$$

$$V_f = \mu \qquad (6.15)$$

where

P_u = Factored compression force at top of pedestal

L_A = Lever arm between centroid of tension loads on bolts and centroid of the compression load on the pedestal. This is a complicated distance to determine exactly. A conservative approximation is to use 2/3 of the bolt circle diameter as the lever arm

μ = Coefficient of friction. For the normal case of grout at the surface of pedestal, $\mu = 0.55$

V_f = Frictional resisting force (factored)

If there is no shear load on the bolts, then

$$V_u \le \varphi V_f$$

where

V_u = Factored shear load at base of vessel, calculated using load factors in load combinations for uplift cases in Table 3.24 for load combination for strength design

Φ = Strength reduction factor = 0.75

6.9.3 Footing Design

The size of spread footings may be governed by stability requirements, sliding, soil-bearing pressure, or settlement. Footings for vertical vessels shall be octagonal or square and sized based on standard available form sizes. When form information is not available, footing dimensions shall be sized with footing faces in 50-mm increments to allow use of standard manufactured wooden forms. Octagons in Table 6.4 are those having faces in 2-in. increments. If extended to the recommended depth specified in the geotechnical report, the pedestal may be adequate without a footing. Footings smaller than 2.0 m in diameter shall be square.

Where a footing is required, the footing thickness shall be a minimum of 300 mm. The footing thickness shall be adequate to develop pedestal reinforcement and satisfy the shear requirements of ACI 318. The footing thickness shall also be checked for top tension without top reinforcement in accordance with ACI 318. If the thickness is not adequate, either a thicker footing or top reinforcing steel is required. Note that increasing the footing thickness is typically more cost-effective for construction than adding a top mat of reinforcing steel except where seismic effects create tensile stresses requiring top reinforcement.

For the first trial, the diameter (D) of an octagonal footing may be approximated by the following formula:

$$D = 2.6(M_{ftg}/q)^{1/3} \tag{6.16}$$

where
M_{ftg} = Nominal overturning moment at base of footing (mt)
q_{all} = Allowable gross soil-bearing (ton/m^2)

A common assumption in the design of soil-bearing footings is that the footing behaves as a rigid unit. Hence, the soil pressure beneath a footing is assumed to vary linearly when the footing is subjected to axial load and moment. The ensuing footing formulas are based on the linear pressure assumption.

Footings shall be designed so that, under sustained loads (operating loads), the total settlement and the differential settlement between footings do not exceed the established limits. The maximum allowable amount of total settlement and differential settlement is typically set by the project's structural engineer based on the sensitivity of the equipment or structure being supported and the piping connected to the equipment. So, there must be agreement by the structural engineer and the piping stress analysis engineer.

6.9.4 Soil Bearing on the Octagon Footing

The octagon shape has its own characteristic when we go through calculating the soil stresses (Tables 6.6 and 6.7; and also as presented in Table 6.10).

TABLE 6.6

Various Dimensions for Octagonal Pedestal

D (m)	B (m)	C (m)	E (m)	Area (m²)	Ze (m³)
1.81	0.53	0.75	1.96	2.72	0.60
1.93	0.57	0.80	2.09	3.09	0.73
2.05	0.60	0.85	2.22	3.49	0.87
2.17	0.64	0.90	2.35	3.91	1.04
2.29	0.67	0.95	2.48	4.36	1.22
2.41	0.71	1.00	2.61	4.83	1.42
2.53	0.74	1.05	2.74	5.32	1.65
2.66	0.78	1.10	2.87	5.84	1.89
2.78	0.81	1.15	3.01	6.39	2.16
2.90	0.85	1.20	3.14	6.95	2.46
3.02	0.88	1.25	3.27	7.54	2.78
3.14	0.92	1.30	3.40	8.16	3.13
3.26	0.95	1.35	3.53	8.80	3.50
3.38	0.99	1.40	3.66	9.46	3.91
3.50	1.03	1.45	3.79	10.15	4.34
3.62	1.06	1.50	3.92	10.86	4.80
3.74	1.10	1.55	4.05	11.60	5.30
3.86	1.13	1.60	4.18	12.36	5.83
3.98	1.17	1.65	4.31	13.15	6.39
4.10	1.20	1.70	4.44	13.95	6.99
4.22	1.24	1.75	4.57	14.79	7.63
4.35	1.27	1.80	4.70	15.64	8.30
4.47	1.31	1.85	4.83	16.53	9.01
4.59	1.34	1.90	4.96	17.43	9.76
4.71	1.38	1.95	5.10	18.36	10.55
4.83	1.41	2.00	5.23	19.31	11.39
4.95	1.45	2.05	5.36	20.29	12.26
5.07	1.48	2.10	5.49	21.29	13.18
5.19	1.52	2.15	5.62	22.32	14.14
5.31	1.56	2.20	5.75	23.37	15.15
5.43	1.59	2.25	5.88	24.44	16.21
5.55	1.63	2.30	6.01	25.54	17.32
5.67	1.66	2.35	6.14	26.66	18.47
5.79	1.70	2.40	6.27	27.81	19.67
5.91	1.73	2.45	6.40	28.98	20.93
6.04	1.77	2.50	6.53	30.18	22.24
6.16	1.80	2.55	6.66	31.40	23.60
6.28	1.84	2.60	6.79	32.64	25.01
6.40	1.87	2.65	6.92	33.91	26.48
6.52	1.91	2.70	7.06	35.20	28.01
6.64	1.94	2.75	7.19	36.51	29.60
6.76	1.98	2.80	7.32	37.85	31.24

TABLE 6.7

More Various Dimensions for Octagonal Pedestal

D (m)	B (m)	C (m)	E (m)	Area (m²)	Ze (m³)
6.88	2.02	2.85	7.45	39.22	32.95
7.00	2.05	2.90	7.58	40.61	34.71
7.12	2.09	2.95	7.71	42.02	36.54
7.24	2.12	3.00	7.84	43.46	38.43
7.36	2.16	3.05	7.97	44.92	40.38
7.48	2.19	3.10	8.10	46.40	42.40
7.60	2.23	3.15	8.23	47.91	44.48
7.73	2.26	3.20	8.36	49.44	46.63
7.85	2.30	3.25	8.49	51.00	48.86
7.97	2.33	3.30	8.62	52.58	51.14
8.09	2.37	3.35	8.75	54.19	53.50
8.21	2.40	3.40	8.88	55.82	55.94
8.33	2.44	3.45	9.02	57.47	58.44
8.45	2.47	3.50	9.15	59.15	61.02
8.57	2.51	3.55	9.28	60.85	63.67
8.69	2.55	3.60	9.41	62.58	66.40
8.81	2.58	3.65	9.54	64.33	69.21
8.93	2.62	3.70	9.67	66.10	72.09
9.05	2.65	3.75	9.80	67.90	75.05
9.17	2.69	3.80	9.93	69.72	78.09
9.29	2.72	3.85	10.06	71.57	81.22
9.42	2.76	3.90	10.19	73.44	84.42
9.54	2.79	3.95	10.32	75.34	87.71
9.66	2.83	4.00	10.45	77.25	91.08
9.78	2.86	4.05	10.58	79.20	94.54
9.90	2.90	4.10	10.71	81.17	98.09
10.02	2.93	4.15	10.84	83.16	101.72
10.14	2.97	4.20	10.98	85.17	105.44
10.26	3.01	4.25	11.11	87.21	109.25
10.38	3.04	4.30	11.24	89.28	113.15
10.50	3.08	4.35	11.37	91.37	117.15
10.62	3.11	4.40	11.50	93.48	121.23
10.74	3.15	4.45	11.63	95.61	125.41
10.86	3.18	4.50	11.76	97.78	129.69
10.98	3.22	4.55	11.89	99.96	134.06
11.11	3.25	4.60	12.02	102.17	138.53
11.23	3.29	4.65	12.15	104.40	143.09
11.35	3.32	4.70	12.28	106.66	147.76
11.47	3.36	4.75	12.41	108.94	152.52
11.59	3.39	4.80	12.54	111.25	157.39
11.71	3.43	4.85	12.67	113.58	162.36

(continued)

TABLE 6.7 Continued

More Various Dimensions for Octagonal Pedestal

D (m)	B (m)	C (m)	E (m)	Area (m²)	Ze (m³)
11.83	3.46	4.90	12.80	115.93	167.44
11.95	3.50	4.95	12.93	118.31	172.61
12.07	3.54	5.00	13.07	120.71	177.90

$$A = \text{Area (m}^2) = 0.8284272D^2$$

$$C = \text{Length of side} = 0.4142136D$$

$$B = C \times \sin 45° = 0.2928932D$$

$$E = \text{Length of diameter} = 1.0823922D$$

$$S_d = \text{Sec. mod. diameter (m}^3) = 0.1011422D^3$$

$$S_f = \text{Sec. mod. flat} = S_d(E/D)$$

$$I = \text{Moment of inertia} = S_d(E/2)$$

Soil-bearing pressure shall be checked for maximum allowable on the diagonal. Soil-bearing pressure used for footing design shall be computed on the flat.

Where the total octagonal footing area is in compression ($e/D \le 0.122$ on the diagonal and $e/D \le 0.132$ on the flat), the soil-bearing pressure shall be computed using the following formulas:

$$q = (P/A) \pm M_{ftg}/s \tag{6.17}$$

$$q \text{ (diagonal)} = P/A \pm M_{ftg}/S_{\text{Diagonal}} \tag{6.18}$$

$$q \text{ (diagonal)} = P/A \left[1 \pm (8.19e/D)\right] \tag{6.19}$$

$$q \text{ (flat)} = P/A \pm M_{ftg}/S_{\text{Flat}} \tag{6.20}$$

$$q \text{ (flat)} = P/A \left[1 \pm (7.57e/D)\right] \tag{6.21}$$

where
D = Distance between parallel sides (m)
q = Toe pressure (ton/m²)
P = Nominal total vertical load including soil and foundation (tons)
A = Bearing area of octagonal footing ($0.828D^2$) (m²)
M_{ftg} = Nominal overturning moment at base of footing (mt)
S = Section modulus (m³)

$$S_{Diagonal} = 0.1011422D^3$$
$$S_{Flat} \quad = 0.10948 \ D^3$$
$$e \qquad = \text{Eccentricity } (M_{ftg}/P) \text{ (m)}$$

The total octagonal footing area is not in compression if the following condition is verified:

$$e/D > 0.122 \text{ on the diagonal}$$

and

$$e/D > 0.132 \text{ on the flat}$$

Soil-bearing pressure shall be computed using Figure 6.10 and the following formula:

$$q = LP/A \qquad\qquad (6.22)$$

where the value of L is obtained from Figure 6.10b.

The e/D ratios for octagon footings may exceed the limits because of the load factors in strength design. L and K values for these conditions are tabulated in Table 6.5 for lateral loads perpendicular to a face. These values shall be used for calculating moments and shears in the footing. They shall not be used to check soil-bearing pressures (Figure 6.11).

From Table 6.8, for (e/D) values greater than or equal to 0.500, assume soil-bearing to be a line load with a length equal to the face dimension of the octagon footing, applied at a distance of (e) from the centerline of the footing.

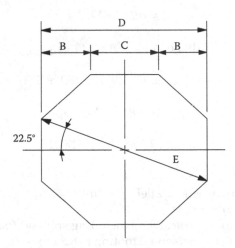

FIGURE 6.10
Octagon propertied dimensions.

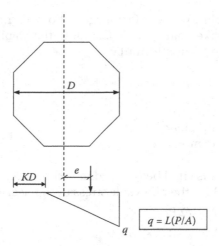

FIGURE 6.11
Pressure distribution under foundation.

6.9.5 Check Stability and Sliding

All foundations subject to *buoyant forces* shall be designed to resist a uniformly distributed uplift equal to the full hydrostatic pressure. The minimum safety factor against flotation shall be 1.20, considering the highest anticipated water level. The stability ratio shall be defined as, The ratio of the resisting moment to overturning moment about the edge of rotation.

TABLE 6.8

Foundation Pressures for Octagon Bases for Large Eccentricities at the Face

e/D	K	L	e/D	K	L	e/D	K	L
0.15	0.05	2.2	0.28	0.45	4.0	0.41	0.76	12.2
0.16	0.09	2.2	0.29	0.48	4.2	0.42	0.79	14.0
0.17	0.13	2.3	0.30	0.49	4.5	0.43	0.81	16.3
0.18	0.16	2.4	0.31	0.52	4.8	0.44	0.84	19.4
0.19	0.19	2.5	0.32	0.55	5.2	0.45	0.86	23.8
0.20	0.22	2.6	0.33	0.57	5.6	0.46	0.89	30.4
0.21	0.25	2.8	0.34	0.60	6.0	0.47	0.92	41.5
0.22	0.28	2.9	0.35	0.62	6.6	0.48	0.94	63.6
0.23	0.3	3.0	0.36	0.64	7.2	0.49	0.97	130.2
0.24	0.34	3.2	0.37	0.67	7.8	0.50	a	a
0.25	0.37	0.40	0.38	0.69	8.7	>0.50	a	a
0.26	0.40	3.6	0.39	0.71	9.6			
0.27	0.42	3.8	0.40	0.74	10.8			

a For e/D values greater or equal to 0.500, assume soil bearing as a line load with a length equal to the face dimension of the octagon footing applied at a distance of (e) from the centerline of the footing.

The minimum stability safety factor against overturning for service loads other than earthquake shall be 1.5. Compute the stability factor of safety ratio, F_S, using the following formula:

$$F_S = M_R + M_{OT} \tag{6.23}$$

where
M_R = resistance moment
M_{OT} = overturning moment

The moment and loads shall be factored in accordance with the load combinations in Table 3.23 for the allowable stress design (service loads).

Eccentricity, $e = \Sigma M_{OT}/\Sigma P$

Resisting moment, $M_R = \Sigma P \times D/2$

D = dimension of footing in the direction of the overturning moment (m)

e = eccentricity (m) = overturning moment at the base of the footing divided by the total vertical load (m)

Hence, the stability factor of safety can be calculated from the following equation:

$$F_S = D/2e$$

6.9.6 Check for Foundation Sliding

The minimum safety factor against sliding for service loads other than earthquake shall be 1.5. The coefficient of friction used in computing the safety factor against sliding for cast-in-place foundations shall be 0.40, unless specified otherwise in a detailed soil investigation. The passive earth pressure from backfill shall not be considered in computing these safety factors.

The minimum overturning "stability ratio" and the minimum factor of safety against sliding for earthquake service loads shall be 1.5. The minimum overturning stability factor of safety for the anchorage and foundations of skirt-supported, vertical vessels classified as SUG III, in accordance with ASCE 7-02, shall be 1.2.

6.9.7 Reinforced Concrete Design

Reinforced concrete design using factored strength design loads shall be in accordance with the relevant design code used in the project. The critical section for moment shall be taken with respect to the face of a square with an area equivalent to that of the pedestal.

Moment shall be checked at the face of the equivalent square. Moment shall be calculated for a 1-m-wide strip as a simple cantilever from the edge of the equivalent square.

Based on ACI 318-08, the punching shear may need to be checked under certain circumstances. The resulting reinforcing steel shall be placed continuously across the entire footing in a grid pattern, the minimum bottom reinforcement being 12-mm bars at 200-mm spacing in each way. The minimum amount of bottom steel shall be according to ACI Code Section 10.5.4, by which this percentage depends on the grade of steel bars.

6.9.7.1 Top Reinforcement

Except where seismic effects create tensile stresses, top reinforcement in the footing is not necessary if the factored tensile stress at the upper face of the footing does not exceed the flexural strength of structural plain concrete, as discussed in Section 6.6.2 and applied in Equation (6.6).

The effective thickness of the footing for tensile stress calculations should be 50 mm less than the actual thickness for footings cast against soil. For footings cast against a seal slab, the actual thickness of the footing may be used for the effective thickness. If the factored tensile stress exceeds the flexural strength of structural plain concrete, top reinforcement should be used if an increase in the footing thickness is not feasible. See the following formulas for footing thicknesses that do not require top reinforcing steel.

For footings cast against soil:

$$t = t_c + 50 \text{ mm} \tag{6.24}$$

For footings cast against a seal slab:

$$t = t_c \tag{6.25}$$

with t_{eff} calculated as follows:

$$t_c = (6M_u/f_t')^{1/2} \tag{6.26}$$

where
  t = Required footing thickness with no top reinforcing steel (m)
  t_c = Calculated footing thickness (m)
  M_u = Factored moment caused by the weight of soil and concrete acting on a 1-m strip in the footing at the face of the equivalent square pedestal (mt/m'), calculated using a load factor of 1.4
  f' = Flexural strength of structural plain concrete (ton/m²)

6.9.7.2 Shear Consideration

Both *wide-beam* and *two-way* action (punching shear) must be checked to determine the required footing depth. Shear, as a measure of diagonal tension, shall be checked at the critical section specified in ACI 318-02, Section 11.1.3.1 (at a distance d from the face of the equivalent square). The shear shall be calculated for a 1-ft-wide strip, as a simple cantilever from the edge of the equivalent square. Beam action assumes that the footing acts as a wide beam with a critical section across its entire width.

Punching shear may need to be checked in some situations in accordance with ACI 318-08. Two-way action (at a distance $d/2$ from the face of the equivalent square) for the footing checks "punching" shear strength. The critical section for punching shear is a perimeter around the supported member with the shear strength computed in accordance with the ACI code.

For footing design, an appropriate depth must be selected so that shear reinforcement is not required. If either permissible shear is exceeded, the thickness of the footing must be increased. The shear strength equations may be summarized as follows based on ACI318-08:

- For beam action

$$V_u \le \phi V_n \le \phi(0.17\sqrt{f_c'}b_w d)$$

where
b_w = Critical width
d = Beam depth

- For two-way action
  $V_u \le$ the least of the following values:

$$0.17(1+(2/\beta))\sqrt{f_c'}b_o d$$

$$0.083(\alpha_s d/b_o + 2)\sqrt{f_c'}b_o d$$

$$1/3\sqrt{f_c'}b_o d$$

where
β = Ratio of long side to short side of the column, concentrated load, or reaction area
α_s = 40 for interior columns
= 30 for edge columns
= 20 for corner columns
B_o = Perimeter of critical section

6.10 Example for Vertical Vessel

The example for vertical vessel present in Figure 6.12 and the foundation plan view and steel detail is presented in Figure 6.13.

6.10.1 Design Data

Vessel Data

Empty weight, D_e = 77.24 tons

Operating weight, D_o = 156.57 tons

Test weight, D_t = 283 tons

Structural Data

Allowable soil-bearing = 1.589 kg/cm^2 = 15.9 ton/m^2

γ = 2.0 ton/m^3

f_c' = 30 N/mm^2

f_y = 420 N/mm^2

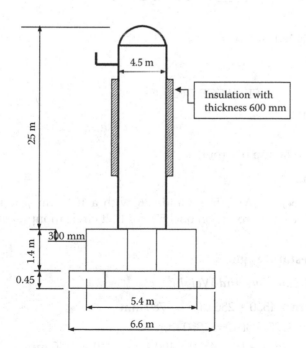

FIGURE 6.12
Sketch for vertical vessel.

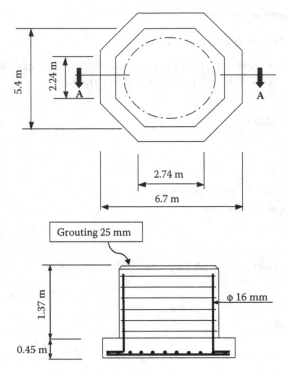

FIGURE 6.13
(a) Plan view. (b) Section A–A.

Wind Load

In accordance with SEI/ASCE 7-02,

V = 115 mph

V = 20.3 tons

M = 262.94 mt (at top of grout)

Anchor Bolts

65-mm, ϕ type, ASTM F1554, Grade 36, with a 100 mm ϕ × 400 mm long sleeve and 350-mm projection on 4.53 m ϕ bolt circle (nonpretensioned).

6.10.2 Pedestal Design

Pedestal Dimensions and Weight

D_{BC} + 250 mm = 4530 + 250 mm = 4780 mm

D_{BC} + 8(B_D) = 4530 + 8(40) = 4850 mm

D_{BC} + S_D + 250 mm − B_D = 4530 + 100 + 250 − 40 = 4840 mm

B_C + S_D + 7 (B_D) = 4530 + 100 + 7 (40) = 4910 mm controls

Use a 4910-mm octagon.
Note: Pedestal diameter had to be increased to 5.40 m to provide a sufficiently large projected concrete failure area to resist the tensile load in the anchor bolts. Alternatively, additional reinforcing steel may be used to transfer anchor bolt forces to concrete.

Pedestal Reinforcement

Pedestal area = 24.13 m², in accordance with Table 6.6

Pedestal weight, $D_p = (24.13 \text{ m}^2)(1.4 \text{ m})(2.5 \text{ ton/m}^3) = 84.5$ tons

M_{ped} = Over turning moment (OMT) at pedestal base = (262.94 mt) + (1.4 m) (20.3 tons) = 291.36 mt

$M_{uped} = 1.6 M_{ped} = 1.6(291.36 \text{ mt}) = 466.176$ mt

Load factors shall be in accordance with Table 3.24, load combination 4

Assume the number of dowels, $N_d = 40$

$D_C = (5.4) - (0.15) = 5.25$ m

$D_e + D_p$ = empty weight of vessel + pedestal weight

$$= 77.24 + 79.51 = 156.75 \text{ tons}$$

$$F_u = 4(M_{uped})/[(N_d)(D_C)] - 0.9(D_e + D_p)/N_d$$

$$= 4(466.176)/[(40)(5.25)] - 0.9(156.75)/40$$

$$= 5.35 \text{ tons}$$

$$A_s = F_u/\varphi f_y$$

$$= (5.35)/(0.9)(0.42) = 141.5 \text{ mm}^2$$

Choose 40 dowels 16 mm in diameter (A_s = 201 mm²) with 12-mm diameters at 350 mm c/c (minimum reinforcement controls).

6.10.3 Anchor Bolt Check

Maximum Tension on Anchor Bolt

$N_u = 4M_u/[(N_b)(B_C)] - 0.9(D_e)/N_b$

$$= 4[(1.6)(262.94)]/[(24)(4.53)] - 0.9(77.24)/24$$

$$= 12.58 \text{ tons}$$

Load factors shall be in accordance with Table 3.24, load combination 4

Maximum Shear on Anchor Bolt

V_u at base top of the grout = 1.6(20.3) = 32.48 tons

Check whether shear load can be taken by friction between the base of the vessel and the top of the grout.

$P_u = M_u/NL_A + 0.9(D_e)$

Choose $L_A = 2/3B_C$ as a conservative measure

$L_A = 2/3 \times 4.53 = 3.0$ m

$P_u = 1.6 \times (262.9/3) + 0.9 \times (77.24) = 209.73$ tons

$V_f = \mu P_u = (0.55)(209.73) = 115.4$ tons

$\varphi V_f = (0.75)(115.4) = 86.55 > 32.48$ tons

Therefore, anchor bolts are not required to resist shear.

Projected Concrete Failure Area

Often, several iterations were required to determine that $D = 4.9$ m would not provide enough projected concrete failure area to resist the maximum tensile load on the anchor, N_u, equal to 12.58 tons, regardless of whatever embedment depth, l_{dh}, was used. To save space, these trial calculations are not shown here. Reinforcing steel either should be added to transfer the tensile load from the anchor bolts to the pedestal or the pedestal diameter should be increased. This second alternative is shown here.

Try increasing D to 5.4 mm.

6.10.4 Footing Design

Select a Trial Octagon Size

$M_{ftg} = \text{OTM at footing base} = (262.94) + (1.85)(20.3 \text{ tons}) = 300.5$ mt

$q_{all} = 15.9$ ton/m²

Trial diameter $= (2.6)(M_{ftg}/q)^{1/3} = 6.6$ m

A 6.7-m octagon area $= 37.1$ m²

Check Required Thickness for Pedestal Reinforcing Embedment

For a 16-mm hooked bar based on ACI 318M-08, the development length will be calculated from the following equation:

$$l_{dh} = [(0.24\Psi f_y)/\lambda(f_c')^{0.5}](d_b) \quad \text{(Section 12.5.2)}$$

where
Ψ = 1.2, if epoxy coated; 1.0, if not coated
λ = 1 for normal concrete
l_{dh} = 308 mm
t_f = 308 + 2 (20 mm bars) + (75 mm cover) = 423 mm

Assume the foundation thickness, $t_f = 450$ mm.

Footing Weights

Weight of pedestal, $D_p = 24.13 \times 1.4 \times 2.5 = 84.5$ tons

The maximum normal force is usually due to the hydrostatic test load if it is water.

Total $(D_s) = 84.5$ tons

$P_t = D_t + D_s = 283 + 84.5 = 367.5$ tons

Check Soil-Bearing and Stability

- Empty + Wind (using Table 3.23, load combination 3)

  $P = P_e = 77.24 + 84.5 = 161.74$ tons

  $M_{ftg} = 291.36$ mt

  $e = M_{ftg}/P = (291.36)/(161.74) = 1.8$ m

  Stability ratio

  $D/2e = (21.73 \text{ ft})/[2(4.48 \text{ ft})] = 2.43 > 1.5 \quad \therefore \text{OK}$

  Stability ratio

  $D/2e = (6.6)/[2(1.8)] = 1.83 > 1.5 \quad \therefore \text{OK}$

  $e/D = (1.8)/(6.6) = 0.27 > 0.122 \quad \therefore L_{diag} = 2.85$

  From Figure 6.13,

  $q = LP/A = (2.85)(161.74)/(37.1) = 12.4 \text{ ton/m}^2 < 15.9 \text{ ton/m}^2 \quad \therefore \text{OK}$

- Operating + wind (using Table 3.23, load combination 2)

  $P = P_o = 241.1$ tons

  $M_{ftg} = 300.5$ mt

  $e = M_{ftg}/P = (300.50/(241.1) = 1.25$ m

  $e/D = (1.25)/(6.6) = 0.189 > 0.122 \quad \therefore L_{diag} = 2.25$

  From Figure 6.13,

  $q = LP/A = (2.25)(241.1)/(36.33) = 14.93 \text{ ton/m}^2 < 15.1 \text{ ton/m}^2 \quad \text{OK}$

- Test + partial wind (using Table 3.23, load combination 6)

  $P = P_t = 367.5$ tons

  Partial wind velocity = 68 mph

  $M_{ftg} = (68 \text{ mph}/115 \text{ mph})^2(300.5) = 105.1$ mt

  $e = M_{ftg}/P = 105.1/367.5 = 0.29$ m

  $e/D = 0.29/6.7 = 0.04 < 0.122$

  $q = P/A[1 + (8.19)(e/D)]$

  $= [(367.5)/(35.2)][1 + (8.19)(0.037)]$

  $= 13.6 \text{ ton/m}^2 < 15.9 \text{ ton/m2} \quad \therefore \text{OK}$

  Choose a 6.6-m octagon.

Bottom Reinforcement

- Operating + wind (using Table 3.24, load combination 3)

$[1.2(D_s + D_o) + 1.6W]$

$P_u = 1.2(241.1) = 289.32$ mt

$M_u = 1.6(300.5) = 480.8$ mt

$e = M_u/P_u = 480.8/289.32 = 1.66$ m

$e/D = 1.66/6.6 = 0.25 > 0.132$

$L = 2.70$ (flat), $K = 0.225$ (flat) (using Table 6.8)

$KD = (0.225)(6.6) = 1.5$

$q = LP/A = (2.7)289.32/36.33 = 21.5$ ton/m^2

Find an equivalent square for pedestal:

Side $2 = 24.13$ m$^2 = 4.9$ m

Projection $= (6.6 - 4.9)/2 = 0.85$ m

On the face of an equivalent square:

$q = 21.5 (5.13 - 0.85)/(5.13) = 17.94$ ton/m^2

$$M = \frac{17.94(0.85)^2}{2} + \frac{(21.5 - 17.9)(0.85)^2}{3} = 7.34 \text{ mt}$$

- Empty + Wind (using Table 3.24, load combination 4)

$[0.9(D_e + D_s) + 1.6W]$

$P_u = 0.9(77.24 + 84.5) = 145.6$ tons

$M_u = 1.6 (300.5) = 480.8$ mt

$e = M_u/P_u = (480.8)/(145.6) = 3.3$ m

$e/D = (3.3)/(6.6) = 0.5 > 0.132$ (flat)

$L = 7.63$ (flat), $K = 0.660$ (flat) (using Table 6.8)

$KD = (0.660)(6.6) = 4.4$ m

$q = LP_u/A = (7.63)(145.6)/(36.33) = 30.6$ ton/m^2

Find an equivalent square for pedestal

On the face of an equivalent square, $q = (30.6)(2.2 - 0.85)/(2.2)$
$= 18.8$ ton/m^2

$$M = \frac{18.8(0.85)^2}{2} + \frac{(30.6 - 18.8)(0.85)^2}{3} = 9.63 \text{ mt}$$

Designing the Concrete Section Based on ACI

For the section, $M = 9.63$ mt.
Choose bars 20-mm in diameter at intervals of 200 mm.

Checking Shear

- Beam shear

  Apply the case of Empty + Wind

  $q = 30.6$ ton/m^2 at the edge

  $q = 18.8$ ton/m^2 at the face of pedestal

  $V_u = 0.85 [(30.6 + 18.8)/2] = 21$ ton/m^2

  $\phi V_n = \phi(0.17)\sqrt{f_c'}b_w d$

  $= 0.75(0.17)(30)0.5(1000)(400)/10,000 = 27.9$ ton/m^2 $> V_u$ OK

- Punching Shear

  The punching shear will be calculated in case of test load and the load combination as shown in Table (3.24) load combination 7.

  $P_u = 1.4(283) = 396.2$

  The length of the new octagon with $D = (4.9 + 0.4) = 5.3$ m

  $b_o = 17.6$ m

  $V_u = 396.2/(17.6)(0.4) = 56.3$ ton/m$^2 = 0.55$ N/mm^2

  $V_u \leq$ minimum of the following

  1. $0.17(1 + (2/\beta))\sqrt{f_c'}$

     $\phi V_n = 0.75(0.17)(1 + (2/1))(30)^{0.5} = 2.1$

  2. $0.083(\alpha_s d/b_o + 2)\sqrt{f_c'}$

     $\phi V_n = (0.75)(0.083)(40\,(400/21,200) + 2)(30)0.5 = 0.94$ N/mm$^2 \leq V_u$ OK

  3. $1/3\sqrt{f_c'}b_o d$

     $\phi Vn = (0.75)(1/3) \times (30)0.5 = 1.35$ N/mm^2

6.11 Pipe Support

The pipe is usually obtained by the pipeline stress analysis engineer, but the following equations and tables provide guidelines for choosing the pipe span. The maximum allowable span depends on the bending stress, vertical deflection, and natural frequency.

The formulation and equation obtained depend on the end conditions. Assuming a straight pipeline with simple support at both ends gives a higher stress and sag and, therefore, results in a conservative span.

$$L = \sqrt{\frac{0.33ZS}{w}} \text{ (stress limitation)}$$

$$L = \sqrt[4]{\frac{\Delta EI}{22.5w}} \text{ (deflection limitation)}$$

For less conservative spans, the end condition may be assumed as between simple support at both ends and fixed condition at both ends.

$$L = \sqrt{\frac{0.4ZS}{w}} \text{ (stress limitation)}$$

$$L = \sqrt[4]{\frac{\Delta EI}{13.5w}} \text{ (deflection limitation)}$$

where
L = Allowable pipe span (m)
Z = Modulus of pipe sections
S = Allowable tensile stress for the pipe materials at design temperature (known as allowable hot stress)
W = Total weight of the pipe, which is equal to the summation of the pipe self-weight, content weight, and insulation weight
Δ = Deflection
I = Moment of inertia
E = Modulus of elasticity of the pipe material at design temperature

Natural Frequency

For most refinery piping, a natural frequency of about 4 cycles per second (cps) is sufficient to avoid resonance in nonpulsating pipelines. However, the natural frequency in cps is related to the maximum deflection, as in the following equation:

$$f_n = \frac{1}{2\pi}\sqrt{\frac{g}{\Delta}}$$

where
g = gravity acceleration
f_n = natural frequency

The deflection is limited to 25 mm. The reason for limiting the deflection is to provide a sufficiently stiff pipe with a sufficiently high natural frequency to avoid large amplitudes under any small disturbing force. From a practical point of view, it seems too low as the natural frequency will be higher for the following reasons:

1. End moments are neglected in the above equation, which will raise the natural frequency by more than 15%.
2. The critical span is usually limited by stress and is rarely reached.
3. The piping weight assumption is often larger than the actual load.

TABLE 6.9

Maximum Spans of Horizontal Pipe Lines (ft)

| Schedule | Limit | Pipe Size (in.) | | | | | | | | | | | | | |
		1	1.5	2	3	4	6	8	10	12	14	16	18	20	24
Schedule 10	L	13	15	17	20	22	25	29	30	32	37	38	39	39	41
	L'	13	16	18	21	24	28	31	34	37	41	42	44	46	48
Schedule 20	L							33	35	36	39	41	42	45	47
	L'							33	37	39	42	44	46	49	52
Schedule 30	L							34	37	39	42	43	46	49	52
	L'							34	38	41	43	45	48	51	55
Standard	L	13	16	18	23	26	31	35	38	41	42	43	44	45	47
	L'	13	16	18	23	26	31	35	38	41	43	45	47	49	52
Schedule 40	L	13	16	18	23	26	31	35	38	41	43	46	49	51	56
	L'	13	16	18	23	26	31	35	38	42	44	45	50	52	57
Schedule 60	L							36	40	43	46	49	52	55	60
	L'							35	39	43	45	48	51	54	59
Extra strong	L	13	17	19	24	27	33	37	41	43	44	46	48	49	51
	L'	13	17	19	23	26	32	36	40	43	44	46	49	51	54
Schedule 80	L	13	17	19	24	27	33	37	42	46	48	52	55	58	63
	L'	13	17	19	23	26	32	36	40	44	46	50	52	55	61
Schedule 100	L							38	43	47	49	53	56	59	65
	L'							37	41	45	47	50	53	56	51
Schedule 120	L					28	34	39	44	48	51	54	57	61	67
	L'					27	32	37	41	45	47	51	54	57	62
Schedule 140	L					28	34	39	44	49	51	54	58	61	67
	L'					27	33	37	42	45	48	51	54	57	62
Schedule 160	L	13	17	20	25	29	35	40	45	49	51	55	58	62	68
	L'	13	17	19	23	27	33	37	42	45	48	51	54	57	63

TABLE 6.10

Maximum Spans of Horizontal Pipe Lines (m)

Schedule	Limit	Pipe Size (in.)

Schedule	Limit	1	1.5	2	3	4	6	8	10	12	14	16	18	20	24
Schedule 10	L	4.0	4.6	5.2	6.1	6.7	7.6	8.8	9.1	9.8	11.3	11.6	11.9	11.9	12.5
	L'	4.0	4.9	5.5	6.4	7.3	8.5	9.4	10.4	11.3	12.5	12.8	13.4	14.0	14.6
Schedule 20	L	0.0	0.0	0.0	0.0	0.0	0.0	10.1	10.7	11.0	11.9	12.5	12.8	13.7	14.3
	L'	0.0	0.0	0.0	0.0	0.0	0.0	10.1	11.3	11.9	12.8	13.4	14.0	14.9	15.8
Schedule 30	L	0.0	0.0	0.0	0.0	0.0	0.0	10.4	11.3	11.9	12.8	13.1	14.0	14.9	15.8
	L'	0.0	0.0	0.0	0.0	0.0	0.0	10.4	11.6	12.5	13.1	13.7	14.6	15.5	16.8
Standard	L	4.0	4.9	5.5	7.0	7.9	9.4	10.7	11.6	12.5	12.8	13.1	13.4	13.7	14.3
	L'	4.0	4.9	5.5	7.0	7.9	9.4	10.7	11.6	12.5	13.1	13.7	14.3	14.9	15.8
Schedule 40	L	4.0	4.9	5.5	7.0	7.9	9.4	10.7	11.6	12.5	13.1	14.0	14.9	15.5	17.1
	L'	4.0	4.9	5.5	7.0	7.9	9.4	10.7	11.6	12.8	13.4	13.7	15.2	15.8	17.4
Schedule 60	L	0.0	0.0	0.0	0.0	0.0	0.0	11.0	12.2	13.1	14.0	14.9	15.8	16.8	18.3
	L'	0.0	0.0	0.0	0.0	0.0	0.0	10.7	11.9	13.1	13.7	14.6	15.5	16.5	18.0
Extra-strong	L	4.0	5.2	5.8	7.3	8.2	10.1	11.3	12.5	13.1	13.4	14.0	14.6	14.9	15.5
	L'	4.0	5.2	5.8	7.0	7.9	9.8	11.0	12.2	13.1	13.4	14.0	14.9	15.5	16.5
Schedule 80	L	4.0	5.2	5.8	7.3	8.2	10.1	11.3	12.8	14.0	14.6	15.8	16.8	17.7	19.2
	L'	4.0	5.2	5.8	7.0	7.9	9.8	11.0	12.2	13.4	14.0	15.2	15.8	16.8	18.6
Schedule 100	L	0.0	0.0	0.0	0.0	0.0	0.0	11.6	13.1	14.3	14.9	16.2	17.1	18.0	19.8
	L'	0.0	0.0	0.0	0.0	0.0	0.0	11.3	12.5	13.7	14.3	15.2	16.2	17.1	15.5
Schedule 120	L	0.0	0.0	0.0	0.0	8.5	10.4	11.9	13.4	14.6	15.5	16.5	17.4	18.6	20.4
	L'	0.0	0.0	0.0	0.0	8.2	9.8	11.3	12.5	13.7	14.3	15.5	16.5	17.4	18.9
Schedule 140	L	0.0	0.0	0.0	0.0	8.5	10.4	11.9	13.4	14.9	15.5	16.5	17.7	18.6	20.4
	L'	0.0	0.0	0.0	0.0	8.2	10.1	11.3	12.8	13.7	14.6	15.5	16.5	17.4	18.9
Schedule 160	L	4.0	5.2	6.1	7.6	8.8	10.7	12.2	13.7	14.9	15.5	16.8	17.7	18.9	20.7
	L'	4.0	5.2	5.8	7.0	8.2	10.1	11.3	12.8	13.7	14.6	15.5	16.5	17.4	19.2

Note: L is calculated with limiting bending stress of S divided by 2. L' is calculated with limiting static deflection 25.4 mm.

Table 6.9 provides a quick reference for values of span, based on the following assumptions:

- Pipe material is A53-grade carbon steel and Table 6.9 (see also Table 6.10) applies conservatively to all steels.
- The temperature ranges from 0°F to 650°F. At 650°F, $S = 12,000$ psi, the modulus of elasticity $E = 25.2 \times 10^6$ from piping code.
- Assumes the fluid is water.
- Assumes the density of insulation is 11 lb/ft^3.
- The thickness of insulation is 1.5 in. for ¼-in pipe, 2.0 in. for 6/14-in. pipe, and 2.5 in. for 16/24-in pipe.

- The pipe is treated as a horizontal beam, supported at both ends, and carrying a uniform load equal to the summation of metal weight, water, and insulation.
- The maximum static deflection was 25 mm.
- The maximum bending stress was equated to the allowable weight stress equal to half the allowable hot stress, *S*.

References

ACI Standard 318-05/318R-05, 2005. *Building Code Requirements for Structural Concrete and Commentary*. Farmington Hills, MI: ACI.

ACI Standard 318-08 M, 2008. *Building Code Requirements for Structural Concrete and Commentary*. Farmington Hills, MI: ACI.

ASCE Standard SEI/ASCE 7–05, 2005. *Minimum Design Loads for Buildings and Other Structures*, ASCE, Ann Arbor, MI.

ASCE 40262. 1997. *Wind Load and Anchor Bolt Design for Petrochemical Facilities*. Ann Arbor, MI: ASCE.

ASTM Standard F1554: Standard Specification for Anchor Bolts, Steel, 36, 55, and 105-ksi Yield, *ASTM Annual Book of Standards*. West Conshohocken, PA: ASTM International.

PIP Standard STC01015, 2006. *Structural Design Criteria*. Austin, TX: PIP.

PIP Standard STE05121, 2006. *Anchor Bolt Design Guide*. Austin, TX: PIP.

7

Steel Structures in Industry

7.1 Introduction

Steel structures are the most common in industrial projects because the characteristics of steel work well for industrial purposes. The following is a sample of steel's advantages to industry:

- Its high strength in tension and compression allows for a small cross section and is, therefore, better for use in long spans.
- Its elasticity makes steel a very good example of a material that complies with Hook's law.
- Its ductility allows a large degree of elongation before breaking, so it can withstand overload without sudden failure.
- Advanced technology is not required to strengthen it.
- Quality controls for steel are usually good because the rolled cross sections are delivered from factories under quality control systems.
- Its properties allow easier and faster erections.
- It is recoverable from scrap.
- It retains value after demolition.

Steel's main disadvantage is that it needs periodic maintenance. It is therefore important that the required maintenance be included in the operating budget.

7.2 Stress–Strain Behavior of Structural Steel

Structural steel is an important construction material. It possesses attributes such as strength, stiffness, toughness, and ductility that are very desirable in modern constructions. Strength is the ability of a material to resist stresses.

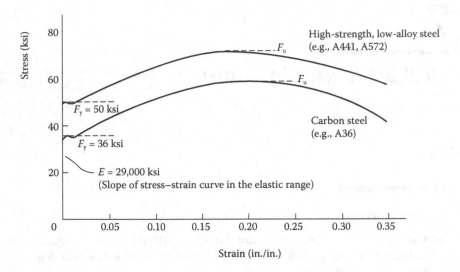

FIGURE 7.1
Uniaxial stress–strain behavior of steel.

It is measured in terms of the material's yield strength, F_y, and ultimate or tensile strength, F_u. For steel, the ranges of F_y and F_u ordinarily used in constructions are 36 to 50 ksi (248 to 345 MPa) and 58 to 70 ksi (400 to 483 MPa), respectively, although higher strength steels are becoming more common. Stiffness is the ability of a material to resist deformation. It is measured as the slope of the material's stress–strain curve. Figure 7.1 shows uniaxial engineering stress–strain curves obtained from tests for various grades of steels. It can be seen that the modulus of elasticity, E, does not vary appreciably for the different steel grades. Therefore, a value of 29,000 ksi (200 GPa) is often used for design.

Toughness is the ability of a material to absorb energy before failure. It is measured as the area under the material's stress–strain curve. As shown in Figure 7.1, most steels (especially the lower grades) possess high toughness that is suitable for both static and seismic applications. Ductility is the ability of a material to undergo large inelastic, or plastic, deformation before failure. It is measured in terms of percent elongation or percent reduction in the area of the specimen, tested in uniaxial tension.

7.3 Design Procedure

Normally, we start by a preliminary design, which is a rapid, approximate, manual method of designing a structure, as opposed to carrying out rigorous analysis and detailed design. The overall aim for a given structure is to identify critical loads, estimate design actions, and select sections. The process is bound by conceptual design, alternative systems, idealization, identification of critical members, and rationalization and depends greatly on the designer's experience and use of appropriate design aids. The term is also often applied to manual design. Preliminary design is needed for the following reasons:

- To obtain sections and weights for cost estimation
- To compare alternative proposals
- To obtain initial sections for computer analysis
- To check a completed design

The need for approximate manual methods is very important. It is necessary to know if the output is right, wrong, or complete nonsense. Methods of preliminary analysis that are not dependent on member sizes are set out for both elastic and plastic theories. Redundant structures are treated as statically determinate by approximately locating points of contraflexure by using subframes or by assigning values of actions at critical positions.

After defining the structure system in the preliminary design, we progress to the detailed engineering by, first, building the computer model and, then, selecting the safe member. Now you are ready to deliver the complete engineering tender package.

7.3.1 Tension Members

Tension members are those structural elements that are subjected to pure tensile forces. The selection of their cross section is one of the simplest and most straightforward problems encountered in steel design. Because stability is of minor concern with tension members, the problem is reduced to selecting a section with sufficient area to carry the design load without exceeding the allowable tensile stress, as determined by the factor of safety.

A structural member is considered a compression member if it is under a compressive load, either alone or with other loadings. Columns, beam-columns, plates, component parts of frames, and the compression flanges of beams and plate girders could all be taken as examples of this category.

The two main differences between tension and compression members are:

- Tension members are held straight by means of tensile loads, while the loads on compression members tend to bend the member out of the plane of loading.

- For riveted and bolted connections, the net area will govern the strength of a tension member, while, for compression members, the rivets are assumed to fill the holes.

For tension members, it is well known that the stability phenomena is not a criterion of the design, but is required to prevent sagging of a tension member if it is too long. It is also utilized to support vibrating equipment; in that case, the slenderness ratio ($\lambda = L/r$), here L is the tension member length and r is the radius of gyration, should be not higher than 300.

7.3.1.1 Slenderness Ratio

The slenderness ratio is obtained from the following equation:

$$\lambda = kL/r$$

where
  L = Slenderness ratio
  k = Buckling length factor

For compression members, k depends on the rotational restraint at the member ends and the means available to resist lateral movements. For tension members:

  $k = 1$

  L = Unsupported length for tension or compression members

  r = Radius of gyration corresponding to the member's effective buckling length (kL)

The buckling factor, k, is considered in Table 7.1. The values in Table 7.1 depend on an ideal case for pure fixation and hinge in the support. Practically, a hinge support will allow some restraint and the fixed support will allow some rotation.

In Table 7.1, for conditions a, b, c, and e, the lower end of the column is assumed to be fixed. In the remaining cases d, and f, the end is assumed to be hinged. Fixed support means that the column is adequately embedded in the concrete footing or connected by a moment-resisting bolted connection.

In the case of a, the top end is considered fixed when the column is in a frame with a heavy girder that is many times more rigid than the column itself. In c and f, the top of the column is allowed to translate, as in the case of columns in unbraced frames with heavy members.

TABLE 7.1

Values of Buckling Factor, *k*, for Different Conditions

	(a)	(b)	(c)	(d)	(e)	(f)
Theoretical *K*, value	0.5	0.7	1.0	1.0	2.0	2.0
Recommended design value when ideal conditions are approximated	0.65	0.8	1.2	1.0	2.1	2.0

Case end conditions		
	Rotation fixed and translation fixed	
	Rotation free and translation fixed	
	Rotation fixed and translation free	
	Rotation free and translation free	

The main method of calculating strength based on the buckling factor is by the Euler formula. All other methods modify the Euler formula and enhance it. The principal of the member strength and its slenderness ratio is shown in Figure 7.2. In general, by decreasing ($\lambda = kL/r$), the strength will increase by using a higher value of radius of gyration (*r*). Theoretically, the steel section can carry the load to infinity, but actually the yield strength is constrained by the section strength as shown in Figure 7.2. There are two formulas to calculate the member strength before and after $\lambda = 133$, which present the maximum load that can be carried by the member without buckling or being deformed.

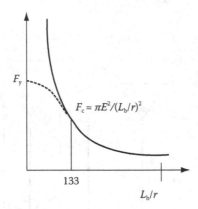

FIGURE 7.2
Relationship between the slenderness ratio and strength.

Example 1: Tension Member by Allowable Stress Design (ASD)

Design the Lower Chord Tension Member
Design force = 30 tons
Geometric length of the member, L_x = 300 cm (using 16-mm bolts for the bolted connection)
For two equal angles, $r = 0.3a$
For one angle, $r = 0.2a$
For star angles, $r = 0.385a$
Solution
$\lambda_{max} = kL/r = 300$
$r = 300/300 = 1$
$a = 1/0.3 = 3.33$ cm
Construction Guideline
$D = 16$ mm (bolt size)
$\phi = 16 + 2$ mm = 18 mm (hole size)
$a - t < 3\phi < 4.8$ cm
$A_{req} = 30/(2 \times 0.6 \times F_y \times 1.2 \times 0.85) = 10.50$ cm²
Choose two equal Angles
$65 \times 65 \times 9$ mm (A = 11.0 cm²)
$A_{net} = 2(11.0 - (1.8 \times 0.9)) = 18.76$ cm²
$f_t = 30/18.76 = 1.595 < 0.6 \, F_y \times 1.2$
LRFD for tension member
When you apply the load resistance force design (LRFD), the following will be the strength of the tension member:
For yielding in the gross section

$$\phi_t = 0.90, \qquad P_n = F_y A_g$$

For fracture in the net section

$$\phi_t = 0.75 \qquad P_n = F_u A_e$$

where

A_e = Effective net area
A_g = Gross area of member
F_y = Specific minimum yield stress
F_u = Specific minimum tensile strength
P_n = Nominal axial strength

Example 2: Previous Example Using Load Resistance Force Design

Design the lower chord tension member
Design force = 30 tons for dead load
L_x = 300 cm (using 16-mm bolts for the bolted connection)
Choose two equal angles
65 × 65 × 9 mm (A = 11.0 cm²)
Solution

1. Load
$P_u = 1.4$ $P_D = 1.4 \times 30 = 42$ tons $\leq f_t P_n$
2. Strength
$\phi_t P_n = 0.9 \times 2.4 \times 22 = 47.52 > P_u$ OK
$\phi_t P_n = 0.75 \times 3.6 \times 18.76 = 50.6 > P_u$ OK

7.3.2 Compression Members

Columns are structural elements subjected to a pure compression force. They are the obvious support for roof trusses or system girders. The main concern for this compression member is buckling.

7.3.2.1 Steps of Preliminary Design

1. Assume the allowable buckling stress from Table 7.2.
2. Calculate the required cross-sectional area.
   $A_g = D_F/F_c$
3. Select the convenient steel section.
4. Generally, check that the section is noncompact and can achieve yield stress without local plate buckling as shown in Tables 7.3 and 7.4.

TABLE 7.2

Approximate Estimates of Critical Stress

D_F (kN)	F_c (N/mm²)
50	60
50–100	70
100–200	80
>200	95

TABLE 7.3

Specified Yield Stress for Steel Compression Members of 36 ksi (248 N/mm²)

kL/r	F_a ksi	F_a N/mm²	kL/r	F_a ksi	F_a N/mm²	kL/r	F_a ksi	F_a N/mm²	kL/r	F_a ksi	F_a N/mm²	kL/r	F_a ksi	F_a N/mm²
1	21.56	148.65	41	19.11	131.76	81	15.24	105.08	121	10.14	69.91	161	5.76	39.71
2	21.52	148.38	42	19.03	131.21	82	15.13	104.32	122	9.99	68.88	162	5.69	39.23
3	21.48	148.1	43	18.95	130.66	83	15.02	103.56	123	9.85	67.91	163	5.62	38.75
4	21.44	147.82	44	18.86	130.04	84	14.90	102.73	124	9.70	66.88	164	5.55	38.27
5	21.39	147.48	45	18.78	129.48	85	14.79	101.97	125	9.55	65.84	165	5.49	37.85
6	21.35	147.20	46	18.70	128.93	86	14.67	101.15	126	9.41	64.88	166	5.42	37.37
7	21.30	146.86	47	18.61	128.31	87	14.56	100.39	127	9.26	63.85	167	5.35	36.89
8	21.25	146.51	48	18.53	127.76	88	14.44	99.56	128	9.11	62.81	168	5.29	36.47
9	21.21	146.24	49	18.44	127.14	89	14.32	98.73	129	8.97	61.85	169	5.23	36.06
10	21.16	145.89	50	18.35	126.52	90	14.20	97.91	130	8.84	60.95	170	5.17	35.65
11	21.10	145.48	51	18.26	125.90	91	14.09	97.15	131	8.70	59.98	171	5.11	35.23
12	21.05	145.13	52	18.17	125.28	92	13.97	96.32	132	8.57	59.09	172	5.05	34.82
13	21.00	144.79	53	18.08	124.66	93	13.84	95.42	133	8.44	58.19	173	4.99	34.40
14	20.95	144.45	54	17.99	124.04	94	13.72	94.60	134	8.32	57.36	174	4.93	33.99
15	20.89	144.04	55	17.90	123.42	95	13.60	93.77	135	8.19	56.47	175	4.88	33.65
16	20.83	143.62	56	17.81	122.80	96	13.48	92.94	136	8.07	55.64	176	4.82	33.23
17	20.78	143.27	57	17.71	122.11	97	13.35	92.04	137	7.96	54.88	177	4.77	32.89
18	20.72	142.86	58	17.62	121.49	98	13.23	91.22	138	7.84	54.05	178	4.71	32.47
19	20.66	142.45	59	17.53	120.87	99	13.10	90.32	139	7.73	53.30	179	4.66	32.13
20	20.60	142.03	60	17.43	120.18	100	12.98	89.49	140	7.62	52.54	180	4.61	31.78

n			n			n			n			n		
21	20.54	141.62	61	17.33	119.49	101	12.85	88.60	141	7.51	51.78	181	4.56	31.44
22	20.48	141.20	62	17.24	118.87	102	12.72	87.70	142	7.41	51.09	182	4.51	31.10
23	20.41	140.72	63	17.14	118.18	103	12.59	86.80	143	7.30	50.33	183	4.46	30.75
24	20.35	140.31	64	17.04	117.49	104	12.47	85.98	144	7.20	49.64	184	4.41	30.41
25	20.28	139.83	65	16.94	116.80	105	12.33	85.01	145	7.10	48.95	185	4.36	30.06
26	20.22	139.41	66	16.84	116.11	106	12.20	84.12	146	7.01	48.33	186	4.32	29.79
27	20.15	138.93	67	16.74	115.42	107	12.07	83.22	147	6.91	47.64	187	4.27	29.44
28	20.08	138.45	68	16.64	114.73	108	11.94	82.32	148	6.82	47.02	188	4.23	29.16
29	20.01	137.96	69	16.53	113.97	109	11.81	81.43	149	6.73	46.40	189	4.18	28.82
30	19.94	137.48	70	16.43	113.28	110	11.67	80.46	150	6.64	45.78	190	4.14	28.54
31	19.87	137	71	16.33	112.59	111	11.54	79.57	151	6.55	45.16	191	4.09	28.20
32	19.80	136.52	72	16.22	111.83	112	11.40	78.60	152	6.46	44.54	192	4.05	27.92
33	19.73	136.03	73	16.12	111.14	113	11.26	77.63	153	6.38	43.99	193	4.01	27.65
34	19.65	135.48	74	16.01	110.39	114	11.13	76.74	154	6.30	43.44	194	3.97	27.37
35	19.58	135	75	15.90	109.63	115	10.99	75.77	155	6.22	42.89	195	3.93	27.10
36	19.50	134.45	76	15.79	108.87	116	10.85	74.81	156	6.14	42.33	196	3.89	26.82
37	19.42	133.9	77	15.69	108.18	117	10.71	73.84	157	6.06	41.78	197	3.85	26.54
38	19.35	133.41	78	15.58	107.42	118	10.57	72.88	158	5.98	41.23	198	3.81	26.27
39	19.27	132.86	79	15.47	106.66	119	10.43	71.91	159	5.91	40.75	199	3.77	25.99
40	19.19	132.31	80	15.36	105.90	120	10.28	70.88	160	5.83	40.20	200	3.73	25.72

TABLE 7.4

Specified Yield Stress for Steel Compression Members of 50 ksi (345 N/mm²)

		F_a			F_a			F_a			F_a			F_a
kL/r	ksi	N/mm²	kL/r	ksi	N/mm²	kL/r	ksi	N/mm²	kL/r	ksi	N/mm²	kL/r	ksi	N/mm²
1	29.94	206.43	41	25.69	177.13	81	18.81	129.69	121	10.2	70.33	161	5.76	39.71
2	29.87	205.95	42	25.55	176.16	82	18.61	128.31	122	10.03	69.15	162	5.69	39.23
3	29.8	205.47	43	25.4	175.13	83	18.41	126.93	123	9.87	68.05	163	5.62	38.75
4	29.73	204.98	44	25.26	174.16	84	18.2	125.48	124	9.71	66.95	164	5.55	38.27
5	29.66	204.50	45	25.11	173.13	85	17.99	124.04	125	9.56	65.91	165	5.49	37.85
6	29.58	203.95	46	24.96	172.09	86	17.79	122.66	126	9.41	64.88	166	5.42	37.37
7	29.5	203.40	47	24.81	171.06	87	17.58	121.21	127	9.26	63.85	167	5.35	36.89
8	29.42	202.84	48	24.66	170.02	88	17.37	119.76	128	9.11	62.81	168	5.29	36.47
9	29.34	202.29	49	24.51	168.99	89	17.15	118.25	129	8.97	61.85	169	5.23	36.06
10	29.26	201.74	50	24.35	167.89	90	16.94	116.80	130	8.84	60.95	170	5.17	35.65
11	29.17	201.12	51	24.19	166.78	91	16.72	115.28	131	8.7	59.98	171	5.11	35.23
12	29.08	200.50	52	24.04	165.75	92	16.5	113.76	132	8.57	59.09	172	5.05	34.82
13	28.99	199.88	53	23.88	164.65	93	16.29	112.32	133	8.44	58.19	173	4.99	34.40
14	28.9	199.26	54	23.72	163.54	94	16.06	110.73	134	8.32	57.36	174	4.93	33.99
15	28.8	198.57	55	23.55	162.37	95	15.84	109.21	135	8.19	56.47	175	4.88	33.65
16	28.71	197.95	56	23.39	161.27	96	15.62	107.70	136	8.07	55.64	176	4.82	33.23
17	28.61	197.26	57	23.22	160.10	97	15.39	106.11	137	7.96	54.88	177	4.77	32.89
18	28.51	196.57	58	23.06	158.99	98	15.17	104.59	138	7.84	54.05	178	4.71	32.47
19	28.4	195.81	59	22.89	157.82	99	14.94	103.01	139	7.73	53.30	179	4.66	32.13

20	28.3	195.12	60	22.72	156.65	100	14.71	101.42	140	7.62	52.54	180	4.61	31.78
21	28.19	194.36	61	22.55	155.48	101	14.47	99.77	141	7.51	51.78	181	4.56	31.44
22	28.08	193.61	62	22.37	154.24	102	14.24	98.18	142	7.41	51.09	182	4.51	31.10
23	27.97	192.85	63	22.2	153.06	103	14	96.53	143	7.3	50.33	183	4.46	30.75
24	27.86	192.09	64	22.02	151.82	104	13.77	94.94	144	7.2	49.64	184	4.41	30.41
25	27.75	191.33	65	21.85	150.65	105	13.53	93.29	145	7.1	48.95	185	4.36	30.06
26	27.63	190.51	66	21.67	149.41	106	13.29	91.63	146	7.01	48.33	186	4.32	29.79
27	27.52	189.74	67	21.49	148.17	107	13.04	89.91	147	6.91	47.64	187	4.27	29.44
28	27.4	188.92	68	21.31	146.93	108	12.8	88.25	148	6.82	47.02	188	4.23	29.16
29	27.28	188.09	69	21.12	145.62	109	12.57	86.67	149	6.73	46.40	189	4.18	28.82
30	27.15	187.20	70	20.94	144.38	110	12.34	85.08	150	6.64	45.78	190	4.14	28.54
31	27.03	186.37	71	20.75	143.07	111	12.12	83.56	151	6.55	45.16	191	4.09	28.20
32	26.9	185.47	72	20.56	141.76	112	11.9	82.05	152	6.46	44.54	192	4.05	27.92
33	26.77	184.58	73	20.38	140.52	113	11.69	80.60	153	6.38	43.99	193	4.01	27.65
34	26.64	183.68	74	20.1	138.58	114	11.49	79.22	154	6.3	43.44	194	3.97	27.37
35	26.51	182.78	75	19.99	137.83	115	11.29	77.84	155	6.22	42.89	195	3.93	27.10
36	26.38	181.88	76	19.8	136.52	116	11.1	76.53	156	6.14	42.33	196	3.89	26.82
37	26.25	180.99	77	19.61	135.21	117	10.91	75.22	157	6.06	41.78	197	3.85	26.54
38	26.11	180.02	78	19.41	133.83	118	10.72	73.91	158	5.98	41.23	198	3.81	26.27
39	25.97	179.06	79	19.21	132.45	119	10.55	72.74	159	5.91	40.75	199	3.77	25.99
40	25.83	178.10	80	19.01	131.07	120	10.37	71.50	160	5.83	40.20	200	3.73	25.72

5. Obtain the value of the slenderness ratio, K.
6. Calculate the slenderness ratio, (kL/r_{min}). It must be less than l_{max}.
7. Based on Table 7.3, get F_c from AISC-89.
8. Compute the actual stress $f_c = D_F/A_g$.
9. If f_c is less than F_c or not more than 3%, that is acceptable.

Example 3: Column Designed by ASD

Design a column to carry a total load of 1000 kN. The column should be hinged at both ends with a buckling length of:
$l_x = l_y = 8000$ mm
Assume $F_c = 95$ N/mm²
$A = (1000 \times 1000/95) = 10{,}525$ mm²
$r = (8000/180) = 44.4$ mm
W10 × 60
$A = 116$ cm², $r_x = 11.2$ cm, $r_y = 6.46$ cm, $t_f = 17.3$ mm, $t_w = 10.7$ mm
$c = (256 - 10.7 - 2 \times 17.3)/2 = 210.7$
$c/t_f = 210.7/17.3 = 12.18 < 14.8$ OK
$d_w = (259.6 - 4 \times 17.3) = 190.4$ cm
$d_w/t_w = 17.8 < 41.3$ OK
$l_y/l_y = 123.8$ $F_c = 67.9$ N/mm² (from Tables 7.5 and 7.6)
$f_c = 1{,}000{,}000 / 11{,}600 = 86.2$ N/mm² $< F_c$ Unsafe
W12 × 65
$A = 12{,}500$ mm², $r_x = 134$ mm, $r_y = 76.4$ mm, $t_f = 15.4$ mm, $t_w = 9.9$ mm
$c = (300 - 9.9 - 2 \times 15.4)/2 = 129.65$
$c/t_f = 129.65/15.4 = 8.41 < 14.8$ OK
$d_w = (300 - 4 \times 15.4) = 238.4$ mm
$d_w/t_w = 238.4/9.9 = 24.08 < 41.3$ OK
$l_y/r_y = 104.575$ $F_c = 87$ N/mm² (from Tables 7.5 and 7.6, 12.47 ksi)
$f_c = 100{,}000/12{,}500 = 80 < F_c$ OK
For load resistance factor design, Tables 7.5 and 7.6 will be used to obtain the allowable stress.

Example 4: Load Resistance Force Design for Column

Design a column to carry a dead load of 50 tons and a live load of 50 tons. The column is hinged at both ends with the following buckling length:
$l_x = l_y = 800$ cm
Load
$P_u = 1.2 \times 50 + 1.6 \times 50 = 140$ tons
Strength
W12 × 65
$A = 12{,}500$ mm², $r_x = 134$ mm, $r_y = 76.4$ mm, $t_f = 15.4$ mm, $t_w = 9.9$ mm
$\lambda = 8000/76.4 = 104.7$
$\phi_c F_c = 17.3 \times 0.07 = 119$ N/mm²
$\phi_c P_n = 119 \times 12{,}500/1000 = 1487.5 > 1400$ kN   OK

TABLE 7.5

Design Stress for Steel Compression Members of 36 ksi (248 N/mm²) with Specified Yield Stress, $\phi_c = 0.85$

kL/r	$\Phi_c F_{cr}$ ksi	$\Phi_c F_{cr}$ N/mm²	kL/r	$\Phi_c F_{cr}$ ksi	$\Phi_c F_{cr}$ N/mm²	kL/r	$\Phi_c F_{cr}$ ksi	$\Phi_c F_{cr}$ N/mm²	kL/r	$\Phi_c F_{cr}$ ksi	$\Phi_c F_{cr}$ N/mm²	kL/r	$\Phi_c F_{cr}$ ksi	$\Phi_c F_{cr}$ N/mm²
1	30.6	211.0	41	28	193.1	81	21.7	149.6	121	14.2	97.9	161	8.23	56.7
2	30.6	211.0	42	27.9	192.4	82	21.5	148.2	122	14	96.5	162	8.13	56.1
3	30.6	211.0	43	27.8	191.7	83	21.3	146.9	123	13.8	95.1	163	8.03	55.4
4	30.6	211.0	44	27.6	190.3	84	21.1	145.5	124	13.6	93.8	164	7.93	54.7
5	30.6	211.0	45	27.5	189.6	85	20.9	144.1	125	13.4	92.4	165	7.84	54.1
6	30.5	210.3	46	27.4	188.9	86	20.7	142.7	126	13.3	91.7	166	7.74	53.4
7	30.5	210.3	47	27.2	187.5	87	20.5	141.3	127	13.1	90.3	167	7.65	52.7
8	30.5	210.3	48	27.1	186.8	88	20.4	140.7	128	12.9	88.9	168	7.56	52.1
9	30.5	210.3	49	27	186.2	89	20.2	139.3	129	12.7	87.6	169	7.47	51.5
10	30.4	209.6	50	26.8	184.8	90	20	137.9	130	12.6	86.9	170	7.38	50.9
11	30.4	209.6	51	26.7	184.1	91	19.8	136.5	131	12.4	85.5	171	7.3	50.3
12	30.4	209.6	52	26.5	182.7	92	19.6	135.1	132	12.2	84.1	172	7.21	49.7
13	30.3	208.9	53	26.4	182.0	93	19.4	133.8	133	12.1	83.4	173	7.13	49.2
14	30.3	208.9	54	26.3	181.3	94	19.2	132.4	134	11.9	82.0	174	7.05	48.6
15	30.2	208.2	55	26.1	180.0	95	19	131.0	135	11.7	80.7	175	6.97	48.1
16	30.2	208.2	56	25.9	178.6	96	18.8	129.6	136	11.5	79.3	176	6.89	47.5
17	30.1	207.5	57	25.8	177.9	97	18.7	128.9	137	11.4	78.6	177	6.81	47.0
18	30.1	207.5	58	25.6	176.5	98	18.5	127.6	138	11.2	77.2	178	6.73	46.4
19	30	206.8	59	25.5	175.8	99	18.3	126.2	139	11	75.8	179	6.66	45.9
20	30	206.8	60	25.3	174.4	100	18.1	124.8	140	10.9	75.2	180	6.59	45.4

(continued)

TABLE 7.5 Continued

Design Stress for Steel Compression Members of 36 ksi (248 N/mm²) with Specified Yield Stress, $\phi_c = 0.85$

kL/r	$\Phi_c F_{cr}$ ksi	$\Phi_c F_{cr}$ N/mm²	kL/r	$\Phi_c F_{cr}$ ksi	$\Phi_c F_{cr}$ N/mm²	kL/r	$\Phi_c F_{cr}$ ksi	$\Phi_c F_{cr}$ N/mm²	kL/r	$\Phi_c F_{cr}$ ksi	$\Phi_c F_{cr}$ N/mm²	kL/r	$\Phi_c F_{cr}$ ksi	$\Phi_c F_{cr}$ N/mm²
21	29.9	206.2	61	25.2	173.7	101	17.9	123.4	141	10.7	73.8	181	6.51	44.9
22	29.8	205.5	62	25	172.4	102	17.7	122.0	142	10.6	73.1	182	6.44	44.4
23	29.8	205.5	63	24.8	171.0	103	17.5	120.7	143	10.4	71.7	183	6.37	43.9
24	26.7	184.1	64	24.7	170.3	104	17.3	119.3	144	10.3	71.0	184	6.3	43.4
25	29.6	204.1	65	24.5	168.9	105	17.1	117.9	145	10.2	70.3	185	6.23	43.0
26	29.5	203.4	66	24.3	167.5	106	16.9	116.5	146	10	68.9	186	6.17	42.5
27	29.5	203.4	67	24.2	166.9	107	16.8	115.8	147	9.87	68.1	187	6.1	42.1
28	29.4	202.7	68	24	165.5	108	16.6	114.5	148	9.74	67.2	188	6.04	41.6
29	29.3	202.0	69	23.8	164.1	109	16.4	113.1	149	9.61	66.3	189	5.97	41.2
30	29.2	201.3	70	23.6	162.7	110	16.2	111.7	150	9.48	65.4	190	5.91	40.7
31	29.1	200.6	71	23.5	162.0	111	16	110.3	151	9.36	64.5	191	5.85	40.3
32	29	199.9	72	23.3	160.6	112	15.8	108.9	152	9.23	63.6	192	5.79	39.9
33	28.9	199.3	73	23.1	159.3	113	15.6	107.6	153	9.11	62.8	193	5.73	39.5
34	28.8	198.6	74	22.9	157.9	114	15.4	106.2	154	9	62.1	194	5.67	39.1
35	28.7	197.9	75	22.8	157.2	115	15.3	105.5	155	8.88	61.2	195	5.61	38.7
36	28.6	197.2	76	22.6	155.8	116	15.1	104.1	156	8.77	60.5	196	5.55	38.3
37	28.5	196.5	77	22.4	154.4	117	14.9	102.7	157	8.66	59.7	197	5.5	37.9
38	28.4	195.8	78	22.2	153.1	118	14.7	101.4	158	8.55	59.0	198	5.44	37.5
39	28.3	195.1	79	22	151.7	119	14.5	100.0	159	8.44	58.2	199	5.39	37.2
40	28.1	193.7	80	21.9	151.0	120	14.3	98.6	160	8.33	57.4	200	5.33	36.7

TABLE 7.6

Design Stress for Steel Compression Members of 50 ksi (345 N/mm²) with Specified Yield Stress, $\phi_c = 0.85$

kL/r	$\Phi_c F_{cr}$ ksi	$\Phi_c F_{cr}$ N/mm²	kL/r	$\Phi_c F_{cr}$ ksi	$\Phi_c F_{cr}$ N/mm²	kL/r	$\Phi_c F_{cr}$ ksi	$\Phi_c F_{cr}$ N/mm²	kL/r	$\Phi_c F_{cr}$ ksi	$\Phi_c F_{cr}$ N/mm²	kL/r	$\Phi_c F_{cr}$ ksi	$\Phi_c F_{cr}$ N/mm²
1	30.6	211.0	41	28	193.1	81	21.7	149.6	121	14.2	97.9	161	8.23	56.7
2	30.6	211.0	42	27.9	192.4	82	21.5	148.2	122	14	96.5	162	8.13	56.1
3	30.6	211.0	43	27.8	191.7	83	21.3	146.9	123	13.8	95.1	163	8.03	55.4
4	30.6	211.0	44	27.6	190.3	84	21.1	145.5	124	13.6	93.8	164	7.93	54.7
5	30.6	211.0	45	27.5	189.6	85	20.9	144.1	125	13.4	92.4	165	7.84	54.1
6	30.5	210.3	46	27.4	188.9	86	20.7	142.7	126	13.3	91.7	166	7.74	53.4
7	30.5	210.3	47	27.2	187.5	87	20.5	141.3	127	13.1	90.3	167	7.65	52.7
8	30.5	210.3	48	27.1	186.8	88	20.4	140.7	128	12.9	88.9	168	7.56	52.1
9	30.5	210.3	49	27	186.2	89	20.2	139.3	129	12.7	87.6	169	7.47	51.5
10	30.4	209.6	50	26.8	184.8	90	20	137.9	130	12.6	86.9	170	7.38	50.9
11	30.4	209.6	51	26.7	184.1	91	19.8	136.5	131	12.4	85.5	171	7.3	50.3
12	30.4	209.6	52	26.5	182.7	92	19.6	135.1	132	12.2	84.1	172	7.21	49.7
13	30.3	208.9	53	26.4	182.0	93	19.4	133.8	133	12.1	83.4	173	7.13	49.2
14	30.3	208.9	54	26.3	181.3	94	19.2	132.4	134	11.9	82.0	174	7.05	48.6
15	30.2	208.2	55	26.1	180.0	95	19	131.0	135	11.7	80.7	175	6.97	48.1
16	30.2	208.2	56	25.9	178.6	96	18.8	129.6	136	11.5	79.3	176	6.89	47.5
17	30.1	207.5	57	25.8	177.9	97	18.7	128.9	137	11.4	78.6	177	6.81	47.0
18	30.1	207.5	58	25.6	176.5	98	18.5	127.6	138	11.2	77.2	178	6.73	46.4
19	30	206.8	59	25.5	175.8	99	18.3	126.2	139	11	75.8	179	6.66	45.9
20	30	206.8	60	25.3	174.4	100	18.1	124.8	140	10.9	75.2	180	6.59	45.4
21	29.9	206.2	61	25.2	173.7	101	17.9	123.4	141	10.7	73.8	181	6.51	44.9

(continued)

TABLE 7.6 Continued

Design Stress for Steel Compression Members of 50 ksi (345 N/mm²) with Specified Yield Stress, $\phi_c = 0.85$

kL/r	$\Phi_c F_{cr}$ ksi	$\Phi_c F_{cr}$ N/mm²	kL/r	$\Phi_c F_{cr}$ ksi	$\Phi_c F_{cr}$ N/mm²	kL/r	$\Phi_c F_{cr}$ ksi	$\Phi_c F_{cr}$ N/mm²	kL/r	$\Phi_c F_{cr}$ ksi	$\Phi_c F_{cr}$ N/mm²	kL/r	$\Phi_c F_{cr}$ ksi	$\Phi_c F_{cr}$ N/mm²
22	29.8	205.5	62	25	172.4	102	17.7	122.0	142	10.6	73.1	182	6.44	44.4
23	29.8	205.5	63	24.8	171.0	103	17.5	120.7	143	10.4	71.7	183	6.37	43.9
24	26.7	184.1	64	24.7	170.3	104	17.3	119.3	144	10.3	71.0	184	6.3	43.4
25	29.6	204.1	65	24.5	168.9	105	17.1	117.9	145	10.2	70.3	185	6.23	43.0
26	29.5	203.4	66	24.3	167.5	106	16.9	116.5	146	10	68.9	186	6.17	42.5
27	29.5	203.4	67	24.2	166.9	107	16.8	115.8	147	9.87	68.1	187	6.1	42.1
28	29.4	202.7	68	24	165.5	108	16.6	114.5	148	9.74	67.2	188	6.04	41.6
29	29.3	202.0	69	23.8	164.1	109	16.4	113.1	149	9.61	66.3	189	5.97	41.2
30	29.3	201.3	70	23.6	162.7	110	16.2	111.7	150	9.48	65.4	190	5.91	40.7
31	29.1	200.6	71	23.5	162.0	111	16	110.3	151	9.36	64.5	191	5.85	40.3
32	29	199.9	72	23.3	160.6	112	15.8	108.9	152	9.23	63.6	192	5.79	39.9
33	28.9	199.3	73	23.1	159.3	113	15.6	107.6	153	9.11	62.8	193	5.73	39.5
34	28.8	198.6	74	22.9	157.9	114	15.4	106.2	154	9	62.1	194	5.67	39.1
35	28.7	197.9	75	22.8	157.2	115	15.3	105.5	155	8.88	61.2	195	5.61	38.7
36	28.6	197.2	76	22.6	155.8	116	15.1	104.1	156	8.77	60.5	196	5.55	38.3
37	28.5	196.5	77	22.4	154.4	117	14.9	102.7	157	8.66	59.7	197	5.5	37.9
38	28.4	195.8	78	22.2	153.1	118	14.7	101.4	158	8.55	59.0	198	5.44	37.5
39	28.3	195.1	79	22	151.7	119	14.5	100.0	159	8.44	58.2	199	5.39	37.2
40	28.1	193.7	80	21.9	151.0	120	14.3	98.6	160	8.33	57.4	200	5.33	36.7

Source: AISC, *Steel Construction Manual: Load Resistance Force Design,* AISC, Chicago, IL, 2001.

7.3.3 Beam Design

Beams are structural members transversely loaded to their lengths and in the plane of their webs. They resist flexure, shear, and torsion stresses. Regardless of the method of design and the design code that you will use, these three aspects must be considered in the design of beams:

- Strength: the cross section of a member must be adequate to resist the applied bending moments and shearing forces.
- Stability: the steel plate elements that compose the cross section must resist the local buckling phenomena. The web should be safe against the buckling created by shearing forces and bending moments. The cross section should have adequate torsion strength to resist lateral torsion buckling (LTB).
- Performance: the deformation of the member must not be excessive under service loads. Limitations of deflections sometimes govern beam design.

Beams are structural elements subjected to bending moment and shearing force only. The normal force is neglected (zero).

The beams are usually used for the following functions:

- Purlines
- Crane track girder (CTG)
- Beam in flooring system, as shown in Figure 7.3
- Girder of rigid beam (Figure 7.4)

The beam design is based on the failure steps shown in Figure 7.5. The load, P, is applied on the center of the beam and is increased gradually. A diagram of the corresponding bending moment is drawn. The beam failure occurs in four steps:

1. The stress on the bottom and top fiber of the steel section is less than the yield strength.
2. The top and bottom fibers of the steel section stresses are equal to the yield strength and the moment is M_y.

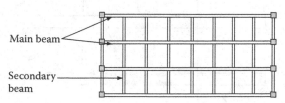

FIGURE 7.3
Floor structure system.

FIGURE 7.4
Photo of roof system.

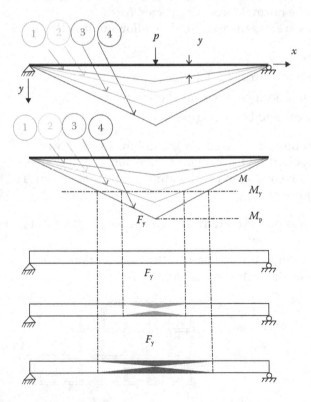

FIGURE 7.5
Simple beam failure stresses in increasing steps.

3. Part of the section has stresses equal to the yield strength.

4. All the cross sections have been affected by stresses equal to the yield strength and so we reach the plasticity limit and the moment is, M_p.

From the AISC, the following are the strengths allowable in the design for steel structure beams:

- For a compact section
  $F_b = 0.66F_y$
- For a noncompact section
  $F_b = 0.60F_y$

Allowable shear stress = $0.4F_y$.

It is important to define whether the section is a compact or noncompact section. A compact section can achieve the plastic moment with local buckling on the web or the flange. A noncompact section can only reach the yields for the lower and bottom fiber without local buckling to the web or the flange. In general, channel, equal, and nonequal angles are always noncompact sections.

7.3.3.1 Lateral Torsion Buckling

The strength of the beam controlled by the second limit state for beams is lateral torsional buckling (LTB). LTB occurs when the compression portion of a cross section is restrained by the tension portion, and the deflection due to flexural buckling is accompanied by torsion or twisting. The way to prevent LTB is to have adequate lateral bracing at adequate intervals along the beam. The limit state is the interval of the bracing. The computed flexural stress, f_b, shall not exceed the allowable flexural stress, F_b, given as follows (in all equations, the minimum specified yield stress, F_y, cannot exceed its limits):

$$\text{For } L_b \le L_c,$$

$$F_b = 0.66F_y \tag{7.1}$$

where L_c = smaller value of

$$200b_f \big/ \sqrt{F_y} \text{ or } 137{,}895/(d/A_f)F_y \text{ for I and channel shapes} \tag{7.2}$$

in which
b_f = Flange width (mm)
F_y = Yield strength (N/mm²)
d = Overall depth of section (mm)
A_f = Area of compression flange (mm²)

For the above sections to be considered compact, in addition to having the width-thickness ratios of their component elements falling within the limiting value, as shown in Table 7.7, the flanges of the sections must be continuously connected to the webs. For $L_b > L_c$, the allowable flexural stress in tension is given by

$$F_b = 0.60F_y \tag{7.3}$$

The allowable flexural stress in compression is given by the larger value calculated from Equation (7.4).

$$F_b = \frac{82737 A_f}{L_d} \tag{7.4}$$

TABLE 7.7

Maximum Width/Thickness Ratio for Compact and Noncompact Sections

	Compact			Noncompact Section		
	St. 52	St. 44	St. 37	St. 52	St. 44	St. 37
Bending						
	$d_w/t_w \le 127/(F_y)^{0.5}$			$d_w/t_w \le 190/(F_y)^{0.5}$		
	66.94	75.91	82	100.15	113.56	122.65
Compression						
	$d_w/t_w \le 58/(F_y)^{0.5}$			$d_w/t_w \le 64/(F_y)^{0.5}$		
	30.56	34.65	37.44	33.73	38.24	41.3
Rolled						
	$c/t_f \le 16.9/(F_y)^{0.5}$			$c/t_f \le 23/(F_y)^{0.5}$		
	8.9	10.1	10.91	12.12	13.74	14.8
Welded						
	$c/t_f \le 15.3/(F_y)^{0.5}$			$c/t_f \le 21/(F_y)^{0.5}$		
	8.06	9.14	9.87	11.07	12.55	13.5
Equal angles						
				$b/t \le 23/(F_y)^{0.5}$		
				8.96	10.15	11.0
Unequal angles						
				$(b + h)/2t \le 17/(F_y)^{0.5}$		
				8.96	10.15	11.0
Tubular member subjected to bending or compression stress						
	$D/t \le 165/F_y$			$D/t \le 211/F_y$		
	45.83	58.92	68.79	58.61	75.35	88.0
T section subjected to compression						
				$b/t \le 30/(F_y)^{0.5}$		
				15.81	17.93	19.36

where

l = Distance between cross sections braced against twist or lateral displacement of the compression flange (mm)

A_f = Compression flange area (mm²)

C_b = $12.5M_{max}/(2.5M_{max} + 3M_A + 4M_B + 3M_C)$

M_{max} = Maximum moment along the unbraced length of the member

M_A = Quarter-point moment along the unbraced length of the member

M_B = Midpoint moment along the unbraced length of the member

M_CD = Three-quarter point moment along the unbraced length of the member

For simplicity in design, C_b can conservatively be taken as unity.

Equation (7.4) is applicable only to I and channel shapes with an axis of symmetry in, and loaded in, the plane of the web. You may check whether the cross section is compact or noncompact regarding lateral tensional buckling by applying:

$$L_c \leq 20b/(F_y)^{0.5}$$

$$L_u \leq (1380A_f/dF_y)C_b$$

Purline design, until now, did not allow the use of covering materials as carrying load with the purlines in design, on the other hand, they have not taken the effect of torsion into consideration.

7.3.3.2 Allowable Deflection

It always depends on the project specifications, but in general the allowable deflection in AISC will be as follows and see also Table 7.8 for British Standards BS5950.

For crane track girder = span/800
For floor and roof with plastering = span/360

TABLE 7.8

Suggested Limits for Calculated Deflection Based on BS5950

(1) Vertical deflection of beam due to imposed load	
Cantilever	Length/180
Beams are not carrying plaster or brittle finished	Length/200
(2) Horizontal deflection of columns due to imposed load and wind load	
Top of columns in single-story building, except portal frames	Height/300
In each building with more than 1 story	Height of that story/300
(3) Crane girders	
Vertical deflection due to static vertical wheel loads from overhead travelling crane	Span/600
Horizontal deflection (calculate)	Span/500

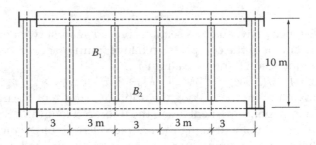

FIGURE 7.6
Floor geometry sketch.

Based on AISC-ASD, beams and girders supporting floors and roofs shall be proportioned with due regard to the deflection produced by the design loads. Beams and girders supporting plastered ceilings shall be so proportioned that the maximum live-load deflection does not exceed 1/360 of the span.

Example 1: Beam Design

The floor system of a shopping center is as shown (Figure 7.6). The floor consists of a 50-mm reinforced concrete (RC) slab over steel corrugated sheets. It is required to design the floor beams B1 and B2.

Loads/m² of the floor:

Corrugated sheet	= 200 N/m²
RC slab	= 1250 N/m²
Sand + tiles	= 1000 N/m²
Live load (LL)	= 4000 N/m²
Total	= 6450 N/m²
Load on the beam	= 6450 × 3.0 = 19,350 N/m'
Assume beam weight	= 1000 N/m'
Total load on beam	= 19,350 + 1000 = 20,350 N/m'

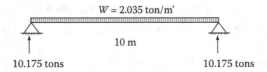

$$M_x = \frac{2.035 \times 10^2}{8} = 254.4 \text{ m kN}$$

$$Q_{max} = R = 101.75 \text{ kN}$$

The compression flange is encased in an RC slab, so the unbraced length is zero.
The cross section is compact in relation to the LTB.
$F_b = 0.66F_y = 0.66(248) = 163$ N/mm²
$Z = 254.4 \times 1,000,000/163 = 1,560,736$ mm³
Choose W18 × 55
$A = 105$ cm²
$h = 460$ mm
$t_w = 9.9$ mm
$t_f = 16$ mm
$S_x = 1620 \times 10^3$ mm³,
$I_x = 37,270 \times 10^4$ mm⁴
$C = 0.5(b - t_w - 2t_f)$, check $c/t_f < 10.91$ OK
$d_w = h - 4t_f$ $d_w/t_f < 82$ OK
The section is compact regarding the local buckling.
Check stresses
$f_{bc} = 254.4 \times 1,000,000/1,620,000 = 157$ N/mm² < 163 N/mm²   (safe)
$Q = 101.75 \times 1000/(460 \times 16) = 13.8$ N/mm² < 0.4 F_y (safe)
Check deflection
$\Delta = $ span/360 = 27.78 mm

$$\Delta = \frac{5wl^4}{384EI}$$

$$\Delta = \frac{5(0.000004 \times 3000)(10,000)^4}{384(205.95)(37,270 \times 10^4)} = 20.4 \text{ mm}$$

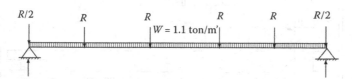

Assume self-weight + wall = 11.0 kN/m'
$M_x = 1225.13$ m kN $Q_{max} = 286.0$ kN
$Z = 1225.13 \times 1000 \times 1000/163 = 7,516,134.97$ mm³
Try W30 × 165
$h = 785$ mm
$b = 268$ mm
$S_x = 7949$ cm³
$I_x = 312,000$ cm⁴
$t_w = 18.5$ mm
$c = 0.5(b - t_w - 2t_f)$, check $c/t_f < 10.91$ OK
$d_w = h - 4t_f$ $d_w/t_f < 82$ OK
$L_u = 3000$ mm
$L_u = 200 \times b/(F_y)^{0.5} = 3870$ mm > 3000 mm   OK
The cross section is compact, regarding the LTB.
$F_b = 0.66F_y = 163.7$ N/mm²
$f_{bc} = 1225.13 \times 1,000,000/7,949,000 = 154.12$ N/mm² < 0.66 F_y (safe)
$Q = 286,000/(785 \times 18.5) = 19.7$ N/mm² < 0.4 F_y (safe)

Check deflection

Δ = span/360 = 41.67 mm W_{eq} = 4 × 6/15 = 1.6 ton/m' = 15.7 N/mm'

$$\Delta = \frac{5(0.0157)(15,000)^4}{384(205.95)(312,000 \times 10^4)} = 16\ mm \quad OK$$

Example 2: Crane Track Girder

Design a crane track girder given the following information:
Span of girder = 19.50 m
Span of crane track girder = 4.90 m
Axis of trolley is at 0.90 m from the centerline of the crane track girder
Distance between the wheels of the crane girder = 2.00 m
Weight of trolley = 1.0 ton
Self-weight of crane track girder = 5.0 tons
Crane capacity = 12.0 tons (Figure 7.7)
Loads
Moving load = weight of trolley + weight to be displaced = 1.0 + 12.0 = 13.0 tons

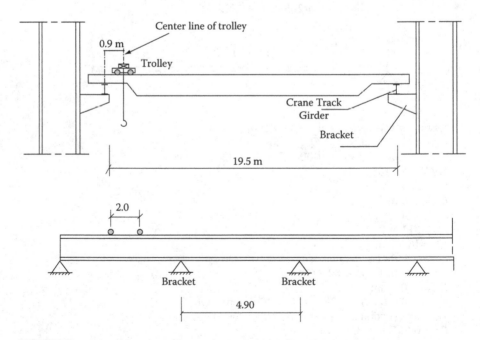

FIGURE 7.7
Crane track girder configuration.

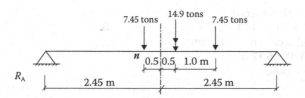

Total load taken by crane track girder = (13(19.5 − 0.9)/19.5) + 2.5 = 14.9 tons
Load on each of the wheels on left crane track girder = 14.9/2 = 7.45 tons
R_A = [7.45(2.45 − 1.5) + 7.45(2.45 + 0.5)]/4.9 = 5.92 tons
$M_x(n)_{LL}$ = 5.92(2.45 − 0.5) = 11.562 mt
M_x (with impact) = 11.562 × 1.25 = 14.452 mt
Assume the self-weight of the crane track girder = 200 kg/m'
M_x = [0.2(4.9)²/8] + 14.452 = 15.052 mt
M_y = 0.10(11.562) = 1.1562 mt

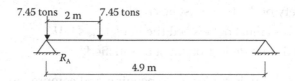

Q_y(LL) = (7.45(2.9/4.9) + 7.45) = 11.859 tons
Q_y(LL) with impact = 11.859 × 1.25 = 14.824 tons, Q_x = 0.1(11.859) = 1.1859 tons
(lateral shock)
Total Q_y = 0.2(4.9)/2 + 14.824 = 15.314 tons
Assume ($Z_x \cong 3Z_y$)
$(M_x/S_x) + (M_y/0.5S_y) > 1.4 × 1.2$ Choose HEB 280
h = 280 mm
b = 280 m
t_f = 18 mm
t_w = 10.5 mm
I_x = 19,270 × 10⁴ mm⁴
I_y = 6595 × 10⁴ mm⁴
S_x = 1376 × 10³ mm³
S_y = 471 × 10³ mm³
Check local buckling
c/t_f and d_w/t_w OK
The section is a compact section with regard to local buckling.
Lateral Torsion Buckling
L_u = 4900 mm
L_u = 20 × $b/(F_y)^{0.5}$ = 3870 mm < 4900   unsafe
The section is noncompact.
F_b = $0.6F_y$
F_{bt} = 15.05 × 100/1376 = 1.09 tons/cm²
F_{bc} = 1.09 + 1.154 × 100/(0.5 × 471) = 1.58 < 1.4 × 1.2   OK
q_y = 15.314/(28 × 1.0) = 0.55 < $0.4F_y$ ton/cm²
q_x = (3/2)(Q_x/A_f) = (3/2)(1.185/28 × 1.8) = 0.035 < $0.4F_y$ ton/cm²
Check LTB
F_{LTB} = 800$A_f C_b/(L_u d)$ = 3.27 > f_{bc} OK

Check deflection (LL only)

$M_{max} = 11.562$ mt $100 \times 11.562 = W_{eq} \times (490)^2/8$

$W_{eq} = 0.0385$ ton/cm' $= 3.85$ ton/m'

$\Delta \leq 4900/800 \leq 6.12$ mm

$\Delta_{act} = 0.66$ unsafe

7.3.4 Design of Beam Column Member (Allowable Stress Design)

Beam columns and structural elements are subjected to compression (normal force) and a bending moment. Safety against the following must be satisfied:

- The buckling compression stress.
- The axial compressor stress $F_c \leq 0.6F_y$.
- The maximum tensile stress $F_t \leq 0.66F_y$.
- The safety of the LTB must be checked.
- Shear stress must not exceed the limit, $t_{max} \leq 0.4F_y$.
- The maximum deflection must be satisfied.

As the beam column is subjected to bending and compression forces, so the compact or noncompact section is defined the flange local buckling as in Table 7.7 but for the web is calculated in case of web capacity can carry the normal force as explained and verified in the following examples 1 and 2.

The value of K will obtained from the next alignment chart as shown in Figure 7.8.

$$G = \frac{\sum (I/L) \text{ column}}{\sum (I/L) \text{ girders}}$$

where $G = 10$ for hinged support and $G = 1$ for fixed support.

$$\frac{f_a}{F_c} + \frac{f_{bx}}{F_{bx}} A_1 + \frac{f_{by}}{F_{by}} A_2 \leq 1.0$$

where

f_a = Actual compression stress due to axial compression

F_c = Allowable compression stress due to buckling

f_{bx}, f_{by} = Actual compression bending stresses on moment about the x and y axes, respectively

F_{bx}, F_{by} = Allowable compression bending stresses for the x and y axes

$$A_1 = \frac{C_{mx}}{\left(1 - \dfrac{f_a}{F_{EX}}\right)}, \ A_2 = \frac{C_{my}}{\left(1 - \dfrac{f_a}{F_{EY}}\right)}$$

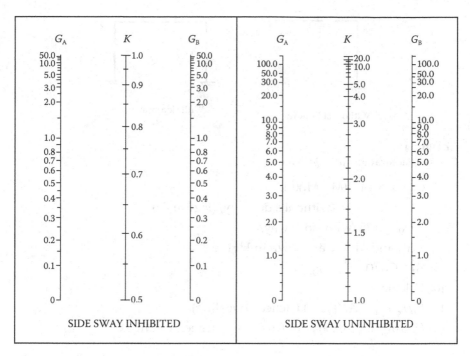

FIGURE 7.8
Alignment charts for buckling length factor, k.

F_{EX}, F_{EY} = Allowable buckling stress in x and y direction, respectively
F_{EX} = $7500/(l_x)^2$
C_M = Moment modification factor
    = 0.85 for frame permitted to sway
    = 1.0 in the case of members with simply supported at the ends

The steps of design the beam column member will be as follows:

1. Assume I_c/I_G = 1.0 (no crane).
    I_c/I_G = 1.5 – 2.0 (with crane).
2. Calculate M_{max}, N_{max}.
3. Assume f = 1.0 ton/cm² for higher N or f = 1.3 ton/cm² for higher M, get approximate cross section.
4. Check compactness (d_w/t_w, c/t_f).
5. Get l_x, l_y. Calculate l_x, l_y.
6. F_c from the maximum (l).
7. F_{EX} = $7500/(l_x)^2$, f_{ca} = N/A, f_{bx} = M_x/Z_x.

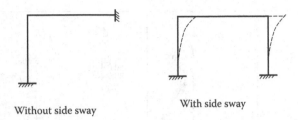

FIGURE 7.9
Cases of side sway and non-side sway.

8. $C_M = 0.6 + 0.4(M_2/M_1)0.4$.

    $0.4 \le C_M \le 0.6$ (without side sway) (Figure 7.9).

    $C_M = 0.85$ (with side sway).

    M_1 and M_2 are as shown in Figure 7.10.

9. $A_1 = C_M/(1 - (f_{ca}/F_{EX}))$.

10. Calculate F_{bx}.

11. $f_{ca}/F_c + f_{bx} \cdot A_1/F_{bc} \le 1.0$ (check buckling).

12. $f_{ca}/0.6F_y + f_{bx}/(0.66\ F_y$ or $0.6\ F_Y) \le 1.0$ (check yielding).

7.3.5 Design of Beam Column Member (LRFD)

The same procedure outlined above will be used but the following equations are applied:

$$P_u/\phi P_n \ge 0.2$$

$$\left(P_u/\phi P_n\right) + 8/9\left[\left(M_{ux}/_f M_{nx}\right) + \left(M_{ux}/_f M_{nx}\right)\right] \le 1.0$$

$$P_u/\phi P_n < 0.2$$

$$\left(P_u/2\phi P_n\right) + \left[\left(M_{ux}/_f M_{nx}\right) + \left(M_{ux}/_f M_{nx}\right)\right] \le 1.0$$

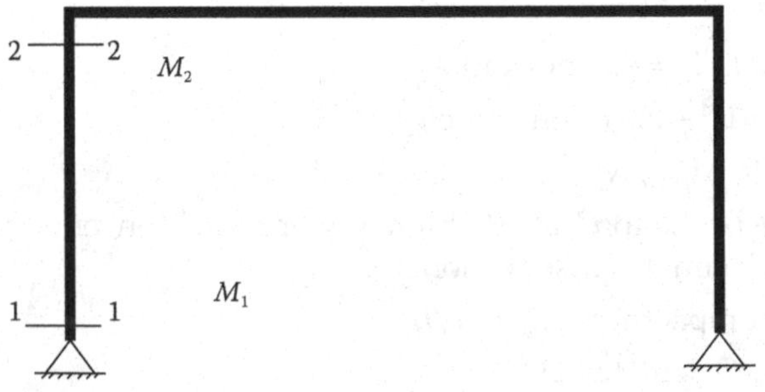

FIGURE 7.10
Portal frame.

Example 1: Frame Design in Case of Side Sway

Design the beam column where the buckling lengths are

$$L_y = 300 \text{ cm} \quad L_x = 1380 \text{ cm}$$

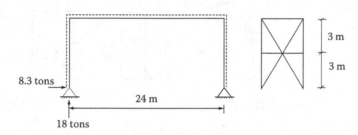

8.3 tons

24 m

18 tons

3 m

3 m

$$M_x = 49.8 \text{ mt} \quad N_{max} = -18.0 \text{ tons}$$

Assume $f_{bx} = 1.30$ ton/cm²
$S_x = 49.80 \times 100/1.3 = 3830.76$ cm³
Choose HEB 500
$F_y = 240$ N/mm²
$t_f = 28$ mm, $t_w = 14.5$ mm, $r_x = 212$ mm, $r_y = 72.7$ mm, $S_x = 4287 \times 10^3$ mm³
$S_y = 842 \times 10^3$ mm³
$A = 23,900$ mm²
Local buckling
Web capacity $= F_y(50 - 2 \times t_f)t_w = 160.95 > 18$ tons
The plastic neutral axis inside the web
$N = 2at_wF_y \qquad a = 2.5$ cm
$ad_w = h/2 + a - 2t_f = 22.5 + 2.5 - 2 \times 2.8 = 19.4$ cm
$d_w = h - 4t_f = 33.8 \qquad a = 0.57 > 0.5$
$d_w/t_f \le (699/(2.4)^{0.5}/(13 \times a - 1) = 69.9$
$d_w/t_f = 33.8/1.5 = 22.53 < 69.9$ OK
Check flange c/t_f OK
The unbraced length
$L = 300$ cm $\qquad L_u = 20b/(F_y)^{0.5} = 387.3 > 300$ OK
$L_x/r_x = 13,800/212 = 65.1 \quad L_y/r_y = 41.27$
F_{EX} (plane of bending) $= 7500/(72.63)^2 = 1.4217$ ton/cm²
$F_c = $ from the table $= 1.13$ ton/cm²
$f_a = 18/239 = 0.075$ ton/cm²
$f_{bx} = 49.8 \times 100/4815 = 1.034$ ton/cm² $= 10.34$ N/mm²
Equivalency and amplification factors
$C_M = 0.85$ (side sway permitted)
$A = 1/(1 - (0.077/1.4217)) = 1.0573 \quad A_1 = C_MA = 0.898 \ge 1.0 \quad A_1 = 1.0$
$(f_{ca}/F_c) + (f_{bx}C_MA/F_{bc}) \le 1.0$
$(0.077/1.13) + (1.034 \times 1.0/0.66F_y) = 0.72 < 1.0$ safe
$f_{ac}/F_c + f_{bx}/F_{bc} \le 1.0$
$(0.075/0.6F_y) + (1.034/0.66F_y) = 0.70 < 1.0$ safe
This section is safe, but not economical. Try smaller section dimensions.

Example 2: Design for Frame without Side Sway

Design the beam column where the buckling lengths are

$$L_y = 500 \text{ cm} \quad L_x = 350 \text{ cm}$$

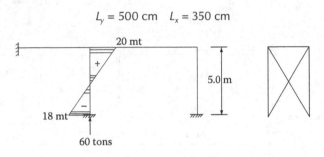

$$M_x = 20 \text{ mt} \quad N_{max} = -60 \text{ tons}$$

Assume $f_{bx} = 1.00 \text{ ton/cm}^2$
$S_x = 20 \times 100/1.0 = 2000 \text{ cm}^3$
Choose HEB 340
$t_f = 21.5 \text{mm}, t_w = 12 \text{ mm}, r_x = 146 \text{ mm}, r_y = 75.3 \text{ mm}, S_x = 2156 \times 10^3 \text{ mm}^3$
$S_y = 646 \times 10^3 \text{ mm}^3$
$A = 17,100 \text{ mm}^2$
Local buckling
Web capacity $= F_y(34 - 2 \times t_f)t_w = 85.54 > 60 \text{ tons}$
The plastic neutral axis inside the web
$N = 2at_wF_y \quad a = 9.61 \text{ cm}$
$ad_w = h/2 + a - 2t_f = 17 + 9.61 - 2 \times 2.1 = 22.31 \text{ cm}$
$d_w = h - 4t_f = 25.4 \quad a = 0.90 > 0.5$
$d_w/t_f \leq (699/(2.4)^{0.5}/(13 \times a - 1) = 41.48$
$d_w/t_f = 25.4/2.15 = 11.81 < 41.48 \quad OK$
Check flange $c/t_f \quad OK$
$L = 500 \text{ cm} \quad L_u = 20b/(F_y)^{0.5} = 387.3 < 500 \quad OK$
The section is noncompact, $F_b = 0.6F_y$
$L_x/r_x = 34.25 \quad L_y/r_y = 66.41$
F_{EX} (plane of bending) $= 7500/(34.25)^2 = 6.39 \text{ ton/cm}^2 = 63.9 \text{ N/mm}^2$
$F_c = 1.18 \text{ ton/cm}^2$ (from Table 7.3)
$f_a = 60/171 = 0.35 \text{ ton/cm}^2$
$f_{bx} = 20 \times 100/2156 = 0.93 \text{ ton/cm}^2$
Equivalency and amplification factors
$f_a/F_c = 0.35/1.18 = 0.29 > 0.15$
$A = 1/(1 - (0.35/6.39)) = 1.05 \quad C_M = 0.6 - 0.4(18/20) = 0.24 < 0.4 \quad C_M = 0.4$
$A_1 = CM \times A < 1.0 \quad A_1 = 1.0$
$C_b = 1.75 + 1.05 (18/20) + 0.3(18/20)^2 = 2.938 > 2.30$, take 2.3
$F_{LTB} = (800 \times 30 \times 2.5) \times 2.3/(500 \times 32) = 6.98 > 0.6 \ F_y$
Checking stresses

- Buckling
  $(f_a/F_c) + (f_{bx}C_MA/F_{bc}) \leq 1.0$
  $(0.35/1.18) + (0.93 \times 1.0/0.6F_y) = 0.95 \quad \text{safe}$

- Yielding
  $f_{ac}/F_c + f_{bx}/F_{bc} \leq 1.0$
  $(0.35/0.6F_y) + (0.99/0.6F_y) = 0.96 < 1.0$ safe

7.4 Steel Pipe Rack Design

The purpose of this section is to provide guidelines and recommend procedures for the analysis and design of steel pipe racks. This design guide defines the minimum requirements for the analysis and design of pipe racks for use in industrial facilities. It covers general design philosophy and also requirements for the analysis and design of pipe racks. Criteria presented here pertain to loads, load combinations, allowable stresses, and superstructure and foundation design.

7.4.1 Pipe Rack Design Guide

Pipe racks are frame structures that support pipes and auxiliary equipment in the process areas of industrial plants. Piping loads can vary greatly from project to project in the same way as loads from wind and earthquake vary. Clearly, it is difficult to define specific criteria for the design of such structures. This guideline sets general requirements that the engineer should incorporate into pipe rack designs, if possible.

This guideline is intended to apply to the following three types of steel pipe racks:

- Strutted main pipe racks
- Unstrutted secondary or miscellaneous pipe racks
- "T" supports

The pipe rack superstructures and foundations shall be designed for the loads and load combinations specified in Chapter 3.

The piping work will be contact with the pipe rack and act as one unit, as shown in Figure 7.11. Therefore, the deflection should be restricted based on the design to avoid higher stress on the piping system. The deflection requirements for pipe rack beams and transverse bents are detailed below.

- The maximum allowable beam deflection, Δ_{max}, due to total load shall be as follows:

$$\Delta_{max} = L/240$$

where L is the span length.

FIGURE 7.11
Photo of a pipe rack.

- The maximum allowable drift limits for a pipe rack shall not exceed

$$H/100$$

where H is the pipe rack height.

7.4.2 Pipe Rack Superstructure Design

The principal structural components of a pipe rack are the transverse beams, the columns, longitudinal struts, and vertical bracing. Design criteria applicable to each of these components are presented below.

In general, the pipe support framing system is designed as a rigid frame with fixed or pinned bases in the transverse direction and as braced frames in the longitudinal direction. Additional longitudinal and/or intermediate transverse beams may be required to support electrical conduit, instrumentation lines, or other small lines. Electrical conduit and cable trays usually must be supported every 3 m, unless the vendor recommends larger spans in the drawings.

Structural components of the pipe rack must be capable of resisting the axial loads, shears, and moments. An elastic analysis shall be used to determine moments and forces in pipe rack members.

7.4.2.1 Structural Steel Expansion

For pipe rack design, provisions shall be made for the thermal expansion of steel, with the structural steel checked for temperature change. Slotted connections (sliding connections) shall be provided in each segment of the pipe rack between vertical bracings to allow for structural steel thermal expansion. The maximum segment for the pipe rack shall be limited to 42 m in length, unless calculations show otherwise.

(1) Transverse beams

- In computing the allowable bending stress, F_b, the unbraced length shall be taken as the span of the beam and the AISC factor, C_b, shall be used to account for end fixity. $C_b = 1.0$ is a very conservative and safe assumption. In no case shall the assumption of lateral support from piping be used in computing, F_b.
- Generally, the horizontal members' depth should not be less than 1/24 of the span.
- If top-flange lateral loads are significant, the transverse beam shall be investigated for bending about the y–y axis and for torsion. This can be estimated by using $M_y \times 2/S_y$.
- In axial load design, the total span of the beam should be used, modified by the appropriate effective length factor for each direction. This factor should be equal to 1.0 for the weak direction of the beam.
- Special consideration shall be given to the design of transverse beams that support large vapor lines to be hydrotested, or that support large anchor or guide forces. Normally, if the local bending stresses are too high, using horizontal bracing locally is required.

(2) Columns

- If using strutted pipe racks, the columns shall normally be designed with pinned or fixed bases, depending on the lateral drift requirements.
- For normal pipe racks, column bases shall be considered pinned in the transverse direction and fixed in the longitudinal direction. The major axis of columns should normally be perpendicular to the longitudinal direction of the pipe rack (i.e., the plane formed by the column web is parallel to the longitudinal direction).
- T support column bases, which are used to carry one or two pipes, shall be considered fixed in both the transverse and longitudinal directions. The major axis of columns may be turned in either direction.
- Column base plates for major and miscellaneous pipe racks and T supports shall be affixed by anchor bolt to the concrete foundation.

(3) Longitudinal struts

- In areas where gravity loading of struts is anticipated, struts shall be designed for axial loads produced by longitudinal pipe loads, plus gravity load moments and shears. Such struts should be designed for the actual load, but not less than 50% of the gravity load of the loaded transverse pipe support beam. This loading requirement will account for the usual piping and electrical conduit.
- Concentrated loads for large pipes shall also be included in the design. Where gravity loading is not anticipated, struts shall be designed for axial loading only. The primary source of axial loads is longitudinal pipe loads.

(4) Vertical bracing

- Vertical bracing may be used to transmit transverse and longitudinal forces to the foundations. K-bracing or X-bracing is usually used for this purpose. In modern research, X-bracing is preferred over K-bracing because of redundancy.
- Braced bays in strutted pipe rack systems should be spaced at 42 m, maximum. Longitudinal bracing should be provided about every fourth bay.
- Compression bracing for steel pipe rack systems shall normally be designed with wide flange shapes. For tension bracing, single angle or double angle are usually used.

Example 1: Pipe Rack Design

Design a typical pipe rack bent in oil plant. The pipe rack configuration shall be as shown in Example 1 in Chapter 3 and the plan layout as shown in Figure 7.12.

Assumptions

Assume in the first trial that the main beams are $W10 \times 33$ for the cable tray support, and $W14 \times 48$ and $W14 \times 53$ for the pipe supports.

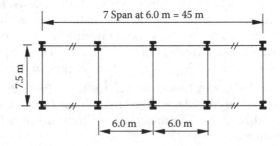

FIGURE 7.12
Pipe rack—plan layout.

The beam levels are 20.00, 25.00, and 20.00. The beams are rigidly connected to the columns (i.e., moment connection). The columns are W14 × 53 and fixed at the base.

The longitudinal struts (W10 × 33) located at levels 17.50, 22.50, and 30.00 act as struts to transfer the thermal load to the vertical bracing of the rack. These levels will be considered as braced in the longitudinal direction.

The loads affect the structure as shown in Figure 7.13. The second step is to build the model in any software package used in structural analysis and then perform the analysis and obtain the stresses and bending moment, as shown in Figure 7.14.

Perform K_z (effective length factor about the column local z-axis) calculations for the columns (Table 7.9).

Refer to detailed calculations in Figure 7.15.

The unity check for the member is illustrated in Figure 7.16. From this figure, it is found that the unity check is lower than the unity, so the chosen member is safe.

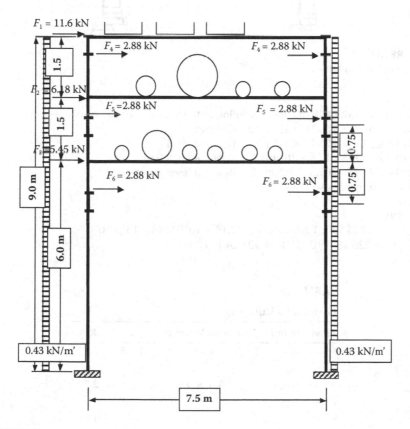

FIGURE 7.13
Sketch for loads on the pipe rack for Example 1.

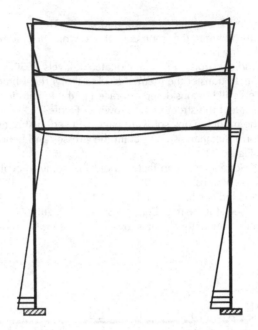

FIGURE 7.14
Sketch of BM diagram.

Check the critical K_z for the governing bottom portion of column:
$W14 \times 53$ $I_x = 225.18 \times 10^6$ m⁴   Column
$W14 \times 53$ $I_x = 225.18 \times 10^6$ m⁴   Beam at level
$W10 \times 33$ $I_x = 71.18 \times 10^6$ m⁴   Beam at level 9 m
$W14 \times 48$ $I_x = 201.5 \times 10^6$ m⁴   Beam at level 7.5 m
$G_B = 1.0$ fixed base
$G_T = \Sigma I_C/L_C$
$\Sigma I_B/L_B$
$\Sigma I_C/L_C = (225.18 \times 10^6/6000) + (225.18 \times 10^6/1500) = 187{,}650$
$\Sigma I_B/L_B = 225.18 \times 10^6/7500 = 30{,}024.133$

TABLE 7.9

G and K Calculation

Elevation (m)	Column Member	G	K_z
9	5, 10	19.78	3.3
7.5		13.09	
	3, 4, 8, 9		2.7
6		7.85	
	1, 2, 6, 7		1.8
0		1.0	

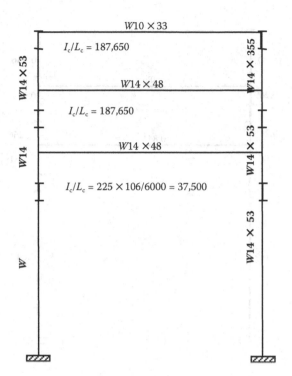

FIGURE 7.15
Sketch of loads on the pipe rack for Example 1.

$G_T = 187,650/30,024.133 = 6.25$
With $G_B = 1.0 + G_T = 6.26K_z = 1.8$
Level 9
$I_B/L_B = 71.18 \times 106/7500 = 9490$
$I_C/L_C = 187,650$
$G = 187,650/9490 = 19.78$
Level 7.5
$I_B/L_B = 201.5 \times 106/7500 = 28,666.7$
$G = 187,650 + 187,865/28,666.7 = 13.09$
Level 6
$G = 37,500 + 187,650/28,666.7 = 7.85$
Pipe rack bent is designed per attached STAAD PRO input and output file.
Check STAAD PRO output for the following:
Unity check
Ensure that unity checks for all structural members are less than 1.0.
Beam deflection
Ensure that maximum beam vertical deflection is less than $L/240$, where L = span length.
Lateral drift
Ensure that maximum lateral drift for the pipe rack is less than $H/100$ for load combinations with wind loads, and $H/100$ for earthquake loads.

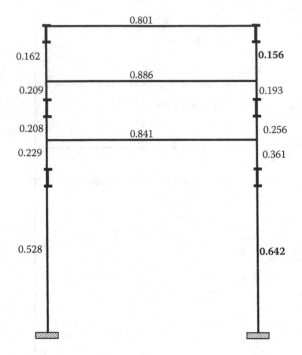

FIGURE 7.16
Sketch of unity check value.

7.5 Stairway and Ladders

The following requirements are summarized from the 1987 edition of *OSHA General Industry Requirements*.

7.5.1 Stairways

Fixed stairs are usually used to access one structural level from another, where operations necessitate regular travel between levels, and for access to the operating platforms of any equipment that requires regular monitoring during operations. Fixed stairs shall also be provided where access to elevations is required daily or at each shift for such purposes as gauging, inspection, regular maintenance, and when one of the following conditions occurs: (1) work may expose employees to acids, caustics, gases, or other harmful substances, or (2) the carrying of tools or equipment by hand is normally required (OSHA 1910.24(b)).

- Every flight of stairs having four or more risers shall be equipped with standard handrails (OSHA 1910.23 (d)(1)).

- Stair Strength—Fixed stairways shall be designed and constructed to carry a load of five times the normal live load anticipated, but never of less strength than required to carry safely a moving concentrated load of 4.4 kN based on OSHA 1910.24 (c).

- Stair Width—A fixed stairway's width is recommended to be not less than 760 mm, according to OSHA 1910.24 (d).

- Angle of Stairway Rise—Fixed stairs shall be installed at angles to the horizontal of between 30° and 50°. Any uniform combination of rise/tread dimensions may be used that will result in a stairway at an angle to the horizontal within the permissible range. Table 7.10 gives rise/tread dimensions that will produce a stairway within the permissible range, giving the angle to the horizontal produced by each combination. However, the rise–tread combinations are not limited to those given in Table 7.9 from OSHA 1910.24 (e).

- Stairway Platforms—Stairway platforms shall be no less than the width of a stairway and a minimum of 760 mm in length measured in the direction of travel, based on OSHA 1910.24 (g).

- Handrails—OSHA 1910.24 (h) states that standard railings shall be provided on the open sides of all exposed stairways and stair platforms. Handrails shall be provided on at least one side of closed stairways, preferably on the right side descending.

- Vertical Clearance—Vertical clearance above any stair tread to an overhead obstruction shall be at least 2.1 m measured from the leading edge of the tread (OSHA 1910.24 (i)).

TABLE 7.10

Stairway Rise and Run Dimensions

Angle to Horizontal	Rise (mm)	Tread Run (mm)
30°35′	165.1	279.4
32°08′	171.45	273.05
33°41′	177.8	266.7[a]
35°16′	184.15	260.35
36°52′	190.5	254
38°29′	196.85	247.65
40°08′	203.2	241.3
41°44′	209.55	234.95
43°22′	215.9	228.6
45°00′	222.25	222.25
46°38′	228.6	215.9
48°16′	234.95	209.55
49°54′	241.3	203.2

[a] Preferred.

7.5.2 Handrails and Railings

Every open-sided floor or platform 1.2 m or more above an adjacent floor or ground level shall be guarded by a standard railing on all open sides, except where there is an entrance to a ramp, stairway, or fixed ladder (OSHA 1910.23 (c)(1)). Many of our clients request guard railings for platforms less than 1.2 m above an adjacent floor or ground. The client's requirements should be discussed. All handrails and railings shall be provided with a clearance of not less than 75 mm between the handrail or railing and any other object, based on OSHA 1910.23 (e)(6).

7.6 Crane Supports

The vertical deflection is very critical in the movement of the crane and it mainly depends on the crane manufacturer. Table 7.11 is a guideline for the maximum allowable deflection of the runway girders under the maximum static load without impact load (where L = the span length).

Vertical deflection of the jib crane support beams shall not exceed $L/225$ (where L = the maximum distance from the support column to the load location along the length of the jib beam) if loaded with the maximum lift plus hoist load(s), without impact.

Lateral deflection of support runway girders for cranes with lateral moving trolleys shall not exceed $L/400$ (where L = the span length) when loaded with a total crane lateral force not less than 20% of the sum of the weights of the lifted load (without impact) and the crane trolley. The lateral force shall be distributed to each runway girder with consideration given to the lateral stiffness of the runway girders and the structure supporting the runway girders.

7.7 Connections

The connection can be welded or bolted, according to the design and fabrication procedure, which is constrained if the connection will be fabricated

TABLE 7.11

Maximum Allowable Girder Deflections

Top-running CMAA Class A, B, and C cranes	$L/600$
Top-running CMAA Class D cranes	$L/800$
Top-running CMAA Class E and F cranes	$L/1000$
Under-running CMAA Class A, B, and C cranes	$L/450$
Monorails	$L/450$

onsite or in the workshop. The following rules should be followed when selecting the weld connection:

- Welding in the field is avoided, if possible, due to welding condition requirements.
- Field welding is not to be performed while it is raining, snowing, or below 0°F.
- In certain ambient temperatures, preheating of the material to be welded is required.
- AWS Code D1.1 (2004b) specifies minimum preheat temperatures that are designed to prevent cracking.

The connections shall conform to the following requirements:

1. Shop connections may be either bolted or welded. Field connections shall be bolted where possible. Connections may be field-welded when conditions are such that a bolted connection is not suitable.
2. Bolted connections for primary members shall utilize high-strength bolts conforming to ASTM A-325-N, bearing-type connections with threads included in the shear plane. However, slip-critical-type connections shall be used in connections subject to vibration or repeated stress reversal.
3. Standard connections shall be designed by the fabricator in accordance with the project construction specifications and loads shown on the drawings. Moment connections and special connections shall be designed by the engineer and shall be shown on the engineering drawings.
4. Moment connections shall preferably be of the bolted end plate type.

In general, the differences between the bolting and welding that will be used in the selection are summarized as follows:

- Bolting is generally a faster operation than welding.
- Bolting does not have the temperature and weather condition requirements that are associated with welding.
- Unexpected weather changes may delay welding operations.

7.7.1 Bolts

Four basic types of bolts are commonly used. They are designated by ASTM as A307, A325, A490, and A449. A307 bolts are called *unfinished* or *ordinary* bolts. They are made from low-carbon steel. Two grades (A and B) are available.

They are available in diameters from 6 to 100 mm and are used primarily for low-stress connections and for secondary members. A325 and A490 bolts are called high-strength bolts. A325 bolts are made from heat-treated medium carbon steel. There are three types of bolts available on the market:

- Type 1 bolts are made of medium-carbon steel.
- Type 2 bolts are made of low-carbon steel.
- Type 3 bolts have atmospheric corrosion resistance and weathering characteristics comparable to A242 and A588 steel.

A490 bolts are made from quenched and tempered alloy steel and thus have a higher strength than A325 bolts. A with A325 bolts, there are three types (Types 1, 2, and 3) available on the market.

Both A325 and A490 bolts are available in diameters from 12 to 38 mm in 1/8-in. increments. They are used for general construction purposes. A449 bolts are available in diameter up to 75 mm and so are used when diameters exceeding 38 mm are needed. They are also used for anchor bolts and threaded rod.

High-strength bolts can be tightened to two conditions of tightness: snug-tight and fully tight. Snug-tight conditions can be attained by a few impacts of an impact wrench, or by the full effort of a worker using an ordinary spud wrench. Snug-tight conditions must be clearly identified on the design drawing and are permitted only if the bolts are not subjected to tension loads and loosening or fatigue due to vibration or load fluctuations are not design considerations. Bolts used in slip-critical conditions (i.e., conditions for which the integrity of the connected parts is dependent on the frictional force developed between the interfaces of the joint) and in conditions where the bolts are subjected to direct tension are required to be fully tightened to develop a pretension force equal to about 70% of the minimum tensile stress F_u of the material from which the bolts are made.

Stresses on the bolt are shown in Figure 7.17. *Fully-tightened* can be accomplished by using the turn-of-the-nut method, the calibrated wrench method, or by the use of alternate design fasteners or a direct tension indicator.

For a bolted connection, the following tools can be used as shown in Figures 7.18 and 7.19.

The common tools used by ironworkers include spud wrenches, pins, and corrections bars of various sizes, as shown in Figure 7.18. Impact wrenches will be needed for certain installations, as shown in Figure 7.19.

Electricity or compressed air is required, depending on the impact wrench being used. A generator as well as an air compressor may be needed. From ECS2008, Table 7.12 presents the size of the bolts, their areas, A_s, and the areas of stress with the torque required, M_a, to achieve the required load. The shearing-force resistance, P_s, is presented in Case I for primary stresses due to dead loads, live loads, or superimposed loads, and the dynamic effects

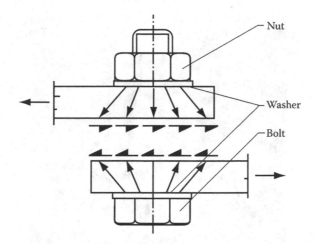

FIGURE 7.17
High-strength bolt stresses.

FIGURE 7.18
Tools for erecting a connection.

FIGURE 7.19
A hydraulic wrench.

and centrifugal forces. Case II presents primary and additional stresses in Case I, plus earthquake loads, wind loads, braking forces, and lateral shocks. Table 7.12 can be applied to grade 8.8 by reducing the values in Table 7.12 by 30%.

The steel detailing guide is shown in Figures 7.20 and 7.21.

TABLE 7.12
Properties and Strengths of High-Strength Bolts (Grade 10.9)

					Permissible Friction Load of One Bolt per One Surface Friction, p_s (tons)			
					St. 37 and 42–44 ($\mu = 0.4$)		St. 50–55 ($\mu = 0.5$)	
d	A (mm²)	A_s (mm²)	T	M_a	I	II	I	II
M12	113	84	5.29	12	1.69	2.01	2.11	2.52
M16	201	157	9.89	31	3.16	3.37	3.95	4.71
M20	314	245	15.43	62	4.93	5.90	6.17	7.36
M22	380	303	19.08	84	6.10	7.27	7.63	9.10
M24	452	353	22.23	107	7.11	8.45	8.89	10.60
M27	573	459	28.91	157	9.25	11.03	11.56	13.78
M30	706	561	35.34	213	11.30	13.48	14.13	16.86
M36	1018	817	51.47	372	16.47	19.64	20.58	24.55

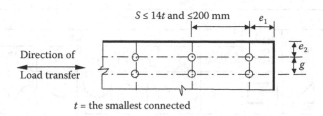

FIGURE 7.20
Spacing in tension or compression members.

7.7.2 Welding

Welded connections are connections whose components are joined together primarily by welds. Welds can be classified by the following criteria:

- Types of welds: groove, fillet, plug, and slot welds
- Position of the welds: horizontal, vertical, overhead, and flat welds
- Types of joints: butt, lap, corner, edge, and tee

Although fillet welds are generally weaker than groove welds, they are used more often because they allow for larger tolerances during erection. Plug and slot welds are expensive to make and they do not provide much reliability in transmitting tensile forces perpendicular to the surfaces. Furthermore, quality control of such welds is difficult because inspection of the welds is rather arduous. As a result, plug and slot welds are normally used just for stitching different parts of the members together.

7.7.2.1 Welding Symbols

A shorthand notation giving important information on the location, size, length, etc., for the various types of welds was developed by the American Welding Society to facilitate the detailing of welds. This system of notation is reproduced in Figure 7.22.

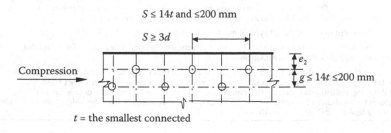

FIGURE 7.21
Staggered spacing in compression.

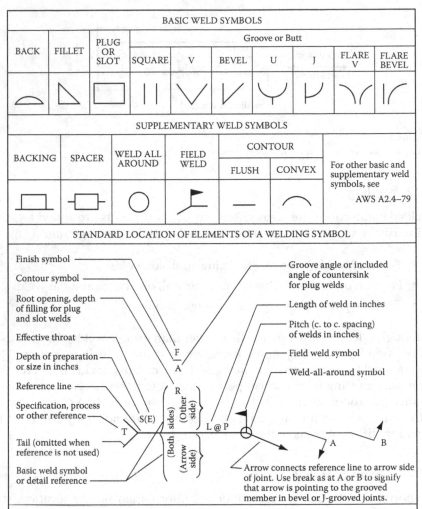

FIGURE 7.22
Welding symbols.

7.7.2.2 Strength of Welds

In ASD, the strength of welds is expressed in terms of allowable stress. In LRFD, the design strength of welds is taken as the smaller of the design strength of the base material and the design strength of the weld electrode. These allowable stresses and design strengths are summarized in Table 7.13. When a design uses ASD, the computed stress in the weld shall not exceed its allowable value. When a design uses LRFD, the design strength of welds should exceed the required strength obtained by dividing the load to be transmitted by the effective area of the welds.

General limitations to the welding will be as follows:

- $S \leq$ min of t_1 or t_2
- Min of t_1 or $t_2 < 20$ mm, $S \geq 5$ mm

  min of t_1 or $t_2 = 20$–30 mm, $S \geq 6$ mm

  min of t_1 or $t_2 = 30$–50 mm, $S \geq 8$ mm

  min of t_1 or $t_2 = 50$–100 mm, $S \geq 10$ mm
- For a building, $S \geq 4$ mm
- L_{min} = largest of ($4S$ or 5 cm)
- $L_{max} = 70S$
- No limit for L for beam-column connections or flange to web welds in welded built-up girders

Example 1: Welded Connection

Figure 7.23 presents an unequal angle, 80 × 200 × 120 connected to the column by a weld with force, $P = 6$ tons.

Calculation
Straining action
$q = 60$ kN $M = 60 \times 45 = 2700$ kN/m
Section 1.1 ($S = 10$ mm)
$q = 60/(2 \times 120 \times 10) = 0.025$ kN/mm^2
$q(M_t) = 2700 \times 60/(2 \times (10 \times (120)^3/12)) = 0.0563$ kN/mm^2
$q(t) = [(0.025)^2 + (0.0563)^2]^{0.5} = 0.061 < 0.2F_u < 0.072$ kN/mm^2 OK
Section 2.2
$I_x = 2 \times (120)^3 \times 10/12 = 2{,}880{,}000$ mm^4
$F_t = 2700 \times (120/2)/2{,}880{,}000 = 0.05625$ kN/mm^2
$F_{eq} = [(0.05625)^2 + 3(0.025)^2]^{0.5} = 0.07 < 0.072 \times 1.1$ OK

TABLE 7.13
Strength of Welds

Types of Weld and Stress	Material	ASD Allowable Stress	LRFD ϕ FBM or ϕ FW	Required Weld Strength Level
Full Penetration Groove Weld				
Tension normal to effective area	Base	Same as base metal	$0.90F_y$	Matching weld must be used
Compression normal to effective area	Base	Same as base metal	$0.90F_y$	Weld metal with a strength level equal to or less than "matching" must be used
Tension of compression parallel to axis of weld	Base	Same as base metal	$0.90F_y$	
Shear on effective area	Base weld electrode	$0.3\times$ nominal tensile strength of weld metal	$0.9(0.6F_y)$ $0.8(0.6FEXX)$	
Partial Penetration Groove Welds				
Compression normal to effective area	Base	Same as base metal	$0.90F_y$	Weld metal with a strength level equal to or less than "matching" must be used
Tension or compression parallel to axis of weld	Base weld electrode	$0.30\times$ nominal tensile strength of weld metal	$0.75(:60FEXX)$	
Shear parallel to axis of weld				
Tension normal to effective area	Base weld electrode	$0.30\times$ nominal tensile strength of weld metal $\leq 0.18\times$ yield stress of base metal	$0.90F_y$ $0.8(0.6FEXX)$	
Fillet Welds				
Stress on effective area	Base weld electrode	$0.30\times$ nominal tensile strength of weld metal	$0.75(:60FEXX)$ $0.9F_y$	Weld metal with a strength level equal to or less than "matching" must be used
Tension or compression parallel to axis of weld	Base	$0.30\times$ nominal tensile strength of weld metal	$0.9F_y$	

Note: See AWS D1.1 for "matching" weld material.
Weld metal one strength level stronger than "matching" weld metal will be permitted.
Fillet welds partial-penetration groove welds joining component elements of built-up members such as flange-to-web connections.
May be designed without regard to the tensile or compressive stress in these elements parallel to the axis of the welds.

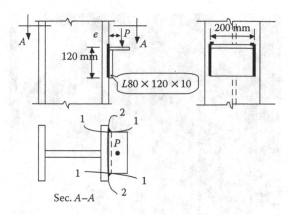

FIGURE 7.23
Welded connection.

7.7.2.3 *Welding in Existing Structures*

Welding to existing structures, as shown in Figure 7.24, to install an extension or during retrofit projects, requires careful consideration of numerous factors:

- Determine weldability and establish a welding procedure by identifying the steel grade
- Select and design the weld—fillet welds are preferred as they avoid overwelding
- Surface preparation—remove contaminants such as paint, oil, and grease
- Loads during retrofit—an engineer should determine the extent to which a member will be permitted to carry loads while heating, welding, or cutting
- Fire hazards—follow all governing codes, regulations, and safety rules to avoid fires

7.7.3 Connection Design

The following are guidelines for connection design in most steel structures used in the process plant, specifically the pipe support and the pipe rack.

FIGURE 7.24
Welding in an existing building.

Example 1: Bracing Connection Design

The normal force is $P_x = 80$ kN, and the bracing consists of two equal angles, $100 \times 100 \times 10$ mm.

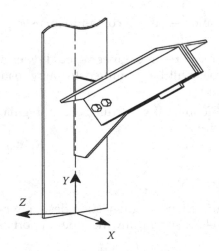

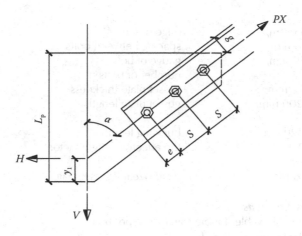

Section Properties

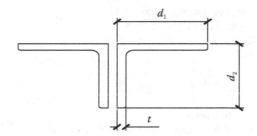

d_1 = 100 mm
d_2 = 100 mm
t = 10 mm
A_g = 1920 mm²

The steel structure member and the connection plates grade is ASTM A36.
F_{us} = 400 N/mm²,   ultimate strength
F_{ys} = 245 N/mm²,   yield strength

Connection Bolts
Choose grade ASTM A325 bolts.
d = 20 mm,   bolt diameter
d_h = 22 mm,   hole diameter
A_b gross = 314 mm²,   bolt gross area
A_b net = 245 mm²,   bolt net area
f_v = 145 N/mm²,   allowable shear stress
f_p = 480 N/mm²,   allowable bearing stress = $1.2F_u$
f_t = 303 N/mm²,   allowable tension stress

Geometry

$e = 35$ mm,	edge distance
$s = 70$ mm,	spacing between bolts
$g = 55$ mm,	gauge of bolts
$N_b = 2$,	number of bolts
$t_p = 12$ mm,	gusset plate thickness
$L_p = 200$ mm,	gusset plate length
$Y_1 = 0$	
$L_1 = 120$ mm,	buckling length
$K = 0.5$,	the effective length factor
$\alpha = 30°$	
$S_w = 8$ mm,	weld size for gusset plate

Check Stress for Bolts

F_s(all) = allowable double shear force per bolt
F_s(all) = A_b gross $\times f_v \times 2$
F_s(all) = 91 kN
F_s(act) actual shear force per bolt = P_x/N_b
F_s(act) = 40 kN
F_s(all) > F_s(act)
F_p(all) = allowable bearing force per bolt
F_p(all) = $f_p \times$ min $(2t, t_p) \times d$
F_p(all) = 115 kN
F_p(all) > F_s(act)

Check Bracing

1. Stress on effective area
   A_{net}, net area for brace $\times$ section = $2(A_g - t \times d_h)$
   $A_{net} = 3400$ mm^2
   A_e = effective area = 0.75 A_{net} = 2550 mm^2
   F_t(act), actual tension stress = P_x/A_e = 31 N/mm^2
   F_t(all) allowable tension stress = 0.5 $\times F_{u1}$ = 200 N/mm^2
   F_t(all) > F_t(act)   OK, safe

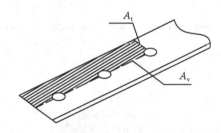

2. Block shear capacity
   Shear area $A_v = 2 \times t[(e + (N_b - 1) \times s - (N_b - 0.5) \times d_h]$
                   $A_v = 1440$ mm^2
   Tension area $A_t = 2 \times t \times [d_1 - g - 0.5 \times d_h]$
                   $A_t = 680$ mm^2

R_{bs}, resistance for block shear
$$R_{bs} = 0.3 \times F_{u1} \times A_v + 0.5 \times F_{u1} \times A_t$$
$$R_{bs} > P_{x}, \quad \text{OK, safe}$$

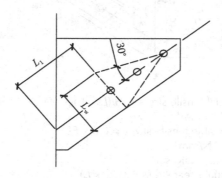

Check Gusset Plate

1. Stress on plate section at plate buckling

$$r = \text{radius gyration} = \sqrt{\frac{I}{A}} = \sqrt{\frac{tp3 \times L/12}{tp \times L}} = tp\sqrt{12}$$

Slenderness ratio, $\lambda = kL_1/r$

$$Cc = \sqrt{\frac{2\pi^2 E}{F_y}}$$

$$kL/r < Cc$$

$$F_a(\text{all}) = [1 - (kL/r)^2/2Cc^2]F_y/[5/3 + 3(kL/r)/8Cc - (kL/r)^3/8Cc^3]$$

$$F_a(\text{act}) = P_x/(L_w \times t_p)$$

$$F_a(\text{all}) > F_a(\text{act})$$

2. Stress on the plate section at point of contact with column
   Distance from point where axial force is acting on the plate length, L_p, center
   $H = P_x \sin a$
   $V = P_x \cos a$
   $M = ecc \times H$

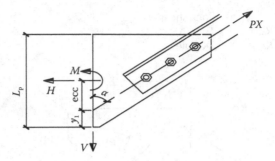

f_t(act), actual tensile stress $= H/(L_p \times t_p) + M/(t_p \times L_p^2/6)$
f_t(act) $= 67$ N/mm^2
f_t(all), allowable tensile stress $= 0.6 \times F_y$
f_t(all) $= 147$ N/mm^2
f_t(all) $> f_t$(atc)   OK, safe
f_v(act), actual shear stress $= V/(L_p \times t_p)$
f_v(act) $= 29$ N/mm^2
f_v(all), allowable shear stress $= 0.4F_y$
f_v(all) $= 98$ N/mm^2
f_v(all) $> f_v$(atc)   OK, safe

3. Block shear capacity

   Shear area $A_v = t_p \times [(e + (N_b - 1) \times s - (N_b - 0.5) \times d_h]$
   $A_v = 864$ mm^2
   Tension area $A_t = t_p \times [d_1 - g - 0.5 \times d_h]$
   $A_t = 408$ mm^2
   R_{bs}, resistance for block shear
   $R_{bs} = 0.3 \times F_{u1} \times A_v + 0.5 \times F_{u1} \times A_t$
   $R_{bs} = 185$ kN
   $R_{bs} > P_x$ OK, safe

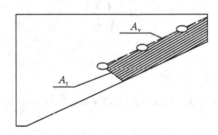

Checking Welds
We will use Electrode E70
$F_w = 0.707 \times 0.3 \times 70$ ksi
$F_w = 14.8$ ksi $= 102$ N/mm^2

Weld Stress Components
$f_1 = H/(2 \times L_p \times s_w) + M/2x(s_w \times L_p^2/6)$
$f_1 = 50$ N/mm^2
$f_2 = V/(2 \times L_p \times s_w)$
$f_2 = 21.7$ N/mm^2

$f_r = 54.5 \text{ N/mm}^2$

$f_r = \sqrt{f_1^2 + f_2^2}$
$f_w > f_r$ OK, safe

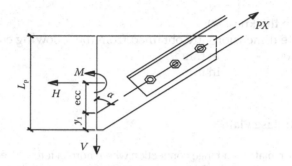

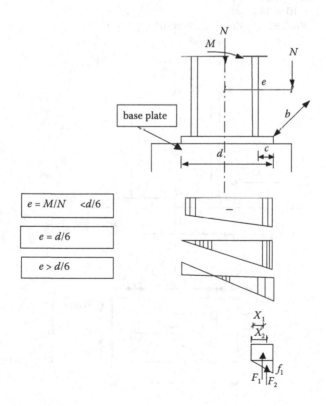

$e = M/N$ $< d/6$

$e = d/6$

$e > d/6$

FIGURE 7.25
Base plate stresses.

7.7.4 Base Plate Design

The stresses on the base plate are shown in Figure 7.25, where d and b are the length and width of the base plate, respectively.

$f_{1,2} = (-N/A) + [6(N_e/bd^2)]$

$M = f_1 X_1 + f_2 X_2$

$f_a = 6M/(bt^2)$

Get base plate thickness (t)

The base plate dimensions are obtained from the following equations:

$D = 0.95d + 100$,   in mm

$B = 0.8b + 100$,   in mm

Example 1: Base Plate

Design a base plate for a hinge connection where normal force, N, equals 300 kN and shear force, Q, equals 30 kN (Figure 7.26). The concrete characteristic strength is 30 N/mm².

$D = 0.95 \times 300 + 100 = 385$ mm

$B = 0.8 \times b + 10 = 340$ mm

$f_c = 300 \times 1000/385 \times 340 = 2.29$ N/mm² $< f_c'$ OK

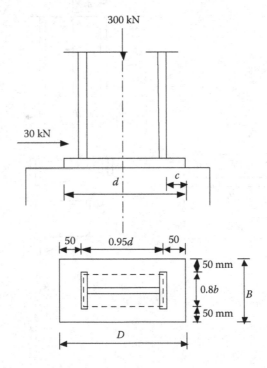

FIGURE 7.26
Base plate design.

$M = 2.29(50)^2/2 = 2862.5$ mm N/mm
$t = (6M/(0.72F_y))^{0.5}$
$t = 9.8$ mm, use 10 mm
If welding
$S = (30/2 \times 0.2F_u \times 300) = 6.9$ mm
Bolts
$A_r = 30/(2 \times 10.8)$

7.8 Anchor Bolt Design

The following are definitions for the different types of anchor bolts.

Cast-in anchor: A headed bolt, anchor rod, or hooked bolt that is installed before the concrete is placed.

Post-installed anchor: An anchor is installed in hardened concrete. Expansion, undercut, and adhesive anchors are examples of post-installed anchors.

Pretensioned anchor bolt: An anchor bolt that is designed to be tensioned to a predefined force to prevent premature failure of the anchor bolt due to fatigue.

Shear lug: A pipe or plate section welded to the bottom of a base plate that is used to resist base shear forces.

The cast-in anchor bolt is the most traditional type in industrial projects; however, the other types are very important for certain facilities, based on fabricator or manufacturer requirements.

7.8.1 Anchor Bolts, Nuts, and Washers

7.8.1.1 Anchor Bolts

For most ordinary structures, ASTM A307 headed bolts, ASTM A36/A36M rods, or ASTM F1554 grade 36 rods should be specified for ordinary strength anchor bolts. Applications requiring high-strength materials should use ASTM A193/A193M grade B7 or ASTM F1554 grade 105 anchor bolts and rods.

Anchor bolt materials and their properties are shown in Table 7.14, while the materials and properties of the anchor bolt nut are presented in Table 7.15.

7.8.1.2 Washers

Because base plates typically have holes larger than oversized holes to allow for tolerances on the location of the anchor rod, washers are usually furnished from ASTM A36 steel plate. They may be round, square, or rectangular, and

TABLE 7.14

Anchor Bolt Materials and Properties

Anchor Material Type		F_y ksi (MPa)	f_{ut} ksi (MPa)	Galvanize?	Ductile?
A307		Not clearly defined	60 (410)	Yes	Yes
A 36 or f1554 grade 36		36 (250)	58 (400)	Yes	Yes
F1554 grade 105		105 (720)	125 (860)	Yes	Yes
A 193 trade and 760 bassed on bolt diameter (d_b)	$d_o \le 64$ mm	105 (720)	125 (860)	No	Yes
	$64 > d_o \le 102$ mm	95 (660)	115 (790)		No	Yes
	$d_o > 102$ mm	75 (515)	100 (690)	No	Yes

generally have holes 2 mm larger than the anchor rod diameter. The thickness must be suitable for the forces to be transferred.

7.8.1.3 Sleeves

The sleeves are very important tools to allow small movements of anchor bolts during the installation of the machine skid after pouring concrete. Sleeves are recommended for use when small movements of the anchor bolt are desired after the bolt is set in concrete. There are two types of sleeves commonly used with anchors. A partial sleeve is primarily used for alignment requirements, while the full sleeve is used for alignment as well as pretensioning.

Sleeves do not affect the tensile capacity of a headed anchor because the tension in the anchor is transferred to the concrete through the head and does not rely on the bond between the anchor and surrounding concrete. Sleeved anchors can only resist shear forces when the sleeve is filled with grout.

A partial-length sleeve is used where precise alignment of anchors is required during installation equipment or where bolt groups or patterns involving six or more anchor bolts with interdependent dimensional requirements. In this situation, the sleeve should be filled with grout after

TABLE 7.15

Anchor Bolt Nut Specifications

Anchor Material Specification	Nut Material Specification
ASTM A307, Grade A	ASTM A563, Grade A
ASTM A36/A36M	ASTM A563, Grade A
ASTM F1554, Grade 36	ASTM A563, Grade A
ASTM A193/A193M, Grade B7	ASTM A563, DH Heavy Hex, or ASTM A194/A194M
ASTM F1554 Grade 105	ASTM A563, DH Heavy Hex, or ASTM A194/A194M

installation is complete. Partial-length sleeves should not be used for base plates on structural steel columns that have oversized holes.

Full-length anchor bolt sleeves are used when anchor bolts will be pretensioned in order to maintain the bolt under continuous tensile stress during load reversal. Pretensioning of anchor bolts requires the bolt to be unbonded over a well-defined, stretched length. When sleeves are used for pretensioned bolts, the top of these sleeves should be sealed and the sleeve should be filled with elastomeric material to prevent grout or water from filling the sleeve.

For vessels of significant height, where the wind effect is high, the forces on the anchor bolt will also be high and the friction between the anchor bolt and concrete is not sufficient to carry the force on the anchor bolt. The placement of a plate, connected to the lower point of the anchor bolt, is recommended. This is illustrated in Figure 7.27.

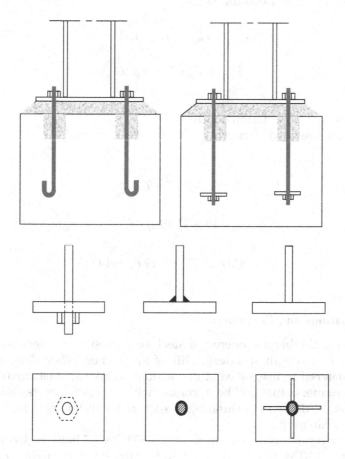

FIGURE 7.27
Anchor bolt detail.

7.8.2 Anchor Bolt Plate Design

The three anchor bolts are connected by the steel plate and the thickness of this plate must be sufficient to withstand the load from the anchor bolts.

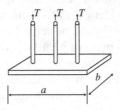

$$b = 3d$$

where d is the anchor bolt diameter.

$$l = l_{min} = l_{RC} - 100$$

$$T_{bond} = F_b \times 3.14 \times d \times l_{min}$$

where
l_{RC} = Concrete foundation depth
F_b = Concrete bond strength

$$T_e = T - T_b$$

$$W = 3T_e/(ba) \le f_c'$$

$$(M/0.72F_y) = (bt^2/6), \text{ get } t$$

7.8.3 Coatings and Corrosion

Corrosion is the biggest enemy of steel, so corrosion is a serious situation affecting the strength and design life of the anchor. When deciding which anchor material to use, or what precautions to take against corrosion, consider the elements that will be in contact with the anchor bolts and the environmental conditions surrounding it, such as the water table, humidity, salt water, and caustic gases.

Galvanizing is a common option for ASTM A307 bolts and ASTM A36/A36M and ASTM F1554 threaded rods. Hot-dip galvanizing is not recommended for ASTM A193/A193M, grade B7, anchor bolts because of the possibility of hydrogen embrittlement that can be caused by the hot-dip galvanizing process.

A practical design for the anchor should consider at least 3 mm as a corrosion allowance in the anchor bolt diameter if it is near highly corrosive environments—such as a building near a coastal area—or if it is washed periodically, as in the case of regular testing of fire-fighting systems.

Pedestal design and anchor arrangement should consider water collection and anchor environment to reduce the amount of contact with corrosive substances or plant wash-down water. Therefore, the tops of pedestals are usually at least 150 mm above the finished paving surface.

Painting the anchor bolts will protect them, but they may require more maintenance. ASTM A193/A193M anchors are required to be painted. If the engineer determines that prolonged contact with a corrosive substance is unavoidable, a metallurgist consultant should determine whether another anchor material or protective options are appropriate. These precautions will be discussed in Chapter 9.

7.8.4 Bolt Types, Details, and Layout

It is traditional to use cast-in-place anchor bolts that come in several configurations. Headed bolts, threaded rods with heavy hex nuts, and J- and L-shaped anchor rods are some common examples. Current design practices do not rely on the concrete bond to develop the tension capacity of smooth rods bent into J and L shapes. While these bolts are still recognized in ACI 318-02, they have no advantage over headed bolts and can be harder to fit into small foundation pedestals.

Determining the number, type, projection, and diameter of anchor bolts should be done as follows:

- Structural steel base plates require a minimum of four anchor bolts for stability during construction per the latest OSHA safety requirements.
- The layout of the anchor bolts and required foundation pedestal size should be established based on the design forces and minimum dimensions in Section 6.3 below.
- The minimum anchor bolt size is 20 mm for most items. However, there are exceptions made for very small pieces of equipment, or when suggested by the manufacturer, or for small miscellaneous steel items such as ladders, stair stringers, small base pipe supports, etc.
- For most structures and equipment, ordinary strength anchor bolts can be used (ASTM A307, A36/A36M, or F1554, grade 36). If the anchor bolt loads require anchor bolts in excess of 50 mm, high-strength anchor bolt materials should be considered.

7.8.4.1 Anchor Bolt Projection

Anchor bolt projection above the top of concrete is computed as follows:

- For single nuts

$$P = t_g + t_p + 1.5d$$

- For double nuts

$$P = t_g + t_p + 2.5d$$

where
  P = Anchor bolt projection above concrete
  t_g = Thickness of grout
  t_p = Thickness of base plate
  d = Diameter of anchor bolt

The thread length required at the top of the anchor bolt must be sufficient to accommodate two nuts and about ½ of the anchor bolt diameter projecting above the top nut. Normally, a thread length is about 3 bolt diameters, to provide some tolerance for errors in the elevation of the anchor bolt placement during construction.

The minimum edge distance shall be in accordance with ACI Standard 318 and ASCE recommendations. Minimum embedment and anchor spacing should be in accordance with the recommendations of the ASCE *Anchor Bolt Report*. Table 7.16 presents the different types of anchor bolt shapes for different pieces of equipment traditionally in industrial facilities.

TABLE 7.16

Types of Anchor Bolts

Equipment Type	Bolt	Anchor Bolt Shape		
		J	K	X
Stair at 45° inclination	M16	×		
Ladders	M12	×		
Steel structure	Variable	×		
Horizontal vessel and heat exchanger	M20/M24		×	
Vertical vessel on steel leg support	Variable			×
Pie racks	M24/M30	×		
Stacks, pumps, and compressor and vertical vessel with skirt	Variable		×	
For high-precision installation	Variable			×

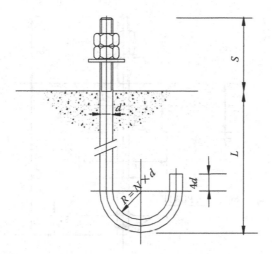

FIGURE 7.28
Anchor bolt type J.

The anchor bolts shown in Figures 7.28, 7.29, and 7.30 reflect the approximate dimensions of these anchor bolts based on concrete characteristic strength 30 N/mm2. These dimensions are used as guidelines and presented in Table 7.17. These dimensions should be verified before use. Table 7.18 presents the height of the threaded parts of different diameters.

7.8.4.2 Edge Distance

The clear distance for anchor bolts or anchor bolt sleeves to the edge of the concrete shall be a minimum of 100 mm. This clear distance is intended to

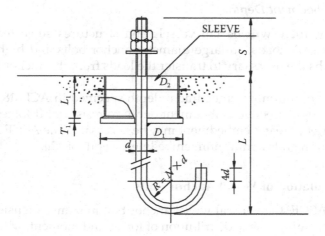

FIGURE 7.29
Anchor bolt, type J.

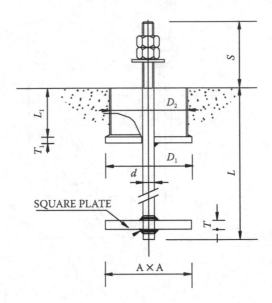

FIGURE 7.30
Anchor bolt, type X.

prevent corrosion and to make sure that the anchor bolts are not in contact with the rebar cage. ACI 318 requires that cast-in headed fasteners that will be torqued have minimum edge distances of $6d$. For constructability reasons, the ASCE *Anchor Bolt Report* recommends a minimum edge distance of $4d$ for ASTM A307, A36, or their equivalent and $6d$ for high-strength bolts. The minimum bolt spacing is $8d$.

7.8.4.3 Embedment Depth

In industry, there will be tall vessels and structures subjected to heavy wind or seismic loads, so large diameter anchor bolts and higher embedment lengths are necessary to transfer the loads from the anchor bolts to the concrete.

There is no minimum embedment depth specified in ACI 318, so long as there is enough effective embedment depth to resist uplift forces. If ductility is required, greater embedment may be necessary. The ASCE *Anchor Bolt Report* recommends a minimum embedment depth of $12d$.

7.8.5 Calculation of Vessel Anchor Bolts

Based on ASCE, the vertical vessel anchor bolt maximum tension is calculated by assuming elastic distribution of forces and moment, which is based on the moment of inertia of the bolt group.

TABLE 7.17

Approximate Dimensions for Anchor Bolt Types J, K, and X (in mm)

d	N	L	H_1	T_1	D_2	D_1	T^a	B^a	L^a
10	2.5	320	100	8	80	110			
12	2.5	360	100	8	80	110			
14	2.5	390	100	8	80	110			
16	2.5	400	100	8	80	110			
18	2.5	400	100	8	80	110			
20	2.5	420	100	8	80	110	15	80	380
22	2.5	460	150	8	80	110	20	80	380
24	2.5	460	150	8	80	110	20	80	380
26	2.5	500	150	10	100	130	20	80	460
28	2.5	500	150	10	100	130	30	80	460
30	3.5	520	150	10	100	130	30	80	430
32	3.5	560	150	10	100	130	30	100	530
34	3.5	560	200	10	100	130	30	100	530
36			200	10	100	130	30	100	540
38			200	14	100	180	30	100	600
42			200	14	150	180	40	100	630
46			200	14	150	180	40	100	690
50			250	14	150	180	40	150	750
52			250	14	150	180	40	150	780
56			250	14	150	240	40	150	840
60			300	14	200	240	45	150	900
64			300	14	200	240	50	150	960

[a] Anchor bolt type X has a bolt size larger than 20 mm.

$$T_b = \frac{4M}{N \times B_C} - \frac{W}{N} \tag{7.5}$$

where
T_b = Tension force on the anchor bolt
M = Maximum moment on vessel
N = Number of anchor bolts
B_C = Bolt circle diameter
W = Minimum weight of vessel

Equation (7.5) is conservative as it assumes that the moment is resisted only by the bolt group. This equation may provide a large embedment length

TABLE 7.18

Height of Threaded Part (mm)

Anchor Bolt Diameter	Height of Threaded Part	
	One Nut and Washer	Two Nuts and Washer
10	15	25
12	20	30
14	25	40
16	25	40
18	30	50
20	30	50
22	35	55
24	35	60
26	40	70
28	40	75
30	45	75
34	50	85
36	55	90
38	60	100
42	65	105
46	70	115
50	75	125
52	80	130
56	85	140
60	90	150
64	95	160

and can control the design in case of tall vessels and stacks. Another sophisticated solution uses the iteration method to define the natural axis location and determine the compressive area and tensile forces in the bolt.

7.8.6 Anchor Bolt Strength Design

A ductile anchorage design can be defined as one where the yielding of the anchor (or the reinforcement or the attachment that the anchor attaches to) controls the failure of the anchorage system. This will result in large deflections, redistribution of loads, and absorption of energy prior to any sudden loss of capacity of the system resulting from a brittle failure of the concrete, according to the ASCE *Anchor Bolt Report*.

Anchors embedded in concrete and pulled to failure fail either by pullout of the concrete cone or by tensile failure of the anchor itself. The former is a brittle failure and the latter is a ductile failure. A brittle failure is sudden and without warning, possibly causing catastrophic results. In contrast, a ductile failure will cause the steel to yield, elongate gradually, and absorb a significant amount of energy, often preventing structures from collapsing. Consequently, when the design of a structure is based upon ductility or energy absorption, one of the following mechanisms for ductility shall be used. The anchors shall be designed to be governed by the tensile or shear strength of the steel and the steel shall be a ductile material.

7.8.6.1 Ultimate Strength Design

When bolts are subjected to both tension and shear, the following equation will apply.

$$\left(\frac{T_u}{\phi_1 P_n}\right) + \left(\frac{V_u}{\phi_2 V_n}\right) \le 1 \tag{7.6}$$

where
  T_u = Factor tensile load per bolt
  ϕ_1 = 0.9, strength reduction factor for tension load
  V_u = Factored shear force per bolt
  ϕ_2 = 0.85, strength reduction factor for shear loads

$$\phi_1 P_n = \phi_t F_t A_s$$

where
  F_t = The smaller of 1.0 times bolt yield stress or 0.9 times bolt tensile strength
  A_t = Bolt tensile stress area

$$\phi_2 V_n = \phi_2 \mu f_y A_s$$

where
  μ = friction coefficient from ACI 349 Appendix B
     = 0.55 when the bottom of the base plate is raised above the concrete surface as on a grout bed
     = 0.7 when the bottom of the base plate is on the concrete surface
     = 0.9 when the top of the base plate is at or below the concrete surface
  F_y = Anchor bolt yield stress
  A_s = Bolt tensile stress area

7.8.6.2 Allowable Stress Design

AISC ASD provides allowable stress design requirements for ASTM A307 and that is valid for A36, A325, A449, and A490 bolts. Based on ASCE, this equation is considered conservative:

$$\left(\frac{f_t}{F_t}\right) + \left(\frac{f_v}{F_v}\right) = 1$$

where

F_t = Calculated tensile stress

F_v = Calculated shear stress

F_t = Allowable tensile stress = $0.33F_u$

F_v = Allowable shear stress

= $0.22F_u$ in case of threaded, excluded from the shear plane

= $0.17F_u$ in case of threaded, included in shear plane

7.8.6.3 Calculate Required Embedment Length

Design an embedment using a straight reinforcing bar welded to an embedment plate as shown in Figure 7.31.

$$l_d = \left(\frac{3}{40}\right)\frac{f_y}{\sqrt{f_c'}}\frac{\alpha\beta\gamma}{\left[(c+k_{tr})/d\right]}d$$

Assume no transverse reinforcement ($k_{tr} = 0$). Use $l_d = 24$ in. with no adjacent anchors or edges ($[c + k_{tr}]/d_b = 2.5$, max.), more than 300 mm of fresh concrete to be cast below the anchor ($\alpha = 1.3$), uncoated anchor ($\beta = 1.0$), and a 19-mm bar ($\gamma = 0.8$).

Determine the required embedment length for the stud to prevent concrete cone failure. The design pullout strength of the concrete, P_d, must exceed the minimum specified tensile strength ($A_s f_{ut}$) of the tensile stress component, as shown in Figure 7.32.

$$P_d > A_s f_{ut}$$

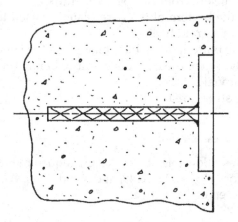

FIGURE 7.31
Reinforcing bar anchor.

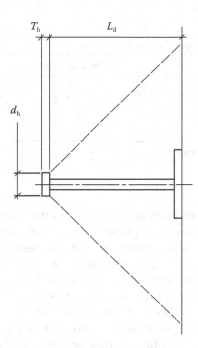

FIGURE 7.32
Embedment stud.

$$P_d = \phi_4 \sqrt{f_c'} A_{CP}$$

$$A_{cp} = \pi \left[(L_d + d_h/2)^2 \right] \phi_4 \sqrt{f_c'} \geq A_s f_{ut}$$

7.8.7 Anchor Design Considerations

If the base plate is designed with oversized bolt holes and there is a shear load in excess of the amount that can be transmitted through friction, it is recommended that:

1. Shear lugs be used.
2. A mechanism to transfer the load from the base plate to the bolt without slippage is incorporated (such as welding washers in place).

If no tensile force is effectively applied to the anchors, the anchors need not be designed for tension. Where the tensile force is adequately transferred to properly designed rebar, there is no requirement to check the concrete breakout strength of the anchor or anchors in tension. If the anchors are welded to the attachment so as to distribute the force to all fasteners, the strength may be based on the strength of the row of anchors farthest from the edge.

TABLE 7.19

Pretension Load and Torque Recommendation

Nominal Bolt Diameter (mm)	Torque (N m)	Pretension Load (kg)
M16	60	3,311
M24	100	7,447
M30	160	18,247
M52	245	37,136

Note: Based on 30 ksi bolt internal stress.

7.8.8 Pretensioning

Pretensioning of anchor bolts can be used to effectively fix the base plate to the foundation. The primary application for pretensioning is for large vibrating equipment, tall process columns, and cantilever stacks. Pretensioning anchor bolts will virtually eliminate fatigue stresses in the anchor bolts and is also effective in eliminating shear forces on the bolts. Proper anchor bolt details, careful installation procedures, and field quality controls are essential to ensure that the anchor bolts are successfully tensioned.

The advantages of the pretension anchor bolt is that it can prevent stress reversals on anchors susceptible to fatigue weakening, may increase dampening for pulsating or vibrating equipment, and will decrease, to some extent, the drift for process towers under wind or seismic load. It will also increase the frictional shear resistance for process towers and other equipment.

The disadvantage of this type of anchor bolt is that it can be a costly process to accurately install them and there is no recognized code or authority that gives guidance on the design and installation of pretensioned anchors. There is little research in this area so specifications depend on the requirements of the manufacturer.

Pretensioning should always be done when recommended by the equipment manufacturer. The manufacturer's instructions should be followed carefully. When not otherwise specified, anchors for turbines and reciprocating compressors should be torqued to the values shown in Table 7.19. When pretensioning is required, high-strength anchors should be used. Pretensioned anchor bolts should also be considered for tall process columns or cantilever stacks, which can develop significant bolt tension under wind loads or seismic loads.

References

Standard HB-17. 2002. *AASHTO Standard Specifications for Highway Bridges*, 17th edition. Washington, DC: American Association of State Highway Transportation Officers.

ACI Standard 318-05/318R-05. 2005. *Building Code Requirements for Structural Concrete and Commentary*. Farmington Hills, MI: ACI.

ACI Standard 350/350R-01. 2001. *Environmental Engineering Concrete Structures*. Farmington Hills, MI: ACI.

ACI Standard 530/ASCE 5. 2002. *Building Code Requirements for Masonry Structures*. Farmington Hills, MI: ACI.

AFPA NDS-2005. 2005. *National Design Specification for Wood Construction*. Washington, DC: American Forest and Paper Association.

AFPA. 2005. *NDS Supplement—Design Values for Wood Construction*. Washington, DC: American Forest and Paper Association.

AISC Standard 316-89. 1989. *Manual of Steel Construction: Allowable Stress Design*. Chicago, IL: AISC.

AISC Standard 316-89. 1989. *Manual of Steel Construction: Load and Resistance Factor Design (LRFD)*. Chicago, IL: AISC.

AISC Standard S334L. 1999. *Specification for Structural Joints Using ASTM A325 or A490 Bolts*. Chicago, IL: AISC.

AISI Standard ANSI/AISC 341-02. 2002. *Seismic Provisions for Structural Steel Buildings*. Washington, DC: American Iron and Steel Institute.

AISI Standard SG 673, Part I. 1987. *Specification for the Design for Cold-Formed Steel Structural Members*. Washington, DC: American Iron and Steel Institute.

AISI Standard SG 673, Part II. 1987. *Commentary on the Specification for the Design for Cold-Formed Steel Structural Members*. Washington, DC: American Iron and Steel Institute.

AISI Standard SG 913, Part I. 1991. *Load and Resistance Factor Design Specification for Cold-Formed Steel Structural Members*. Washington, DC: American Iron and Steel Institute.

AISI Standard AISI SG 913, Part II. 1991. *Commentary on the Load and Resistance Factor Design Specification for Cold-Formed Steel Structural Members*. Washington, DC: American Iron and Steel Institute.

API Standard 650. 1998. *Welded Steel Tanks for Oil Storage*, 10th edition. Washington, DC: API.

ASCE Standard 7-02. 2002. *Minimum Design Loads for Buildings and Other Structures*. Ann Arbor, MI: ASCE.

ASCE Standard SEI/ASCE 37-02. 2002. *Design Loads on Structures during Construction*. Ann Arbor, MI: ASCE.

ASCE. 1997. *ASCE Guidelines for Seismic Evaluation and Design of Petrochemical Facilities*. Ann Arbor, MI: ASCE.

ASCE Standard 40262. 1997. *Wind Load and Anchor Bolt Design for Petrochemical Facilities*. Ann Arbor, MI: ASCE.

ASCE. 1997. *ASCE Design of Blast Resistant Buildings in Petrochemical Facilities*. Ann Arbor, MI: ASCE.

American Society of Mechanical Engineers (ASME). A17.1. 2007. *Safety Code for Elevators and Escalators*. New York, NY: ASME.

ASTM Standard A36/A36M. 2001. *Standard Specification for Carbon Structural Steel, ASTM Annual Book of Standards*. West Conshohocken, PA: ASTM International.

ASTM Standard ASTM A82. 2007. *Standard Specification for Steel Wire, Plain, for Concrete Reinforcement*. West Conshohocken, PA: ASTM International.

ASTM Standard ASTM A185. 2007. *Standard Specification for Steel Welded Wire Fabric, Plain, for Concrete Reinforcement*. West Conshohocken, PA: ASTM International.

ASTM Standard A193/A193M. 2005. *Standard Specification for Alloy-Steel and Stainless Steel Bolting Materials for High-Temperature Service.* West Conshohocken, PA: ASTM International.

ASTM Standard A307-04. 2004. *Standard Specification for Carbon Steel Bolts and Studs, 60,000 psi Tensile Strength.* West Conshohocken, PA: ASTM International.

ASTM Standard A325. 2007. *Standard Specification for Structural Bolts, Steel, Heat Treated, 120/105 ksi Minimum Tensile Strength.* West Conshohocken, PA: ASTM International.

ASTM Standard A325M. 2007. *Standard Specification for High-Strength Bolts for Structural Steel Joints [Metric].* West Conshohocken, PA: ASTM International.

ASTM Standard A354. 2007. *Standard Specification for Quenched and Tempered Alloy Steel Bolts, Studs, and Other Externally Threaded Fasteners.* West Conshohocken, PA: ASTM International.

ASTM Standard 490/490M. 2004. *Standard Specification for Heat-Treated Steel Structural Bolts, 150 ksi Minimum Tensile Strength.* West Conshohocken, PA: ASTM International.

ASTM Standard A615/A615M. 2008. *Standard Specification for Deformed and Plain Billet-Steel Bars for Concrete Reinforcement.* West Conshohocken, PA: ASTM International.

ASTM Standard A706/A706M. 2008. *Standard Specification for Low-Alloy Steel Deformed Bars for Concrete Reinforcement.* West Conshohocken, PA: ASTM International.

ASTM Standard A992/A992M. 2006. *Standard Specification for Steel for Structural Shapes for Use in Building Framing.* West Conshohocken, PA: ASTM International.

ASTM Standard F1554. 2007. *Standard Specification for Anchor Bolts, Steel, 36, 55, and 105 ksi Yield Strength.* West Conshohocken, PA: ASTM International.

AWS Standard AWS D1.1. 2006. *Structural Welding Code—Steel.* Miami, FL: American Welding Society.

BS5950. 2000. *Structural Use of Steelwork in Building, Part 1.*

CMAA No. 70. 2000. *Specifications for Top Running Bridge and Gantry Type Multiple Girder Overhead Electric Traveling Cranes.* Crane Manufacturers Association of America (CMAA).

CMAA No. 74. 2004. *Specifications for Top Running and Under Running Single Girder Overhead Electric Traveling Cranes Utilizing Under Running Trolley Hoist.* Crane Manufacturers Association of America (CMAA).

Lui, E. M. 1999. Structural steel design. In *Structural Engineering Handbook.* W. F. Chen, Ed. Boca Raton: CRC Press LLC.

OSHA 29 CFR 1910. *Occupational Safety and Health Standards.*

OSHA 29 CFR 1926. *Safety and Health Regulations for Construction.*

PCI MNL 120. 2004. *Design Handbook—Precast and Prestressed Concrete.* Chicago, IL: Precast/Prestressed Concrete Institute.

PIP Standard 686. *Recommended Practices for Machinery Installation and Installation Design.* Austin, TX: PIP.

Steel Joist Institute (SJI). 2006. *SJI Standard Specifications and Load Tables.* Forest, VA: SJI.

8

Assessment of Existing Structures

8.1 Introduction

Typically, the main cause of deterioration of reinforced concrete structures in oil and gas facilities is corrosion. Most of the oil and gas delivered to these onshore facilities comes from offshore, so they are usually located near the sea and, therefore, affected by chloride from seawater.

When concrete structure shows signs of corrosion, we want to repair it. Before we do that, however, we must accurately evaluate the building and identify the underlying environmental conditions. After performing a structural risk assessment, we can identify the parts of the concrete elements that need to be repaired and determine the method of repair. When the assessment method is correct, and accurately determines the degree of building risk, the repair process will also be correct, adequate, and precise. In medicine, if the diagnosis of a disease is correct, the selected medicine will also be right. The process of building assessment is the preliminary step in diagnosing the causes of defects in the structure as a result of corrosion.

We will think in two directions: determine the method of repair and defining reasonable protection methods to protect the steel bars from corrosion. The process of assessing a structure often occurs in two stages. The first stage, the initial assessment of the building, involves defining the problem and developing a valuation of the structure. In the second stage, a detailed assessment of the building carefully defines the problems and the reasons for them. At this stage, the detailed inspection of the whole building will be carried out. A basis has been established for the inspection and evaluation of buildings through Technical Report 26 of the Concrete Society (1984), as well as the standards of the American Concrete Institute.

There are many reasons for the deterioration of reinforced concrete structures, from the presence of cracks to falling concrete. Among those reasons are increasing stresses, plastic shrinkage, frost, cracked concrete (in the case of plastic concrete), movement of the wood form during construction, and, in the case of an aggregate, contamination of the aggregate by alkalinity.

There are also some structures, such as reinforced concrete pipelines in sanitary projects and underground concrete structures (e.g., tunnels), in which the surface contacts the soil directly, where the carbonate present in the water attacks the concrete. This is a type of deterioration that is important to consider. We will also discuss corrosion resulting from the exposure of a concrete structure to the atmosphere.

As stated previously, there are several types of cracks. Therefore, the engineer performing the evaluation must be highly experienced in that area so that he can specify precisely the cause of the deterioration in the concrete. A wrong diagnosis will result in the wrong repair, which will cause heavy financial losses and a negative impact on structural safety.

Evaluation of the structure involves a structural assessment, assessment of the state of corrosion, and evaluation of the rate of corrosion and the possibility of the concrete collapsing. From the information gathered, we can decide if the concrete member can withstand loads. A case study will also be performed to assess if there is deflection occurring that is more than the allowable limit. If so, this will cause cracks or, in case of reduction of the cross-section area, falling concrete cover.

8.2 Preliminary Inspection

In this inspection, it is important to identify clearly the geographic location of the structure, the nature and circumstances of weather conditions surrounding the structure, the method of construction of the structure, its structural system, and the method of loading.

This assessment is often performed preliminarily by visual inspection, concentrating on cracks and falling concrete. Data are collected about the thickness of the concrete cover and the nature of the structure. This includes assessment of the quality of the concrete, quality of construction, and the nature of the structural system, whether it consists of a beam and column system, a slab on load-bearing masonry, or prestressed or precast concrete.

All such information is preliminary. It is necessary to perform some simple measurements. For example, to determine the depth of chloride in the concrete, it is necessary to take a sample from the concrete that fell and perform laboratory tests on it. The safety of the structure must be calculated precisely, especially after reduction of the cross-sectional area of steel due to corrosion, as well as after the fall of the concrete cover, which reduces the total area of the concrete member. These two areas are the main coefficients that directly affect the capacity of the concrete member to carry loads.

8.2.1 Collecting Data

The following data must be collected before the visual inspection can be performed and must be stated in the structure assessment report:

- Construction year
- The contractor and engineering office
- The structure system
- The drawings and project specifications
- The construction method
- The concrete tests, if applicable
- The environmental conditions (near sea or chemical factory)

All of the data must be provided to you by the owner, if you are a third-party expert. In most cases, all of the data cannot be collected, as you will be inspecting a structure that may be more than 20 or 30 years old and the drawings and specifications will not be available, as there were no software drawing programs at that time. All the hardcopy drawings, in most cases, will be lost or in bad condition. If the owner has a big organization with an engineering department, there may be a document control system and you can find the drawings if the building is an office or industrial structure. For residential buildings, you will be very lucky if the owner still has these drawings.

The construction year is very critical information. From this you can define the codes and specifications that were applied in that year, depending on your experience. By knowing the construction year, the contractor who constructed the building, and the engineering office that provided the engineering design, one can imagine the condition of the building based on the reputation of the contractor and the engineering office.

The construction system can be determined by visual inspection, through drawings, or by talking with the users of the building and the engineering office that reviewed the engineering design documents or supervised the construction. The information about environmental conditions will be provided to you by the owner. You can also observe the conditions onsite and collect data about the environmental condition from other sources, taking into consideration all the data—temperatures in the summer, winter, morning, and night, as well as the wind and wave effect.

If you are an owner, you must be sure that all the necessary data are delivered to the consultants who do the inspection, as any information is valuable. Engineering work is like a computer—good in, good out, and, of course, rubbish in, rubbish out. So you must verify all the data that you deliver to the expert performing the building assessment.

In an industrial facility plant, the conditions and mode of operations are not clear to the consultant engineer, so the report that you must deliver to him about the vibration of the machines, the heat of liquid in pipelines, and so on, has a major important impact on the concrete structure assessment result. Take into consideration when collecting data from a process plant or factory that the mode of operations may change from time to time, so an inclusive history must be delivered to the engineers who perform the evaluation process. Those persons are usually structural engineers and not familiar with these situations.

If the data of wind and waves do not exist or are not known, you must obtain it from a third party specialized in metocean criteria. This information is very important in the case of offshore structures or marine structures where the assessment depends on these data if wind and waves represent the loads most affecting the structure.

Concrete compressive strength tests are usually performed in any building by cylinder specimens (in American code) or cube specimens (in the British standard), but it is often a problem to find these data after 20 years or more, depending on the age of the building. If these data are not available, a test must be performed using the ultrasonic pulse velocity test, Schmidt hammer, or core test.

8.2.2 Visual Inspection

Visual inspection is the first step in any process of technical diagnosis, and it starts by generally viewing the structure as a whole and then concentrating afterward on the general defects that define precisely the corrosion deterioration and the extent of corrosion on the steel reinforcement.

As we have stated before, the process of assessing the building must be performed by an expert, as the cracks in the concrete structure may not be the cause of the corrosion. Corrosion is not the only factor that can cause cracks; however, it is the main factor in major deterioration of structures. Figure 8.1a, b, and c illustrates the various forms and causes of cracks in beams, slabs, and columns.

In general, cracks in concrete can have many causes. They may affect appearance only or they may indicate significant structural distress or a lack of durability. Cracks may represent the total extent of the damage or they may point to problems of greater magnitude. Their significance depends on the type of structure, as well as on the nature of the cracking. For example, cracks that are acceptable for building structures may not be acceptable for water-retaining wall structures.

The proper repair of cracks depends on knowing the causes of the cracks and selecting the repair procedures that take these causes into account; otherwise, the repair may only be temporary. Successful long-term repair procedures must prevent the causes of the cracks, as well as the cracks themselves.

Cracks may occur in plastic concrete or hardened concrete. Based on ACI code, examples of plastic concrete cracks include plastic shrinkage cracking, settlement cracking, and, after hardening, dry shrinkage cracking. The Concrete Society's *Technical Report 54* (2000) mentions that different types of cracks occur at different times in the life of a concrete element, as shown in Table 8.1. So, along with recognition of a crack pattern, knowledge of the time

(a)

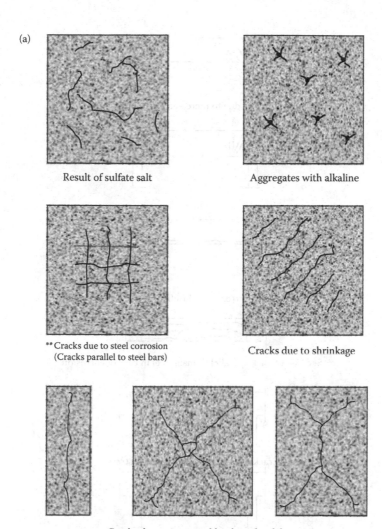

Result of sulfate salt

Aggregates with alkaline

** Cracks due to steel corrosion
(Cracks parallel to steel bars)

Cracks due to shrinkage

Cracks due to increased load on the slab

FIGURE 8.1
(a) Types of cracks in reinforced concrete slabs. (continued)

of the first appearance of the cracks is helpful in diagnosing the underlying cause.

8.2.2.1 Plastic Shrinkage Cracking

Plastic shrinkage cracking in concrete occurs most commonly on the exposed surfaces of freshly placed floors and slabs and other elements with large surface areas when they are subjected to a very rapid loss of moisture caused by low humidity and wind, high temperatures, or both. Plastic shrinkage usually occurs before final finishing, before the start of curing. When

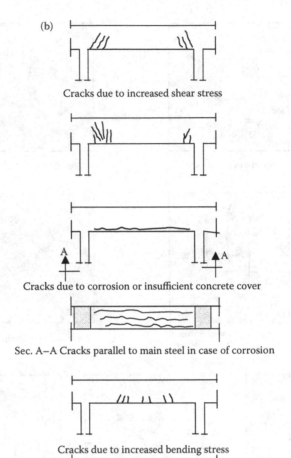

(b)

Cracks due to increased shear stress

Cracks due to corrosion or insufficient concrete cover

Sec. A–A Cracks parallel to main steel in case of corrosion

Cracks due to increased bending stress

Cracks due to compression failure

FIGURE 8.1 (continued)
(b) Types of cracks in reinforced concrete beams. (continued)

moisture evaporates from the surface of freshly placed concrete faster than it is replaced by bleed water, the surface concrete shrinks.

Due to the restraint provided by the concrete on the drying surface layer, tensile stresses develop in the weak, stiffening plastic concrete, resulting in shallow cracks that are usually not short and run in all directions. In most cases, these cracks are wide at the surface. They range from a few millimeters to many meters in length and are spaced from a few centimeters to as much as 3 m apart. Plastic shrinkage cracks may extend the full depth of elevated structural slabs.

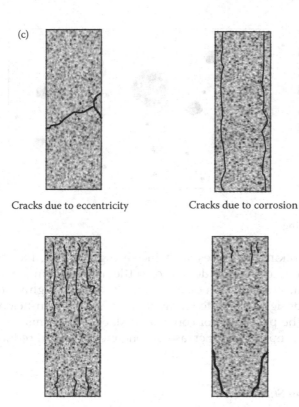

Cracks due to eccentricity   Cracks due to corrosion

Cracks due to increased column load

FIGURE 8.1 (continued)
(c) Types of cracks in reinforced concrete columns.

8.2.2.2 Settlement Cracking

After initial placement, vibration, and finishing, concrete has a tendency to continue to consolidate. During this period, the plastic concrete may be locally restrained by reinforcing steel, a prior concrete placement, or formwork. This local restraint may result in voids or cracks adjacent to the restraining element, as shown in Figure 8.2. When associated with reinforcing steel,

TABLE 8.1

Typical Times for Appearance of Defects

Type of Defect	Typical Time of Appearance
Plastic settlement cracks	10 min to 3 h
Plastic shrinkage cracks	30 min to 6 h
Crazing	1–7 days, sometimes much longer
Early thermal contraction cracks	1 to 2 days until 3 weeks
Long-term drying shrinkage cracks	Several weeks or months

Note: From the Concrete Society Technical Report 54.

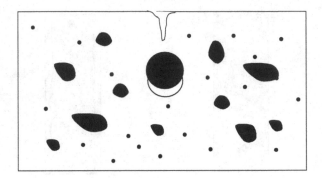

FIGURE 8.2
Settlement cracking.

settlement cracking increases with increasing bar size, increasing slump, and decreasing cover. The degree of settlement cracking will be magnified by insufficient vibration or the use of leaking or highly flexible forms. Proper form design, adequate vibration, provision of a sufficient time interval between the placement of concrete in slabs and beams, use of the lowest possible slump, and an increase in concrete cover will reduce settlement cracking.

8.2.2.3 Drying Shrinkage

The common cause of cracking in concrete is drying shrinkage. This type of shrinkage is caused by the loss of moisture from the cement paste constituent, which can shrink by as much as 1% per unit length. Unfortunately, aggregate provides internal restraint that reduces the magnitude of this volume change to about 0.05%. On wetting, concrete tends to expand.

These moisture-induced volume changes are a characteristic of concrete. If the shrinkage of concrete could take place without any restraint, the concrete would not crack. It is the combination of shrinkage and restraint, usually provided by another part of the structure or by the subgrade, that causes tensile stresses to develop. When the tensile stress of concrete is exceeded, it will crack. Cracks may propagate at much lower stresses than are required to cause crack initiation.

In massive concrete structural elements, as in the case of a foundation under a heavy machine, tensile stresses are caused by differential shrinkage between the surface and the interior concrete. Larger shrinkage at the surface causes cracks to develop that may, with time, penetrate deep into the concrete. The magnitude of the tensile stress is influenced by a combination of the following list of factors:

1. The amount of shrinkage
2. The degree of restraint

3. The modulus of elasticity
4. The amount of creep

The amount of drying shrinkage is influenced mainly by the amount and type of aggregate and the water content of the mix. The greater the amount of aggregate, the smaller is the amount of shrinkage. The higher the stiffness of the aggregate, the more effective it is in reducing the shrinkage of the concrete. This means that concrete that contains sandstone aggregate has a higher shrinkage, by about twice, than that of concrete containing granite, basalt, or limestone. The higher the water content, the greater the amount of drying shrinkage.

Surface crazing on walls and slabs is an excellent example of drying shrinkage on a small scale. Crazing usually occurs when the surface layer of the concrete has a higher water content than the interior concrete. The result is a series of shallow, closely spaced, fine cracks. Shrinkage cracking can be controlled by using properly spaced contraction joints and proper steel detailing. Shrinkage cracking may also be controlled using shrinkage-compensating cement.

8.2.2.4 Thermal Stresses

Temperature differences within a concrete structure may be due to cement hydration, changes in ambient temperature conditions, or both. These temperature differences result in differential volume changes. The concrete will crack when the tensile strains due to the differential volume changes exceed their tensile strain capacity.

The effects of temperature differentials due to the hydration of cement are normally associated with mass concrete such as large columns, piers, beams, footing, retaining walls, and dams, whereas temperature differentials due to changes in the ambient temperature can affect any structure. The concrete rapidly gains both strength and stiffness as cooling begins. Any restraint of the free contraction during cooling will result in tensile stress. Tensile stresses developed during the cooling stage are proportional to the temperature change, the coefficient of thermal expansion, the effective modulus of elasticity, and the degree of restraint. The more massive the structure, the greater the potential for temperature differential and higher degrees of restraint. Procedures to help reduce thermally induced cracking include reducing the maximum internal temperature, controlling the rate at which the concrete cools, and increasing the tensile strain capacity of the concrete.

Hardened concrete has a coefficient of thermal expansion that may range from 7 to $11 \times 10^{-6}/C$ with an average of $10 \times 10^{-6}/C$. When one portion of a structure is subjected to a temperature-induced volume change, the potential for thermally induced cracking exists. The designers must give special consideration to structures in which some portions are exposed to temperature changes, whereas other portions of the structure are either partially or completely protected. A drop in temperature may result in cracking in the exposed element, whereas increases in temperature may cause cracking in the protected portion of the structure. So the designer must allow movement in the structure by recommending the use of contraction joints and providing the correct details for it.

Structures that have a high difference in temperatures are usually built in areas near the desert, where there are greater differences in temperature between afternoon and midnight. Moreover, in countries with high temperatures, they usually use air conditioning inside the building, so there will be a high probability of cracks due to the difference in temperatures inside and outside the building. So, the designer should take these stresses into consideration.

8.2.2.5 Chemical Reaction

As shown in Figure 8.1a, cracks resembling a star on the concrete surface are an indication of chemical reaction. This reaction occurs with the presence of aggregate containing active silica and alkalis derived from cement hydration, admixtures, or external sources such as curing water, ground water, or alkaline solutions stored or used in the finished structure. The alkali–silica reaction results in the formation of a swelling gel, which tends to draw water from other portions of the concrete. This causes local expansion and accompanying tensile stresses and may eventually result in the complete deterioration of the structure.

Groundwater that has sulfate is a special durability problem for concrete. Because the sulfate penetrates hydrated cement paste, it comes in contact with hydrated calcium aluminates. Calcium sulfoaluminate will form, and the accompanying increase in volume results in high-tensile stresses that cause cracking. Therefore, it is recommended to use Portland cement types II and V, which contain low tricalcium aluminates and will reduce the severity of the cracks.

As noted in the following figures, in case of corrosion in the steel reinforcement, the cracks you will see will be parallel to the steel bars of any concrete members such as beams, slabs, and columns. Moreover, brown spots on the concrete surface are an indication of corrosion in the steel reinforcement. At this stage, we rely solely on the experience of the engineers who perform the inspection. This is a disadvantage of the visual inspection—it is completely dependent on the experience of the engineer performing the inspection, as there are some cracks that can have more than one cause, with or without the corrosion effect.

8.3 Detailed Inspection

The purpose of the detailed inspection is to determine, accurately whenever possible, the degree of seriousness of the deterioration of the concrete.

Therefore, the number of collapses that have occurred, the cause of the deterioration in the concrete, and the amount of repair that will be needed must be defined precisely at this stage, as such quantities will determine the amount placed in the tender that will be put forward to contractors for repair.

At this stage, we need to be familiar with the reasons for the deterioration, in detail, and the capability to perform a failure analysis technique. Initially, we will use the method of visual inspection, maybe in conjunction with the use of a small hammer. We will use other measurements to determine the depth of the transformation of carbon in the concrete and the degree of corrosion of the steel in concrete and how far it extends in the steel bars. Moreover, at this stage, it is important to define the degree of the concrete's electric resistivity to predict the corrosion rate. All of these required measurements will be discussed in this chapter. As you should know, the weather conditions that affect the building are the main factor that affects the measurement readings and also affect the selection of the method of repair.

8.3.1 Methods of Structure Assessment

There are several methods by which to assess the structure, as mentioned in terms of the degree of extension of corrosion in the steel reinforcement bars and its impact on the whole structure. The first and most important method is the visual inspection, as it is a cost-effective and easy-to-use method.

This is followed by other methods that require capabilities and are often used in the case of structures of special importance with a high cost of repair. These require the use of sophisticated technology to identify the degree of corrosion of the steel reinforcement in concrete with high accuracy as it affects the total cost of the state of structure. Concrete bridges and special buildings, such as parking garages, tunnels, and other special concrete structures, are examples of such structures.

Each of the measurement tools that are used has a specific accuracy with advantages and disadvantages. In general, the process of assessing the structure from corrosion attack must be conducted by a person who has an acceptable amount of experience with the process of corrosion. Experience is key here and plays a major role in a successful assessment. However, there are some devices that rely on experience only for their use and for the knowledge of environmental factors that affect the equipment readings and how to overcome them.

Table 8.2 summarizes every method and its disadvantages, as well as the user's ability to work with that method and the performance rate. This might help to estimate the cost of the inspection used to evaluate the building, as well as the performance rates that will assist us in constructing the inspection time schedule.

These measurements will determine the cause of corrosion and the degree of the passive protection layer for the steel bars, as well as the expected corrosion rate for every part of the building. From this information, one can determine

TABLE 8.2

Practical Methods to Evaluate Concrete Structures

Methods	Inspection	User	Approximate Performance Rate
Visual inspection	Surface defects	General	1 m²/s
Chain or hammers	Void behind cover	General	0.1 m²/s
Concrete cover measurement	Distance between steel bars and concrete surface	General	1 reading every 5 min
Phenolphthalein	Carbonation depth	General	1 reading every 5 min
Half cell	Evaluate corrosion risk	Expert	1 reading/5 s
Linear polarization	Corrosion rate	Expert	1 reading in 10–30 min
Radar	Defects and steel location	Expert	1 m/s[a] by using a car or 1 m² in 20 s

[a] Add more time during scheduled plan preparation.

what kind of repair the structure needs and its method of construction and also calculate the repair required to strengthen the concrete member.

8.3.2 Concrete Test Data

The following are tests that should be performed to obtain the data for the concrete strength, if this information is not available. There are variations between these tests such as the accuracy, effect on the building, and the cost. Therefore, you have the responsibility to decide which of these tests is suitable for the building that you need to evaluate.

Moreover, there are some precautions required for these tests. For example, when you take the samples, the location must be based on the visual inspection results that were done before. Moreover, when you take the samples or do the loading test, this must be accompanied by a suitable temporary support to the structure and the adjacent members. These temporary support locations and strengths should be able to withstand the load in case of weakness occurring in the structural member that you test. These tests will follow ACI228-89 R1 and BS1881.

8.3.2.1 Core Test

This test is considered one of the semidestructive tests. This test is very important and popular for the study of the safety of structures as a result of changing the loading system or a deterioration of the structure as a result of an accident such as fire, weather factors, or when there is a need for temporary support for repair and there is no any accurate data about concrete strength.

This test is not too expensive and it is the most accurate test of the strength of concrete that can be carried out. Core testing is done by the cutting of cylinders from a concrete member, which could affect the integrity of the structure.

Therefore, the samples must be taken in strict adherence to the standard. This will ensure accuracy of the results without weakening the structure.

For the structure with deterioration due to corrosion of the steel bars, the structure has lost most of its strength due to the reduction of the steel cross-section area, so more caution must be exercised when performing this test. To perform this test, it is important to select a concrete member that will not affect the whole building safety. The codes and specifications provide some guidance to the number of cores to test and these values are as follows:

- For volume of concrete member, $V \leq 150$ m³, take 3 cores.
- For volume of concrete member, $V > 150$ m³, take $(3 + (V - 150/50))$ cores.

The degree of confidence of the core test depends on the number of tests that you can take, which must be as few as possible. The relationship between number of cores and confidence will be given in Table 8.3.

Before you choose the location of the sample, you must first define the location of the steel bars to assist you in selecting the location of the sample. The sample must be taken away from the steel bars to avoid taking a sample containing steel reinforcement bars, if possible. We must carefully determine the places to preserve the integrity of the structure; therefore, this test should be performed by an experienced engineer. Conducting such an experiment should be undertaken with proper precautions. To determine the responsibility of individuals, review the nondestructive testing that has been conducted in an accurate manner.

Figure 8.3 presents the process of taking the core from the reinforced concrete bridge girder.

8.3.2.1.1 Core Size

Note that the permitted diameters are 100 mm in the case of a maximum aggregate size of 25 mm and 150 mm in the case of a maximum aggregate size not exceeding 40 mm. It is preferable to use 150-mm diameter core size whenever possible, as it gives more accurate results, as shown in Table 8.4. Table 8.4 represents the relationship between the dimensions of the sample

TABLE 8.3

Number of Cores and Deviation in Strength

Number of Cores	Deviation Limit between Expected Strength and Actual Strength (Confidence Level 95%)
1	+12%
2	+6%
3	+4%
4	+3%

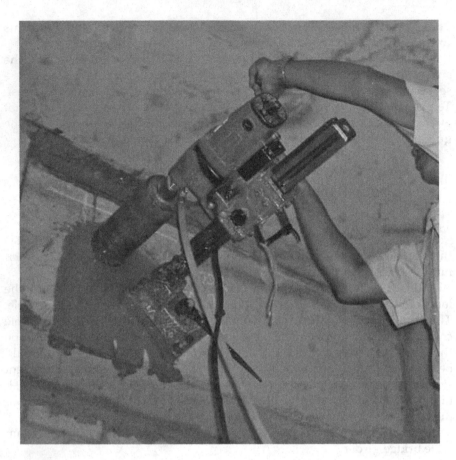

FIGURE 8.3
How to take a core sample.

TABLE 8.4

Core Size with Possible Problems

Test	Diameter (mm)	Length (mm)	Possible Problem
First	150	150	May contain steel reinforcement
Second	150	300	May cause more cutting depth to concrete member
Third	100	100	Not allowed if the maximum aggregate size is 25 mm
			May be cut with a depth less than required
Last	100	200	Less accurate data

and potential problems, so this table should be discussed when choosing the reasonable core size.

Some researchers have stated that the core test can be done with a core diameter of 50 mm in the case of a maximum aggregate size of not more than 20 mm overall. It was noted that small cores sizes give different results than large sizes. Because of the seriousness of the test and to avoid taking a high number of samples, there needs to be close supervision when the sample is taken. Moreover, the laboratory test must be certified and the test equipment must be calibrated with a certificate of calibration from a certified company.

Sample extractions are performed using pieces of a cylinder equipped with a special alloy mixture containing diamond powder to feature pieces in the concrete during the rotation of the cylinder through the body. The precautions taken should be an appropriate match for the sampling method and consistent pressure should be applied carefully depending on the experience of the person doing the test, depending on the expertise of the technical.

After that, the core will be filled with dry concrete of suitable strength or with grouting, which is a popular method. Other solutions depend on using epoxy and injecting it into the hole and then inserting a concrete core of the same size to close the hole. In any case, the filling must be done soon after making the cutting so the material for the filling will be standing by next to the technician who did the cutting, as this core affects the integrity of the structure.

Every core should be examined and photographed, and noted gaps must be classified within the core as a small void (between 0.5 and 3 mm), void (between 3 and 6 mm), or a big void (more than 6 mm). It must also be examined whether the core of nesting and determine the shape, type, and color gradient of aggregates any apparent qualities of sand as well. In the laboratory, the dimensions, weight of each core, density, steel bar diameter, and distance between the bars will be measured.

8.3.2.1.2 Sample Preparation for Test

After cutting the core from the concrete element and processing the sample, test by leveling the surface of the core and then taking the sample, which has a length of not less than 95%, and not more than double, the diameter. To level the surface, use a chainsaw with a spare concrete or steel cutting disk, and after that, prepare the two ends of the sample by covering them with mortar or sulfide and submerging them in water at a temperature of $20 \pm 2°C$ for at least 48 h before testing the sample.

The sample is put into a machine test and, gradually, the influence load is introduced at the regular and continuous rate of 0.2 to 0.4 N/mm^2, until it reaches the maximum load at which the sample is crushed. To determine the estimated actual strength of a cube by obtaining the crushing stress from the test, use the following equation, as λ is the core length divided by the core diameter.

- For a horizontal core, the strength calculation will be as follows:

  Estimated actual strength of the cube = $2.5/(1/\lambda) + 1.5 \times$ core strength

  where λ = core length/core diameter.
- For a vertical core, the strength calculation will be as follows:

$$\text{estimated actual strength for cube} = 2.3/(1/\lambda) + 1.5 \times \text{core strength} \qquad (8.1)$$

For the existing steel in the core perpendicular to the core axis, the previous equations will be multiplied by the following correction factor:

$$\text{correction factor} = 1 + 1.5(S\varphi)/(LD) \qquad (8.2)$$

where
L = Core length
D = Core diameter
S = Distance from steel bar to edge of core
φ = Steel bar diameter

Cores should be free of steel, preferably. If steel is found, you must use the correction factor, taking into account that it is used only in the event that the value obtained ranges from 5% to 10%. We must use the results to core, but if the correction factor is more than 10%, the results of the tests cannot be trusted and you should take another sample.

When evaluating the test results, the following points must be taken into account:

- Once the sample is submerged in water, there is a decrease in strength of up to 15% from that of dry concrete.
- The equation to calculate the expected actual concrete strength does not take into account any differences in direction between the core and the standard cube direction.
- It is stated that the concrete is acceptable if the average strength of the cores is at least 75% of the required strength and the calculated strength for any core is greater than 65% of the required strength.

In case of prestressed concrete, the concrete strength is acceptable if the average strength of the cores is at least 80% of the required strength and the calculated strength for any core is not less than 75% of the required strength.

8.3.2.2 Rebound Hammer

This is a nondestructive method of testing, so it is useful in determining the estimated concrete compressive strength. This is the most common method because it is easy to do and inexpensive, compared with other tests. However, it gives less precise data results. This test relies on concrete strength which is done by measuring the hardening from the surface. It identifies the compressive strength of the concrete member by using calibration curves of the relationship between reading the concrete hardening and concrete compressive strength.

There are different types of rebound hammers and they most commonly give an impact energy of 2.2 N/mm. There is more than one way to show results, based on the manufacturer. In some cases, the reading will be an analog or digital number or the readings may be stored on a drive.

Inspect the device before using it through the calibration tools attached to the device when you purchased it. The calibration should be within the allowable limit based on the manufacturer's recommendation. The first and most important step in the test is cleaning and smoothing the concrete surface at the sites that will be tested by hone in an area of about 300 × 300 mm. It is preferable to test on a surface that has no change after cast or a surface that has not had any smoothing done during the casting process.

On the surface to be tested, draw a net of perpendicular lines in both directions between 20 and 50 mm apart. The intersection points will be the points to be tested. The test points in any case must be at least 2 cm from the edge. Figure 8.4 illustrates that the surface must be cleaned before the test is performed and that the rebound hammer is perpendicular to the surface as shown.

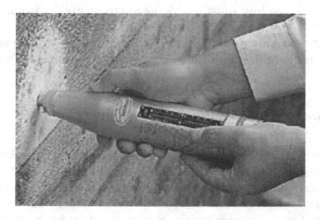

FIGURE 8.4
Testing and reading the number.

The following recommendations should be followed during the test:

- For any conditions, the hammer must be perpendicular to the surface that will be tested because the direction of the hammer affects the value of the rebound number, due to the weight of the hammer.
- The wet surface gives a reading from the rebound hammer of up to 20% less than the reading from a dry surface.
- The tested concrete member must be fixed and must not vibrate.
- You must not use the curves for the relationship between concrete compressive strength and rebound number given by manufacturer directly. You must calibrate the hammer by taking the readings on concrete cubes and crushing the concrete cubes to obtain the calibration of the curves. It is important to perform this calibration from time to time, as the spring inside the rebound hammer loses some stiffness over time.
- You must use one hammer only when making a comparison between the quality of concrete at different sites.
- The type of cement affects the readings, as in the case of concrete with high-alumina cement, which can yield higher results than concrete with ordinary Portland cement by about 100%.
- Concrete with sulfate-resistant cement can yield results of about 50% less than those of ordinary Portland cement.
- Concrete with a higher cement content gives a lower reading than concrete with a lower cement content, with a gross error of only 10% in any case.

The number of reading must by high enough to give reasonably accurate results. The minimum number of the reading is 10, but we usually take 15 readings. The extreme values will be excluded and the average taken for the other remaining values. From this, the concrete compressive strength will be known.

8.3.2.3 Ultrasonic Pulse Velocity

This test is one of the nondestructive types (ACI 228-89-R1 1989; BS 1881, 1971a, 1971b; Bungey, 1993); its concept is to measure the speed of transmission of ultrasonic pulses through the construction member. By measuring the time required for the transmission of impulses and by knowing the distance between the sender and receiver, the pulse velocity can be calculated.

The calibration of these velocities is performed by knowing the concrete strength and its mechanical characteristics. Then, you may use it for any other concrete with the same procedure of identifying compressive strength, the dynamic and static modulus of elasticity, and the Poisson ratio.

The equipment must have the capability to record time for the tracks, with lengths ranging from 100 to 3000 mm accurately, + 1%. The manufacturer should define methods for using the equipment in different temperatures and humidity. There must be an available power transformer sender and receiver of natural frequency vibrations between 20 and 150 kHz, bearing in mind that the frequency appropriate for most practical applications in the field of concrete is 50 to 60 kHz.

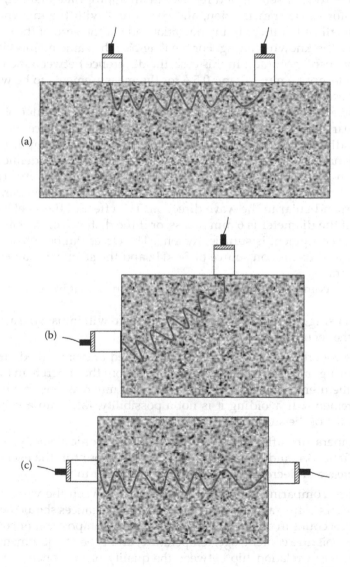

FIGURE 8.5
(a) Surface transmission. (b) Semidirect transmission. (c) Direct transmission.

There are different methods of wave transmission and surface transmission, which are illustrated in Figure 8.5a. Semidirect transmission is shown in Figure 8.5b, and direct transmission is shown in Figure 8.5c.

The ultrasonic test (UT) equipment consists of two rods of metal with lengths of 250 and 1000 mm. The first is used in the determination of zero for the measurement, and the second is used in the calibration. In both cases, the time of the passage of waves through each rod is known. Connect the ends of the rods of sender and receiver in an appropriate manner, measure the time for pulse transmission, and compare it with the known reading. For the small rod, if there is any deviation, adjust the zero of the equipment to provide the known reading. For the long bar, the same method is used to define the result accuracy. In this case, the difference between the two readings should not be more than ±0.5% for the measurements to be within the required accuracy.

The wave transmission velocity value of steel is twice the value of concrete, so steel bars in a concrete member being tested will influence the accuracy of the reading because the value will be high for the wave impulse velocity. To avoid that, the location of the steel reinforcement must be defined beforehand with respect to the path of the ultrasonic pulse velocity. The pulse velocity should be corrected based on the location of the steel bar, if parallel or perpendicular to the wave direction. The effect of the steel bar can be ignored if the diameter is 6 mm or less, or if the distance of the bar from the end of the equipment is sufficiently long. The effect can be considered zero if we use a transmission source of 54 kHz and the steel bar diameter is less than 20 mm.

The most common errors in taking this measurement include:

1. Not using the reference bar to adjust to zero will impact the accuracy of the results.

2. The concrete surface, which is well-leveled and smoothed after the pouring, may have properties different from the concrete in the core of the member and, therefore, avoid it as much as possible in measurements. If avoiding it is not a possibility, take into account the impact of the surface.

3. Temperature affects the transmission ultrasonic velocity, as seen in Table 8.4, and that must be taken into account in the event of an increase or decrease in temperature of 30°C or more.

4. When comparing the quality of concrete between the various components of the same structure, similar circumstances should be taken into account in all cases with regard to the composition of concrete, the moisture content, age, temperature, and type of equipment used. There is a relationship between the quality of the concrete and the speed as shown in Table 8.5.

TABLE 8.5

Relation between Concrete Quality and Pulse Velocity

Pulse Velocity (km/s)	Concrete Quality Degree
>4.5	Excellent
4.5–3.5	Good
3.5–3.0	Fair
3.0–2.0	Poor
<2.0	Very poor

8.3.2.4 Inherent Variations in In Situ Strength

There has been much research performed in this area, and tests on in situ concrete based on BS6089 indicate the following:

1. In situ strength can vary within a structural member, both randomly and, often, in an ordered fashion.

2. The magnitude of variations of in situ strength within structural members varies from one member to another in a random fashion.

3. With the height of a concrete lift, in situ strength decreases toward the top, even for slabs, and can be 25% less at the top than in the body of the concrete. Concrete of lower strength is often concentrated in the top 300 mm or 20% of the depth, whichever is less.

4. At 28 days after casting, columns can have a mean in situ strength of 65% of the mean standard cube strength, with strength in the individual columns carrying from 50% to 80%. The little evidence available for floor slabs suggests that the mean in situ strength may only be 50% of the mean standard cube strength.

5. The gain in in situ strength from 28 days onward is not consistent. At 6 months, the increase in mean strength can vary from 0% to 25% and, at 1 year, from 0% to 35% of the 28-day strength.

Thus, the normal variation of in situ strength within and between structural members has to be considered when evaluating in situ results. Examination of individual test results will identify whether variations between the results are excessive. Further tests may be required to establish whether certain results are rogue values or not.

The results of the test yield the estimated in situ cube strength, which is needed to evaluate whether the design strength is acceptable or not. The design strength is given by f_{cu}/g_m, where $\gamma_m = 1.5$ for ultimate strength.

Based on BS6089, the level of in situ cube strength that may be considered acceptable in any particular case is a matter of engineering judgment but should not normally be less than 1.2 times the design strength. The

TABLE 8.6

Comparison between Different Test Methods

Test Method	Probable Damage	Precaution Requirement
Overload test	Possible member loss	Member must be isolated or allowance of distributed the load to the adjacent members
		Extensive safety precaution just in case collapse failure
Cores	Holes to be made good	Limitations of core size and numbers
		Safety precaution for critical member
Ultrasonic	None	Need two smooth surfaces
Rebound hammer	None	Need a smooth surface

particular strength level selected should include an allowance for possible future deterioration of the strength of the concrete that may result from chemical attack, weathering, vibration, or some unforeseeable impact or circumstances. So, apply the following equation in case of $\gamma_m = 1.5$ to accept or refuse the concrete:

$$\text{estimated in situ cube strength} \geq 1.2(f_{cu})/1.5 \geq f_{cu}/1.25$$

8.3.2.5 Comparison between Different Tests

From the above discussion of the different methods of determining the hardened concrete strength, it is apparent that all have advantages and disadvantages. These are summarized in Table 8.6.

Moreover, these tests are differentiated by their costs, their treatment of the concrete member, and their accuracy of measured strength to actual concrete strength. This comparison is illustrated in Table 8.7.

After you obtain the results of the tests, you will find that the data indicates strength lower than the concrete strength specified in the drawings or project specifications. The value of concrete strength in the drawing is the standard cube (cylinder) compressive strength after 28 days. The cube or the cylinder was poured and compacted based on the standard and the curing process of

TABLE 8.7

Performance Comparison between Different Methods

Test Method	Damage to Concrete	Representative to Concrete	Accuracy	Speed of the Test	Cost
Overload test	Variable	Good	Good	Slow	High
Cores	Moderate	Moderate	Good	Slow	High
Ultrasonic	None	Good	Moderate	Fast	Low
Rebound hammer	Unlikely	Surface only	Poor	Fast	Very low

TABLE 8.8

Variation between Standard Cube Strength and Actual Strength on Site

| Concrete Member | Actual Strength/Standard Cube Strength after 28 Days | |
	Average	Range
Column	0.65	0.55–0.75
Wall	0.65	0.45–0.95
Beam	0.75	0.6–1.0
Slab	0.5	0.4–0.6

28 days, so the conditions were different from what will happen on the site. For example, the curing will not be in 28 days and you can not immerse the concrete slabs or columns in water continuously for 28 days. The temperature onsite is different and varies from that in the laboratory. In addition to that, the concrete pouring and compaction method is different for the cubes or cylinder than that in the actual member size. All these variations must be taken into consideration within the parameters of the design codes, because when you make the test, you measure the strength after complete hardening of the concrete. Table 8.8 is a comparison between a standard cube test and an in situ test for different reinforced concrete members.

8.3.3 Sources of Concrete Failure

Corrosion is not the only source of failure. There are many other factors that cause failure of reinforced concrete structures that must be considered when you perform the inspection. Therefore, it is important to know them very well. These sources of failure are as follows:

1. Unsuitable materials
   (a) Unsound aggregate
   (b) Reactive aggregate
   (c) Contaminated aggregate
   (d) Wrong type of cement
   (e) Cement manufacturer error
   (f) Wrong type of admixture
   (g) Substandard admixture
   (h) Contaminated admixture
   (i) Organically contaminated water
   (j) Chemically contaminated water
   (k) Wrong type of reinforcement
   (l) Steel bar size error

2. Improper workmanship
   (a) Faulty design
   (b) Incorrect concrete mixture (low or high cement content or incorrect admixture dosage)
   (c) Unstable formwork
   (d) Misplaced reinforcement
   (e) Error in handling and placing concrete (segregation, bad placement, and inadequate compact)
   (f) Curing incomplete
3. Environmental
   (a) Soil alkali
   (b) Seawater or sewage
   (c) Acid industry
   (d) Freezing and thawing
4. Structural
   (a) Load exceeds design
   (b) Accident such as ballast load or dropped object
   (c) Earthquake load

8.4 Test Methods for Corroded Steel in Concrete

During the corrosion process, the volume of the steel bar is increased and the internal stresses caused by that increase cause the concrete to crack. These cracks may be vertical or horizontal and increase in length and width until the fall of the concrete cover. To determine the parts of concrete element that cause the separation between concrete cover and concrete itself need to be known exactly.

8.4.1 Manual Method

The deteriorated area can be defined by using a hammer to hit the concrete cover and listening to the sound. If the sound indicates air behind the surface of the concrete cover, this means that the steel bar is corroded and the possibility exists of a concrete cover separation.

For a large area of concrete surface, you can use a steel chain and let it move over the surface to define the separated cover and damaged parts. There are other more complicated methods than using a hammer or steel chain such as infrared or ultrasound.

By the above inspection methods, using the manual hammer and steel chain, you can define the area that is expected to be corroded and predict concrete cover spall within a short time so that it may be addressed in the next repair. This method is faster, easier to use, and lower in cost than methods such as radar, ultrasound, or infrared ray and can be used in big structures or when evaluating huge areas such as decks of bridges. The use of the hammer is often associated with the process of visual inspection, where it is easily carried, low-cost, effective at locating defects, and dirty in that it makes marks at the places where the sound of a vacuum behind the concrete surface is heard.

Generally, these methods, whether manual, radar, or infrared, need professionally trained workers, with sufficient experience and competence, to read the measurement and define the defected area accurately.

There are precautions and limits on the use of these methods, in the case of the existence of water in the cracks or where the separation is deep inside the concrete, as well as in the case of the steel bars embedded deeply within the concrete, as it will be difficult to define the separate areas. Moreover, the existence of water and the depth of separation is something that affects the readings in the case of radar and infrared.

The execution of the repair process should remove an area larger than the defective area defined before by the different methods, because there is time difference between structural evaluation and the start of the repair due to administration work, defining the funds, and receiving approval, and this duration is sufficient to increase the volume of concrete at risk of separation. It should be noted that a disadvantage of infrared is that appropriate weather conditions must be present to provide reasonable accuracy and so it is rarely used in industrial projects.

8.4.2 Concrete Cover Measurements

The thickness of the concrete cover is usually measured in modern construction to ensure that the thickness of the concrete cover conforms with the specifications discussed in Chapter 5 to protect the steel from corrosion. Measurements of the thickness of the concrete cover in structures involve noting where the corrosion begins. The lack of concrete cover thickness increases the corrosion rate as a result of chlorides or carbonation, which increase the rate of steel corrosion. The lack of cover also increases exposure to moisture and oxygen, which are the primary components of the corrosion process. The measurement of the concrete cover thickness explains the causes of corrosion and identifies areas that have the capability to corrode faster.

The measurement of the thickness of the concrete cover needs to define axes y, x in order to determine the thickness of the concrete cover at every point on the structure. The equipment that measures the thickness of the concrete cover is both simple and of advanced technology, as you can obtain the measurement reading as numbers. Figure 8.6 presents the shape of the electromagnetic cover meter and illustrates its method of reading. A radiograph

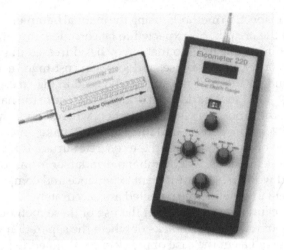

FIGURE 8.6
Shape of concrete cover thickness measurement machine.

may be used on bridges, but at a very high cost (Cadry and Gamnon, 1992; Bungey, 1993).

The magnetic cover method is simple to perform but can be affected by the distance between steel bars, and the thickness of the concrete cover has a significant influence on the readings. This method depends on a supply of electricity through a 9-V battery and measures the potential voltage envelope when completed by the buried steel bars, as shown in Figure 8.7.

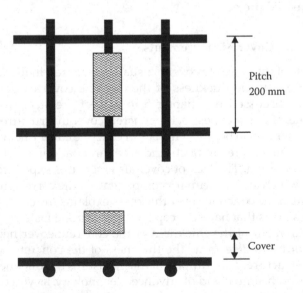

FIGURE 8.7
Device location on concrete surface.

Maximum signal   Minimum signal

FIGURE 8.8
The correct device location.

The British standard is the only standard that is concerned with the measurement of the concrete cover after construction. This is addressed in BS1881, Part 204. In 1993, Alldred studied the accuracy of the measurement of the cover when there are more steel bars close together and he suggested using more than one head of measurement, as that will increase the accuracy of the reading and the small heads have an impact on the accuracy of the equipment (Figure 8.8).

The problem with this method is that dense steel reinforcement in the concrete section will give inaccurate data. In this case, the equipment should be calibrated based on the existing steel bars, as the reading is also affected by the type of steel. The people who work on the equipment must be competent and aware of any thing that can affect the reading such as bolts, steel wires, etc.

8.4.3 Half-Cell Potential Measurements

This measurement unit is traditionally used in oil and gas plants to monitor the cathodic protection system for the pipeline and the facilities. You will find that a well-trained corrosion team can assist you in these measurements.

This system is used to determine the bars that have lost passive protection. The bars in the concrete lose the passive protection when exposed to carbonating or chlorides. This chemical reacts with the alkaline around the steel and destroys the passive layer.

The half-cell equipment, defined by ASTM C876, is a rod of metal in a solution of the same metal ions. For example, a copper rod in a saturated solution of copper sulfate is a very common piece of equipment. There are other types such as a bar of silver in a silver chloride solution, which can be linked through the voltmeter or other potentiometer connected to the steel bars as shown in Figure 8.9.

If the steel bars have a passive protection layer, the potential volts will range from 0 to 200 mV copper/copper sulfate. If there has been a breakdown of the passive protection layer, and some quantity of steel melting results in the movement of ions, the potential volts will be around 350 mV and higher. When the value is higher than 350 mV, it means that the steel has already started to corrode.

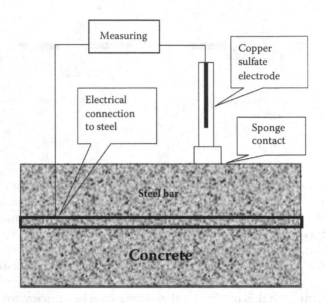

FIGURE 8.9
Half-cell equipment.

It is recommended to use half-cell equipment consisting of silver in a silver chloride solution or mercury in a mercury oxide solution. Copper in a copper sulfate solution can be used, but it is not recommended because it requires regular maintenance, thus, there will likely be some mistakes in readings.

The measurement of the steel reinforcement bars by the half-cell is to define the probability of corrosion to the steel bars. The ASTM C867 specifications present a method to clarify the data obtained from measuring the potential volts. Table 8.9 shows these values for the copper/copper sulfate and the silver/silver chloride equipment and explains the condition of corrosion in each case of corresponding value.

TABLE 8.9

ASTM Specification for Steel Corrosion for Different Half Cells

Silver/Silver Chloride	Copper/Copper Sulfate (mV)	Standard Hydrogen Electrode (mV)	Calomel (mV)	Corrosion Condition
>–106	>–200	>+116	>–126	Low (10% risk of corrosion)
–106 to –256	–200 to –350	+116 to –34	–126 to –276	Intermediate corrosion risk
<–256	<–350	<–34	<–276	High (<90% risk of corrosion)
<–406	<–500	<–184	<–426	Severe corrosion

The inaccuracy of such measurements is a result of the presence of water, which increases the negative values without corrosion in steel, as wetting columns and walls give high negative values for potential volts regardless of the degree of steel corrosion. The negative voltage difference underwater in the offshore structures, despite the lack of oxygen, reduces the rate of corrosion.

8.4.4 Electrical Resistivity Measurement

The process of steel corrosion is the process of an electrical and chemical reaction as the electrical resistance of concrete is a major factor in the rate of steel corrosion. The corrosion rate in the steel reinforcement depends on the movement of ions from the anode to the cathode, causing corrosion. The four-probe resistivity operation is used to measure the resistance of the soil to electricity and then make some changes to the system to measure the electrical resistance of concrete.

The resistance of concrete is measured by the four probes by passing electric current between two outside probes and measuring the potential volts between the inner two probes, as illustrated in Figure 8.11. This measures the electrical resistivity of the concrete. Some equipment is needed to make a hole in the concrete to fix the probes, but new equipment puts it on the surface only.

The electrical resistance is affected so much by the moisture in the porous concrete and it is significantly affected by the quality of the content, the water-to-cement ratio in the concrete mix, the curing, and the use of additives. The level of chlorides has less impact on the electrical resistance of concrete.

It must be taken into account during the measurement that there should be greater distance between the probes with the larger sizes of aggregate to avoid the measurement of the electrical resistance of a piece of aggregate only. It is important to measure away from the steel bar, so the readings must be taken perpendicular to the steel reinforcement, and Figure 8.10 illustrates the appropriate position from which to take readings.

It is possible to take the measurement of electrical resistance of the concrete cover with the use of only two probes by making the steel reinforcement act as the first probe with the second probe moving over the concrete surface.

The rate situation of corrosion of the steel can be distinguished through comparison of the results obtained from the electrical resistance of concrete according to Broomfield et al. (1987, 1993). This is demonstrated by the following values:

- $> 120 \text{ K}\Omega \text{ cm}$ low corrosion rate
- $10–120 \text{ K}\Omega \text{ cm}$ low to medium corrosion rate
- $5–110 \text{ K}\Omega \text{ cm}$ high corrosion rate
- $< 5 \text{ K}\Omega \text{ cm}$ very high corrosion rate

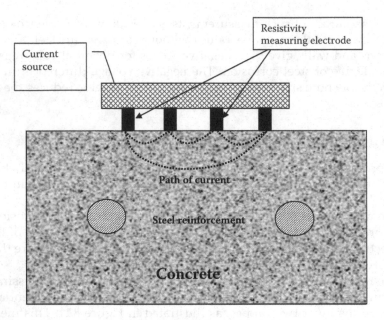

FIGURE 8.10
Function of electrical resistivity measurement equipment.

Many researchers have studied the relationship between electrical resistivity and corrosion rate by using electron probes. The following notes reflect some of their conclusions:

- > 100 K Ω cm   cannot differentiate between passive and active steel reinforcement
- 50–100 K Ω cm   slow corrosion rate
- 50–10 K Ω cm   a medium to high corrosion rate with active steel corrosion
- < 10 K Ω cm   electrical resistivity has no impact on the corrosion process

From the previous relations, this method is considered an effective method of determining the corrosion rate in steel reinforcement. The electrical resistance is expressed in the corrosion rate of steel reinforcement and is a good representation of concrete quality control and the density of concrete, which depends on the compaction method onsite.

According to a study from the University of Liverpool, if the angle of measurement is perpendicular to the steel, the reading error can be reduced to the smallest possible value. It also found electrical resistance when more than 10 K Ω cm cannot distinguish between active and passive protection of reinforcing steel and, if the electrical resistance is 10 K Ω cm or less, the resistance is not effective in the process of corrosion.

8.4.5 Measurement of Carbonation Depth

The carbonation depth is formulated in the propagation of carbon dioxide inside the cover of the concrete, which reduces the concrete alkalinity value. The concrete alkalinity value is expressed by pH values equal to 12–13. With carbon dioxide propagation, a reaction reduces the alkalinity. This will affect the steel surface and damage the passive protection layer around the steel bars, which allows the corrosion process to start. Therefore, it is very important to define the depth of the concrete transformed by carbonation and how far away it is from the steel bars.

This test is performed by spraying the surface of broken concrete or breaking it with special tools to obtain the carbonation depth by phenolphthalein dissolved in alcohol. This solution becomes pink on contact with the surface of concrete with an alkalinity pH value of 12 to 13.5, and the solution turns gray or blue if the concrete's alkalinity pH value is less than 9. In this case, steel bars lose the passive protection layer.

To ensure the accuracy of the test, it is important to make sure that the breaking of the part occur at the same time as testing, whether beams, columns, or slabs. After measuring the carbonation depth of the concrete cover and the distance between the carbonation depth and the steel bar, it is easy to evaluate the corrosion risk of the steel reinforcement.

The best solution is to mix phenolphthalein with alcohol and water to a strength of 1 mg phenolphthalein to 100 mm of alcohol and water (50:50 mixing ratio) or more alcohol than water, if the concrete is completely dry to humidifying the surface with water (Building Research Center, 1981; Parrott, 1987).

If the thickness of carbonation is between 5 and 10 mm, it is going to be necessary to break the passive layer to the steel bar from 5 mm from the presence of color change on the surface of the concrete. Some of the aggregate and some concrete mixtures have a dark color, which makes reading phenolphthalein difficult, and therefore, we must take into account whether the surface is clean when testing. This test must be done in the areas accessible to the breakers, as well as facilitate the work of repairing the part that was broken.

8.4.6 Chloride Test

The chlorides test (AASHTO T260-84, 1984; ASTM D1411-82, 1982) relies on analysis of samples of the concrete powder to determine the quantity of chloride. This test is done by making a hole with a drilling machine inside the concrete and extracting concrete powder out of the hole or through the broken part of the concrete.

Several separate samples must be taken at different depths. The depth of the hole will be varied to increase the accuracy and always taken from 2 to 5 mm. The first 5 mm should be taken from the surface of concrete where the chloride concentration is often very high, particularly in structures exposed to sea water and chlorides.

The holes are made by special devices to collect concrete powder product in the process of drilling the hole. The concrete powder, which will be taken at every drilling depth, will be added to a solution of acid, which will be determined by the amount of the chloride concentration. There are two principal ways of measuring in situ (Quanta Strips) and a method for determining the electron ion (specific ion electron). In the experiments regarding chloride concentration, it is important to have professionally trained workers.

When it is assembled, the readings of chloride concentration at various depths shows the shape of chloride concentration from the surface of concrete into the interior of the concrete. From the general figure, we can determine whether chlorides are within the concrete or from air and environmental factors affecting the concrete from the outside. Figure 8.11 shows the difference between the two cases. It can be seen from Figure 8.11 that the impact of chlorides from the outside is, in the beginning, high and then gradually decreases with depth, until it reaches a constant value.

On the other hand, if the chlorides exist inside the concrete from the concrete mix (salt water and aggregate have high chloride counts), it can be seen from Figure 8.11 that, near the surface, the chloride concentration is minimal and increases with depth until it reaches a constant value.

The chlorides test method is done by melting a sample in acid and calibration work afterward based on the specifications of BS1881, Part 124. In

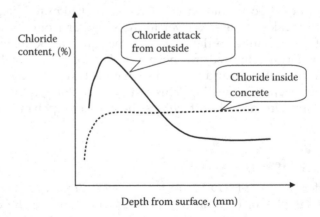

FIGURE 8.11
Comparison between chloride inside the concrete and chloride attack from the outside.

addition to the acid-melting method, this can be done in water following the specifications of ASTM D1411 (1982) and AASHTO T260 (1984).

If chloride concentration exceeds 0.6 of hydroxyl concentration, it causes failure in the passive layer around the steel reinforcement. This ratio is approximately 0.2% to 0.4% chlorides by weight of cement or 1 lb/yd of cubic concrete.

By obtaining the distribution of chloride concentration with depth from Figure 8.11, you can determine the degree of impact of chlorides on the steel reinforcement and whether that impact is from inside or outside the concrete.

8.5 Structure Evaluation Technique

The structure evaluation technique can be best illustrated by applying all the tools and techniques we have discussed to the following real-world examples of structural evaluation.

8.5.1 Case Study One: Structural Evaluation

Our first case study involves an administration building located near the red sea in Egypt. The building is a steel structure constructed around 1975 and consists of two stories. The ground floor is a waiting area for employees to wait for their flights and the first floor is an administration office consisting of an aviation and airport control room. It is a steel structure composed of a beams and columns system, with a reinforced concrete slab on a steel corrugated sheet.

This building is located about 200 m from the Red Sea. The first impression upon visual inspection is that there has been no maintenance performed for a long time, as the corrugated sheets outside are corroded to 300 mm above-ground. The painting condition is generally good, but not the original structural painting, so there has been periodic maintenance. Discussions with the owner, a very important step in any assessment, yield the information that there has been no maintenance for approximately 5 years due to a shortage of funds.

Near the bathroom, there are signs of water on the false ceiling, so this is inspected in detail. Upon removal of the false ceiling, it is found that the corrugated sheet of the floor beneath the first floor is corroded, as shown in Figure 8.12.

The challenge is to define whether the corroded corrugated sheet is load-bearing or not. If it is, it needs to be replaced. For this type of building, replacing this part involves removing the slab and fixing a new composite section.

FIGURE 8.12
Corrosion in a corrugated sheet.

This process requires a special contractor and is very expensive. Destroying the building and rebuilding may be a better solution from an economic point of view.

On the other hand, if it is not load-bearing, the corrugated sheet may be used as shuttering and after pouring the concrete, there is no need to remove it. This method was common during that period. The engineering and construction company was a specialized one from the United States.

The drawing and specifications no longer exist, so it is necessary to collect information to confirm the structural system and whether either of these two methods of construction was used. If the slab is a composite section, the steel reinforcement will be light and there will always be a steel wire mesh for shrinkage. If it is only shuttering, there will be steel bars sufficient to carry the load, so it will have been designed as reinforced concrete slab. By collecting information from the previous engineers that reviewed the design and others who supervised the construction of the building and by making a hole to inspect the steel bars, it is confirmed that there are steel bars 13 mm apart, and therefore, this corrugated sheet is not load-bearing but was used as shuttering. As a result, the building may be repaired and not demolished.

In summary, evaluating thoroughly and do not jump to any conclusions from first inspection. Collect data from all available sources. Even if the information does not seem valuable, it may improve the data you already have. This decision is critical; a building will either be destroyed or live another 30 years. You can imagine the consequences if you make a wrong decision.

8.5.2 Case Study Two: Structural Assessment

There are cracks on the ring beam around steel tanks carrying oil mixed with a high percentage of water. Fortunately, the tanks, constructed in 1983,

FIGURE 8.13
(a) Cracks in ring beam. (b) Spalling of concrete cover. (c) Reduction in stell cross-sectional area.

are not located in an area of high environmental importance. Upon observation, a leak is discovered in the tanks containing hydrocarbon and water that is affecting the reinforced concrete beam. As shown in Figure 8.13a, the corrosion is parallel to the stirrups, every 200 mm, and is also on the corner steel bars. The cause and the shape of the cracks reveal that there is corrosion on the steel bars. Figure 8.13b and c shows that there are some areas with spalling of the concrete cover and the reduction of the steel bar cross-sectional area is more than 20%; therefore, it needs complete repair. From a structural point of view, the steel bars have a tension force, as the ring beam is always under tension.

8.5.3 Case Study Three: Structural Assessment

As we mentioned, the first step of assessment is always to collect data. This site is located in the desert with no great effects from temperature. There are two reciprocating gas lift compressors on foundations. There was one originally; the second was installed after one year with similar load and design dimensions. After installing the second machine, the site engineers observed a large vibration on the administration building. This administration building, constructed around 1975, is about 100 m away from the machine. The cracks are presenting between the reinforced concrete columns and the bricks.

After collecting data, it becomes apparent that the cracks are due to resonance: the vibration wave is transferring from the machine to the building. There may be rocks underneath the foundations. The civil and structural solution is to put rubber around the tanks or use some other method to cut the wave propagation to the building. In cooperation with the mechanical

FIGURE 8.14
Concrete foundation cracks.

engineers, it was decided to change the machine speed by reducing its revolution per minute (rpm) in order for one of the machines to avoid the resonance. This solution is the right one, as the vibration was decreased and the repair of the cracks was successfully completed. It should be noted that the solution to a problem does not always rely solely on the structural engineer, especially in the oil and gas industry.

8.5.4 Case Study Four: Structural Assessment

In this case study, we will address the foundation shown in Figure 8.14. Before we are able to discover a solution, we must first collect data. This plant is near the Mediterranean Sea and, after a review of the soil data, based on our experience, we determine that this area has a bad soil type and find that the soil report mentions clearly that the settlement is predicted to be about 70 mm along 20 years. Moreover, the plant is about 100 m from the sea and the relative humidity is very high.

The probability of settlement and of a corrosion problem is very high. The foundation that is shown in the photo is one of three foundations using a steel I-beam section to carry a manifold. The problem is that there is settlement in this foundation and no tie beam. It seems clear that the solution is to increase the foundation size and connect them by a beam, if the steel is in good condition. If it is not in good condition, we may need to install another foundation.

Now, let us stop and reflect on whether these supports are really necessary, as they do not seem to be in the original drawings. After performing the piping stress analysis, it is found that these supports are not needed and that these foundations require only a minor change in the piping configuration. So, again, do not run too quickly toward a structural solution.

8.6 Structural Assessment

The tests that have been discussed in the preceding sections are only tools to assist in the process of evaluating structures. The main factor in this process is the experience of the engineer examining the structure. The evaluation of the building will determine whether the structure should be subject to repair or if the repair process will be nearly the cost of demolition and reconstruction. The decision-making process will be based on the assessment of the building in terms of safety and repair cost; therefore, the evaluation of the building is vital.

The scope of the repair work depends on the accuracy of the test results and the experience of the skilled labor conducting the tests. When determining the method of repair and the necessary cost value of the repair process, it is important not to forget the methods required to protect the structure from corrosion after repair.

References

AASHTO Standard T260-84. 1984. *Standard Method of Sampling and Testing for Total Chloride Ion in Concrete Ratio Materials*. Washington, DC: American Association of State Highway Transportation Officers.

ACI Standard 318-97. 1997. Standard building code requirements for reinforced concrete, *ACI Manual of Concrete Practice*. Farmington Hills, MI: ACI.

ACI Standard ACI222R-89. 1990. *Corrosion of Metals in Concrete*, Report by ACI Committee 222. Farmington Hills, MI: ACI.

ACI Standard 228-89-R1. 1989. In-place methods for determination of strength of concrete, *ACI Manual of Concrete Practice*. Farmington Hills, MI: ACI.

Alldred, J. C. 1993. Quantifying the losses in cover-meter accuracy due to congestion of reinforcement. *Proceedings of the Fifth International Conference on Structural Faults and Repair*, Vol. 2. Edinburgh: Engineering Techniques Press, pp. 125–130.

ASTM Standard C876. 1991. Standard Test Method for Half-Cell Potentials of Uncoated Reinforcing Steel in Concrete, *ASTM Annual Book of Standards*. West Conshohocken, PA: ASTM International.

ASTM Standard D1411-82, 1982: Standard Test Methods for Water Soluble Chlorides Present as Admixes in Graded Aggregate Road Mixes, *ASTM Annual Book of Standards*. West Conshohocken, PA: ASTM International.

ASTM Standard C 42-90m. 1990. Standard Test Method for Obtaining Strength and Testing Drilled Cores and Sawed Beams of Concrete, *ASTM Annual Book of Standards*. West Conshohocken, PA: ASTM International.

Broomfield, J. P., Langford, P. E., and McAnoy, R. 1987. Cathodic protection for reinforced concrete: Its application to buildings and marine structures. In *Corrosion of Metals in Concrete, Proceedings of Corrosion/87 Symposium*, Paper 142, Houston, TX: NACE, pp. 222–325.

Broomfield, J. P., Rodriguez, J., Ortega, L. M., and Garcia, A. M. 1993. Corrosion rate measurement and life prediction for reinforced concrete structures. *Proceedings of the Fifth International Conference on Structural Faults and Repair*, Vol. 2. Edinburgh: Engineering Techniques Press, pp. 125–130.

British Standard Institution. BS 1881-6. 1971. *Methods of Testing Concrete, Analysis of Hardened Concrete*. London: BSI.

British Standard Institution. BS 8110. 1985. *Structural User of Concrete Part 1: 1985 Code of Practice for Design and Construction*. London: BSI.

British Standard Institution. BS 1881-5. 1983. *Methods of Testing Concrete Part 5: Methods of Testing Hardened Concrete for Other Than Strength*. London: BSI.

British Standard Institution. BS 1881-120. 1983. *Part 120 Method for Determination of the Compressive Strength of Concrete Cores*. London: BSI.

Bungey, J. H., ed. 1993. Non-destructive testing in civil engineering. *Proceedings of the Fourth International Conference on Non-Destructive Testing in Civil Engineering*, Liverpool, April 8–11, 1997.

Cady, P. D., and E. J. Gannon. 1992. *Condition Evaluation of Concrete Bridges, Relative to Reinforcement Corrosion, Vol. 1, State of the Art of Existing Methods*, Washington, DC: National Research Council, SHRP-S-330.

Concrete Society. 1984. Technical Report 26: *Repair of Concrete Damaged by Reinforcement Corrosion*. Crowthorne, UK: The Concrete Society.

Concrete Society. 1987. Technical Report 11: *Concrete Core Testing for Strength*. Crowthorne, UK: The Concrete Society.

Concrete Society. 2000. Technical Report 54: *Diagnosis of Deterioration in Concrete Structures*. Crowthorne, UK: The Concrete Society.

El-Reedy, M. A. 2009. *Advanced Materials and Techniques for Reinforced Concrete Structure*. Boca Raton, FL: CRC Press.

Parrott, L. J. 1987. *Review of Carbonation in Reinforced Concrete*. Slough, UK: British Cement Association.

9

Methods of Protecting Foundations from Corrosion

9.1 Introduction

The minimum requirements of various codes are often insufficient to ensure the long-term durability of reinforced concrete in severe exposures, such as those found in marine splash zones such as jetties and harbors. Some of the newer structures such as oil and gas projects built in marine areas, but not in splash zones, are experiencing corrosion problems due to airborne chlorides. Marine structures in warmer climates prevalent in the Middle East, Singapore, Hong Kong, South Florida, etc., are especially vulnerable because of the high temperatures, which not only increase the rate of chloride ingress, but also the corrosion rate once the process is initiated.

In this section, a brief description of supplemental corrosion protection measures is given for structures especially at risk. Good-quality concrete is considered the primary protection method, but various combinations of this with supplements are necessary to reach the desired design life of the structure.

One of the most effective means to increase corrosion protection is to extend the time until chloride or a carbonation front reaches the steel reinforcement. The minimum code requirements allowing the use of concrete with a water/cement ratio (w/c) of less than 0.45 and concrete cover thickness more than 40 mm are totally inadequate for the structures and environmental conditions outlined above, if a design life of 40 years or more is expected. In many applications, designs complying with minimum code requirements would not provide as much as 10 years of repair-free service.

Prevention is better than cure. This idea is well known and is a good expression of the benefits of protecting reinforced concrete structures, as protecting a concrete structure is easier and less expensive than repairing one. The reason is that the cost of repair and renovation of some reinforced concrete members is very expensive, such as the case of foundation repair. In reality, the protection of the structure from corrosion is the protection of the investment along the structural lifetime.

In this chapter, our aim is to discuss the different methods used internationally to protect reinforced concrete structures from corrosion. Recently, there has been more research and development toward providing economic and effective methods to protect the steel reinforcement, as this subject has billion dollars of investment in construction worldwide.

The first method of protection is to perform good quality control in design and construction and take into consideration all precautions to avoid cases of corrosion, as stated in the codes and specifications of different countries. These specifications vary with different weather conditions the structure is exposed to and the function of the building, such as an offshore underground structure.

The second line of defense from corrosion are the external methods such as using galvanized steel bars, epoxy-coated steel bars, stainless steel, or using additives added to the concrete during the pouring process. Other methods include the use of external membranes to prevent water permeability or cathode protection. All of these methods have advantages and disadvantages. In this chapter, all these methods will be discussed and provide the ways of application for each type of protection.

9.2 Corrosion Inhibitor

There are two types of corrosion inhibitors. The first type is called an anodic inhibitor and the second type is the cathodic inhibitor. The anodic inhibitor depends on a passive protection layer on the steel reinforcement bars. The cathodic protection is based on preventing the propagation of oxygen in the concrete. It has been observed that the most effective protector is the anodic inhibitor and it is commonly used in practice. We will explain these protections to clarify the advantages and disadvantages of both.

9.2.1 Anodic Inhibitors

The most common anodic inhibitor is calcium nitrate. It is well known as it is compatible with the process of pouring concrete at the site where there is no adverse impact on the properties of concrete, whether it is fresh or in the hardening state.

Other types, such as sodium nitrate and potassium nitrate, are very efficient at preventing corrosion but are not used with aggregates containing alkaline, as they react with cement and cause extensive damage to the concrete. Broad calcium nitrate has been widely used since the mid-1970s (Arnon et al., 1997). The use of calcium nitrate accelerates the concrete setting time. Broomfield (1995) mentioned that retarder must be added at the mixing plant.

The mechanism by which these compounds inhibit corrosion is associated with the stabilization of the passivation film that tends to be disrupted when chloride ions are present at the steel level, largely due to the interference with the process of converting the ferrous oxide to the more stable ferric oxide.

We note that the anodic materials are used in concrete exposed to chlorides directly, as in contact with sea water, where the corrosion inhibitor reacts with chlorides and increases the amount of chloride concentration necessary to cause corrosion based on the tests described in Table 9.1.

The amount of calcium nitrate to add is determined based on the amount of chloride exposed to concrete; this can be done in practice or through knowing the quantity of chloride from previous experience. The addition of corrosion inhibitors does not diminish the importance of attention to the quality control of concrete and maintaining the appropriate thickness of the concrete cover. In the case of high-strength, high-quality concrete, choosing and constructing the right concrete cover thickness and appropriate density of concrete, according to specifications, may obviate the need for a corrosion inhibitor for 20 years—although it is used when structures are exposed to chlorides directly (e.g., offshore structures).

9.2.2 Cathodic Inhibitor

A cathodic inhibitor is usually added to the concrete during mixing. There is a new type that is painted on the concrete surface after hardening and it propagates to the steel bars through the concrete porosity and provides an isolation to reduce the quantity of oxygen, as it is the most important driver of the corrosion process.

There are many tests performed on the corrosion inhibitor based on ASTM G109-92 and their purpose is to define the effects of the chemical additives on the corrosion of the steel reinforcement embedded in concrete. The tests indicate that the corrosion inhibitor enhances the protection of the steel bars rather than only controlling the concrete quality. The tests also indicate that a cathodic inhibitor is less effective than an anodic inhibitor.

TABLE 9.1

Amount of Required Calcium Nitrate to Protect Steel
Reinforcement from Corrosion due to Chlorides

Calcium Nitrate Amount (kg/m³, and 30% solution)	Amount of Chlorides Ions on Steel (kg/m³)
10	3.6
15	5.9
20	7.7
25	8.9
30	9.5

From these tests, we find that it is essential to add a large quantity to the concrete mix to obtain more effective corrosion protection for the reinforced steel bars. On the other hand, adding cathodic inhibitors such as Ameen, phosphates, and zinc retard the setting time by significant amounts especially when we add a large quantity. So, when we decide to use cathodic inhibitor, we must take the setting time into consideration.

One can conclude from this information that the anodic inhibitor is more effective than the cathodic inhibitor, so it is customary to use the anodic inhibitor. On the other hand, if we use the cathodic inhibitor, we must increase the setting time.

9.3 Epoxy Coating of Steel Reinforcement

It is important to paint the steel bars using certain types of epoxies that can protect the steel from corrosion. This method yields positive results, especially in steel exposed to seawater. This study has been performed by the Federal Highways Association (FHWA), which has been evaluating the use of epoxy to coat steel reinforcement exposed to chloride attack.

Other research, such as that done by Pike et al. (1972), Cairns (1992), and Satak et al. (1983), demonstrates the importance of the use of painted steel reinforcement. Epoxies have been used for painting reinforced steel for bridges and offshore structures since 1970.

From a practical point of view, there are some shortcomings that must be avoided through the use of some precautions, such as avoiding any friction between the bars, which would affect the coating layer and reduce effectiveness. Also, it is difficult to use methods for measuring the corrosion rate such as polarization or half cell, so it is difficult to predict the steel corrosion performance or measure the corrosion rate.

Painting the steel reinforced bars is done extensively in the United States and Canada and, over 25 years, more than 100,000 buildings used coated bars, which is equal to 2 million tons of epoxy coated bars. The coated steel bar must follow ASTM A 775M/77M-93, which set the allowable limits as follows:

- The coating thickness should be in the range of 5–12 mils, where 1 mil = 1/1000 in.
- Bending of the coated bar around a standard mandrel should not lead to formation of cracks in the epoxy coating.
- The number of pinhole defects should not be more than 6/m.
- The damaged area on the bar should not exceed 2%.
- These deficiencies cited by the code are the result of operation, transportation, and storage.

There are some precautions that must be taken into consideration in these phases to avoid cracks in the paint. Andrade et al. (1994) and Gustafson and Neff (1994) define methods of storage, steel reinforcement bending, carrying the steel, and pouring concrete. Using painted steel reinforcement will reduce the bond between the concrete and steel; therefore, we must increase the development length of steel bars to overcome this reduction in bond strength. According to the ACI code (ACI Committee 318), the increase of the development length should be from 20% to 50%.

ACI Committee 318 (1988) found that when painting steel bars, it is necessary to increase the development length by 50%, in the case of concrete cover less than 3 times the steel bar diameter or a distance between the steel bars that is less than 6 times the bar diameter. In other circumstances, increase the development length by 20%.

The Egyptian code does not take painted steel into account. The study conducted by El-Reedy et al. (1995) found that the equation mentioned in the Egyptian code for the calculation of the development length can be applied in the case of painting steel bars with epoxies without increasing the development length. However, the thickness of the paint coating must be governed by not more than the value cited by the American code, which is 300 μm. Under any circumstances, it is strictly prohibited to paint mild steel, as the bond strength in the smooth bars is a result of friction and you will lose all the bond strength if they are coated.

Care must be taken not to increase the thickness of the paint coating, which should not in any case be more than 12 mils. Some research states that when steel reinforcement with a coating of 14 mils was used for the main steel reinforcement in concrete slabs, testing found extensive cracks, which led to separation between steel bars and the concrete.

There is a comparison between steel bars without coating and other bars coated by epoxy where both are exposed to water from the tap and then placed in water containing sodium chloride and sodium sulfate. The corrosion rate between them was compared and Table 9.2 shows the rate of corrosion for the coated and uncoated steel bars. Table 9.2 shows that the corrosion rate is much slower in the coated steel bars than in the uncoated steel bars.

This method is cheap and widely used by workers and contractors in North America and the Middle East. The coating of reinforcement steel by epoxy does not negate the necessity of using high-quality concrete for maintaining

TABLE 9.2

Corrosion Rate for Coated and Uncoated Steel Bars

Case	Corrosion Rate (mm/yr)	
	Tap Water	$NaCl$ 1% + Na_2SO_4 0.5%
Coated	0.0678	0.0980
Uncoated	0.0073	0.0130

the reasonable concrete cover. In some countries, the steel manufacturers can deliver coated steel bars. This is a better option than coating the bars onsite, as you cannot control the thickness of the coating without special tools to measure the coating thickness.

9.4 Galvanized Steel Bars

Some researchers in the United States recommend the use of galvanized steel in reinforced concrete structures. An FHWA report recommends that the age of galvanized steel should be limited to 15 years in the case of high-quality concrete under attack by chlorides, according to the research of Andrade et al. (1994). A galvanized bar is used effectively in structures exposed to carbonation. Accelerated depletion of the galvanizing occurs if galvanized bars are mixed with ungalvanized bars.

The process of galvanization involves the addition of a layer of zinc. A steel rod is immersed in a zinc solution at a temperature of 450°C and then cooled. As a result, a zinc coating forms on the bars consisting of four layers. The outer layer is a layer of pure zinc and the other layers are mixtures of steel and zinc.

Like most metals, zinc will corrode with time. The rate of corrosion under different weather conditions can be calculated from the amount of corrosion and the amount of time that has passed. The relationship between the thickness of the zinc layer and its lifetime is presented linearly in Figure 9.1.

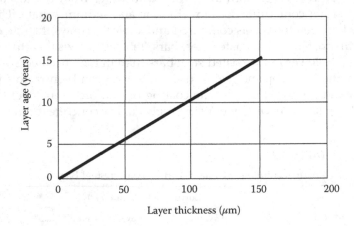

FIGURE 9.1
Relation between zinc layer thickness and expected life.

The galvanized coating must be tested after the bending of the steel bars. Care must be ensured when welding. The maximum zinc cover is about 200 μm and must follow ASTM A767/A767A M-90.

The stability of zinc essentially depends on the stability of the pH value of the concrete. The passive protection layer forms at a pH of 13.3, but when that value increases to more than 13.3, the zinc will dissolve until it vanishes completely. This situation is illustrated in Figure 9.2.

We find that the process of galvanization is totally dependent on the proportion of pH in the concrete pores and the alkalinity of the Portland cement. If the galvanization pH is equal to 12.6, the reduction in coating thickness coverage will be 2 μm. If the pH is equal to 13.2, the shortage in the galvanized layer will reach 18 μm, with the observation that this is happening before the passive protection layer.

Through laboratory tests using different types of Portland cement with different alkalinities, it has been determined that, if you assume that the rate of corrosion is equal and the thickness of the cover is 60 μm, the lifetime will be about 200 years in the case of a low-alkalinity cement and 11 years in the case of a high-alkalinity cement. This is according to building research done in the UK in 1969.

Therefore, it is noted that the thickness of the cover must be more than 20 μm. The American specifications (ASTM 767/A767 M-90) for galvanized steel bars identifies two types of galvanization: Class I and Class II, which have a cover thickness of more than 1070 and 610 g/m^2, respectively, which is the equivalent of 85 and 150 μm, respectively.

In 1969, Building Research, UK, recommended a maximum zinc cover thickness of about 200 μm, as an increase in the thickness of cover will reduce the bond between the steel and concrete (Building Research Establishment, 1969). If the pH is between 8 and 12.6—note that zinc is more stable in this case than in the case of an increasing pH value—the zinc gives high efficiency even when exposed to carbonation, because this reduces the pH, as we have discussed before.

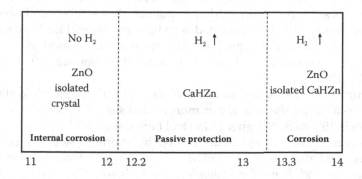

FIGURE 9.2
Galvanized steel behavior in solution with pH from 11 to 14.

However, in the case of attack from chlorides, the galvanization will not prevent corrosion, but will significantly reduce the rate. Galvanization can increase the value of chlorides necessary to cause corrosion from 150% to 200%. This will increase the time required to cause corrosion by four times, according to a study by Yeomans (1994). The specifications of the Institute of Equatorial Concrete were released in 1994 and provide guidelines that should be followed when using galvanization. These include:

- Galvanization increases the protection of steel corrosion but does not compensate for the use of concrete with appropriate concrete cover.
- We must avoid putting galvanized steel with nongalvanized steel, as this will quickly cause the galvanic layer to corrode.
- We must examine the galvanic layer of the steel reinforcement after bending and fabrication and increase the bending diameter.
- Precautions should be taken when using welding with galvanized steel

9.5 Stainless Steel

Stainless steel is not used in industrial structures, but it is recommended for the foundations of critical equipment under severe environmental conditions. *Critical equipment* may be defined as unique equipment that will cause a plant shutdown if it should stop.

A steel bar coated with a layer of stainless steel with a thickness from 1 to 2 mm is considered more expensive in comparison with other steel bars. The same precautions mentioned for galvanized steel should be taken into account for stainless steel. For example, do not put stainless steel or steel bars coated with stainless steel next to uncoated steel bars, as this will lead to corrosion.

In 1995, the design of some balconies used as stainless steel reinforcing bars adjacent to normal steel and the process of erosion was quick, as found by Miller (1994).

The high cost of stainless steel limits its applications. Stainless steel costs about 8 to 10 times the cost of commonly used steel. The cost of using stainless steel is 15% to 50% higher than steel bars coated with epoxy.

The specifications of steel bars coated by stainless steel are now being prepared and discussed by ASTM under the title *Standard Specifications for Deformed and Plain Clad Stainless Steel Carbon Steel Bars for Concrete Reinforcement*, which will provide specifications for the installation of stainless steel–coated bars and will follow specifications A 480/480 M, for types

304, 316, or 316 L. Three levels of yield stress are defined: 300, 420, and 520 MPa.

In the commercial market, there is reinforcing steel covered by stainless steel with a yield strength of about 500 MPa and maximum tensile stress of about 700 MPa. The available diameters are 16, 19, 22, 25, and 32 mm, and some factories are now manufacturing bars with diameters of 40 mm.

The use of stainless steel is not recommended with welding, but if it is necessary, the welding should be done by Tungsten Inerges and the welding wire should be of the same materials as the stainless steel.

9.6 Fiber Reinforcement Bars

In the past decade, different studies have applied extensive theoretical and practical research to the replacement of steel reinforcement bars with filler-reinforced polymer (FRP), as this material is not affected by corrosion and so will be very economical along a structure's lifetime.

Figure 9.3 illustrates that rolled steel sections made from glass-reinforced polymer (GFRP) are more expensive than traditional steel sections, but there are no associated costs, such as for painting and maintenance work, as GFRP is not affected by corrosion. Moreover, GFRP is lightweight with a density of 2.5 g/cm^3, compared with the density of steel, which is about 7.8 gm/cm^3.

From its weight, which is very low, and its resistance, which is higher than steel, GFRP will be worth using for adding additional floors to existing

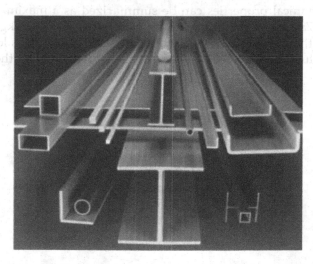

FIGURE 9.3
GFRP structural sections.

buildings. Now, it is used to replace traditional steel gratings, handrails, and ladders on offshore structure platforms in the oil industry. For all the reasons discussed here, bars of this material are being developed and manufactured as an alternative to steel bars.

Figure 9.4 illustrates how GFRP bars are manufactured in a certain way so that protrusions on the surface allow it to bond with concrete. It is now manufactured and used on a small scale while application and marketing research is conducted. There is much ongoing research into the performance of these bars over time.

As a result of limited production, the cost of these bars is very high; however, the cost of maintenance is very low. Moreover, it will reduce the value of the dead load to the structure, which will reduce the total cost of the structure.

Some structures in Canada were built using bars of GFRP, such as a marine structure at the port of Wharf. This structure is a thick wall of pre cast concrete designed to have a maximum strength of 450 kg/cm². The structure is exposed to different temperatures ranging from 35°C to –35°C.

GFRP has also been used to build bridges in Quebec that are regularly exposed to salt, as it is used to dissolve ice when it forms. One such bridge was built in Antherino and another on Vancouver Island. Vancouver has been asked to perform tests on all of these facilities exposed to different weather conditions. Tests conducted include sampling and x-ray examination. These structures are between ages 5 and 8 years.

The results of that research found no impact from the various factors on the GFRP bars and found that the cycles of wet and dry did not affect the bars. By microscopic examination, it was found that a strong bond exists between the bars manufactured from GFRP and the concrete.

The mechanical properties can be summarized as a maximum strength of about 5975 kg/cm² and a maximum bond strength of 118 kg/cm², but it was found that the modulus of elasticity was about five times less than the steel modulus. To overcome problems that might arise from the creep, the

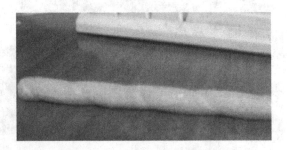

FIGURE 9.4
Illustration of the shape of GFRP reinforcement bars.

American specifications ACI 440 suggest that tensile stress must not be less than 20% less than the maximum tensile strength.

9.7 Protecting Concrete Surfaces

A concrete surface needs to be protected from the permeability of water and the propagation of chloride, such as in offshore structures exposed to sea waves or bridge decks where salt is used to melt ice. It is important to prevent permeability to protect the concrete surface from carbonation and to prevent the propagation of carbon dioxide. The other factors that increase the corrosion rate are humidity and oxygen, which should also be prevented from propagating inside the concrete to decrease the corrosion rate.

There are two popular methods of protecting concrete surfaces. They are using liquid materials applied by spraying or brush-painting and using sheets and membranes of rubber, plastic, or textile immersed in bitumen. The latter is usually used in small projects.

Isolation from water is also a problem for epoxy-coated reinforcement steel. As we have discussed, it is critical to prevent any damage or erosion to the coating due to external mechanical factors such as tearing, exposure to a heat source, or other factors. It is important to follow any required specifications during execution. There are some specifications that define execution methods additional to supplier recommendations.

There are different types of surface protection that use liquid materials, but the function is the same: to prevent the propagation of water inside the concrete through concrete voids, as shown in Figure 9.5.

9.7.1 Sealers and Membranes

Sealers and membranes have traditionally been used for providing protection to concrete structures exposed to severe chemical attacks. However,

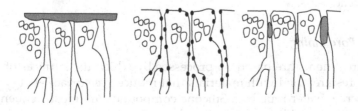

FIGURE 9.5
Sketch presenting different coatings.

with the increase in frequency of durability problems generated by corrosion of steel in regular reinforced concrete structures, they are being increasingly used as a means of mitigating this durability problem. Membranes and sealers can provide protection in the following two ways:

1. Eliminating or slowing down the penetration of chlorides and carbonation to maintain steel passivity
2. Reducing moisture movement into the concrete to keep it dry and slow the propagation of corrosion reactions

Sealers and membranes can be classified into several types and each type represents a family of materials with a different chemical composition. We will now discuss each of these different types.

9.7.1.1 Coating and Sealers

These consist of continuous films applied to the concrete surface with a thickness in the range of 100–300 mm. The films are composed of a binder and fillers. The performance depends on the compounding of the fillers. Compositions having a similar binder may be quite different in their performance.

The coatings or sealers are obtained in a liquid form, which is brushed or sprayed onto the concrete surface in an operation that is very much like painting. The successful performance of coatings and sealers depends not only on the quality of the materials used, but also on the application. The concrete surface should be clean and sound. Weak and cracked concrete should be removed, holes should be filled, and, if necessary, a leveling coat should be applied.

If the membrane applied is polymeric, the surface should be dry. Silane treatments should be done on wet concrete. This is needed to facilitate better penetration of the liquid compound into the pores. Coatings should usually be applied in two layers to obtain a continuous film without pinholes. The service life of the membranes and sealers ranges between 15 and 20 years. Thus, continuous maintenance and follow-up is necessary, including additional treatments.

9.7.1.2 Pore Lining

These treatments are based on processes by which the surfaces of the concrete pores are lined with materials that reduce the surface energy, making the concrete water repellent. Silicone compounds are most frequently used for this purpose. Silicone resin can be dissolved in an organic liquid that deposits a film of the resin on the pore surface as it evaporates.

9.7.1.3 Pore Blocking

These treatments are based on materials that penetrate into the pores and react with some of the concretes constituents. The resulting products are insoluble and are deposited into the pores to block them. The most common materials used for this purpose are liquid silicates and liquid silico-fluorides.

Experience tells us that a membrane used to prevent water permeability has a lifetime of about 15 years and, after that, must be replaced. There are some problems with proactive maintenance, as in the case of bathrooms.

The maintenance is very expensive, as it requires removing all of the tiles and plumbing accessories and starting from scratch. In residential buildings, maintenance will usually be a corrective action, which is prone to failure. In any mature, reinforced concrete structure, most of the problems are due to isolation, as the main steel of the slab is lower and the primary signs of corrosion will be cracks parallel to the main steel bars. These signs will be visible on the lowers floors, but the user of the bathroom will not see them. Therefore, repairs will not be made until there is falling concrete cover and serious corrosion problems in the steel bars.

When using a membrane, care must be taken in execution to avoid any tear in the membrane. Bad executions where specifications and supplier recommendations are not followed will lead to literally paying for nothing, as the membrane is like William Shakespeare's famous line, "To be or not to be." Any defect in the membrane will cause permeability of water and start corrosion.

There are two well-known methods of application of isolations for foundations, swimming pools, or tanks. These methods isolate surfaces directly exposed to water and surfaces without direct exposure to water. These types of isolation must be executed by a professional contractor with experience in this type of work, as the chemicals and membranes require competent workers and a specialized contractor that you trust. Recently, some research found that there are problems with using isolation by membrane in cases of cathodic protection, as there are some gases that accumulate in the anode and need to escape, but the membrane will prevent this and thus the completion of the electrical circuit.

9.7.2 Cathodic Protection by Surface Painting

The corrosion of steel in concrete occurs as a result of attacking chlorides, carbonization of the concrete surface. Incursion into the concrete that reaches the steel will reduce the concrete alkalinity and the presence of moisture and oxygen will drive corrosion until a complete deterioration of the concrete occurs. This has been discussed in Chapter 2.

There are some materials that are painted onto the surface of concrete to saturate it. These materials will propagate through the concrete to reach the steel at speeds from 2.5 to 20 mm/day, by capillary rise phenomena, such as water movement, by penetrating the concrete with water.

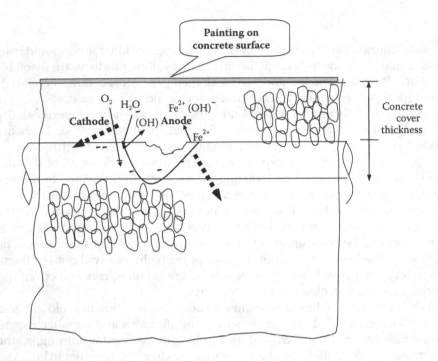

FIGURE 9.6
Painting concrete surface to provide cathodic protection.

Therefore, when those materials reach the surface of steel, an isolation layer around the steel bar surface will reduce the oxygen in the cathode area to the surface on the cathode and reduce the melting of steel in the water in the anode area, thereby delaying the process of corrosion and reducing its rate. Figure 9.6 presents the influence of this cathodic protection coating on the surface on protecting the steel bars.

These new advanced materials are used in the case of new construction, existing structures where corrosion of the steel bars has started, or, in cases where the presence of corrosion is clear and require complete repair to the damaged concrete surface and surface-painting.

9.8 Cathodic Protection System

This method is the most expensive method of protection; it is used usually in the protection of pipelines in the petroleum industry. It is intended not for use in reinforced concrete structures recently, but for special structures because of its higher cost and the need for special studies, design, execution, and monitoring.

Cathodic protection, if applied properly, can prevent the corrosion of steel in concrete and stop corrosion that is already in progress. It accomplishes this by making the steel bar a cathode through the use of an external anode. Electrons are supplied to the reinforcing bar from the anode through the ionically conductive concrete. The current supplied should be sufficiently high so that all of the local cells are inhibited and the entire steel surface becomes anodic. The external current can be supplied by connecting the steel to a metal that is higher in the electrochemical series (e.g., zinc). It serves as the anode relative to the cathodic steel. In this method, the anode gradually dissolves as it oxidizes and supplies electrons to the cathodic steel. This type of cathodic protection is called a *sacrificial anode*.

An alternative method for cathodic protection is based on supplying electrons to the reinforcing steel from an external electrical power source. The electrical power is fed into an inert material that serves as the anode and is placed on the concrete surface. This method is referred to as *impressed current anodic protection*. The anode is frequently called a *fixed anode*.

9.8.1 Cathodic Protection

The principle behind the use of a sacrificial anode was discovered in 1824 by Sir Humphrey Davey. The discovery was used to protect the metal parts of boats completely submerged under the water from corrosion.

At the beginning of the twentieth century, the use of that technology for the protection of pipelines buried underground began when it was discovered that the soil resistance to electricity allows very high cathodic protection with constant direct current.

Cathodic protection is also used in the construction of modern structures with higher importance levels and special nature. The first practical application in the area of reinforced concrete is in a bridge on the highway in a mountainous area north of Italy. Then, the use of cathode protection is more widespread with the development of research and technology. As stated by Broomfield (1995), after a survey of consultants in England who work in the area of cathode protection of concrete structures, cathodic protection has been established to protect against cathodic disabilities of about 64,000 m² in about 24 structures in Spain, the United Kingdom, and the Middle East.

The largest manufacturer of anodes has supplied enough anodes to cover about 400,000 m² around the world, including a garage for 60 cars and 400 different forms of bridges and tunnels.

Cathodic protection is used in certain types of constructions because their high cost or specific nature of construction requires a system of monitoring. It is used most often in structures exposed to chlorides that exist inside the concrete mix or because of penetration of chlorides within the concrete as a result of surrounding environmental conditions.

Chlorides influence on concrete has a special nature in that when it starts, corrosion results. It is necessary to completely remove the part of the concrete that is contaminated by chlorides. Sometimes the work of repair is very difficult and involves structural issues.

Such repair in a case where chlorides are in the concrete mix is considered impossible. Experience has found that the limited availability of electrical protection is more effective in stopping the process of corrosion, as mentioned in some previous studies, than the traditional method in the case of concrete pollution by chlorides.

The U.S. FHWA has stated that the only method of repair that has been proved to stop corrosion in concrete bridges with salts is protection by electrical current, irrespective of the content chlorides concrete. The use of cathode protection specifically to prevent the corrosion of steel reinforcement in concrete or to stop a corrosion process that has already begun relies on making the steel a continuous cathode through the use of an external anode. This is illustrated in Figure 9.7.

As seen in Figure 9.7, the form on the surface with the steel part will be the cathode with negative electrons and the other part will work as the anode and, from this, the corrosion process is formulated as explained in Chapter 4. The electrons will be generated on the steel reinforcement surface by placing the anode on the concrete surface. In this case, the concrete will be the electrical conductive between it and the steel reinforcement, so the cathodic protection is formulated on the steel reinforcement. As shown in Figure 9.7, this method of cathodic protection is called sacrificial protection and the anode in this case is called a sacrificial anode.

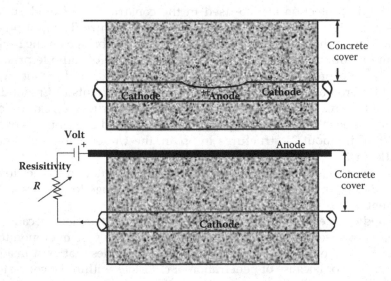

FIGURE 9.7
Cathodic protection method.

There are two main methods for cathodic protection. The first method depends on using a sacrificial anode, as discussed, and is called *sacrificial* protection. In this case, the anode will be zinc, and the zinc will corrode instead of the steel reinforcement. This method is used in submerged structures because the electron movement will be small, and as will the potential volts between the two materials, so the cathodic protection can be maintained for a long time.

The second method of cathodic protection generates electrons on the steel reinforcement in concrete by using an outsource of electricity. This source will be conducted by the anode, which will be put inside the concrete. This is called a *fixed anode*. This process is shown in Figure 9.8.

An example of a fixed anode is a wire mesh inserted into the concrete at the surface, where it works as an anode. The conductivity among the anode, the steel reinforcement, and the batteries is achieved by using cables as shown in Figure 9.8. The source of energy will be normal batteries.

The RILEM report addressed how to make cathodic protection more valuable:

- The electrical conductivity to the steel reinforcement must be continuous.
- The concrete between the steel reinforcement and the anode must have the capability to conduct electricity.
- The alkaline aggregates must be avoided.

9.8.2 Cathodic Protection Components and Design Consideration

The cathode protection system consists of an impressed current, anode, and the electric conductive, which is the concrete in this case. This depends on the concrete's relative humidity, which has a great effect on the electrical cables and the negative pole system that is on the steel reinforcement. The

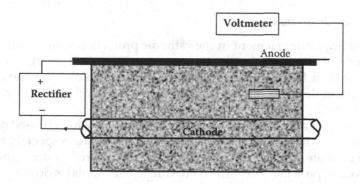

FIGURE 9.8
Cathodic protection method.

last component of this system is the wiring, which will conduct between the anode and the source of direct current. We must not forget one more very important element, which is the control and measurement system.

The most important and expensive element in the cathodic protection system is the anode, which should be capable of resisting chemical, mechanical, and environmental effects through the structure's lifetime. In general, it is preferable to be the structure's lifetime rather than the lifetime of the coating layer. The anode surface must be large enough to have a little density for the electric current so as to allow any deterioration to the anode due to its deterioration.

The cathodic protection of concrete structures is different from any other application, as concrete may have pores containing water in some areas, while other parts may be dry or have water on the surface only. It is different than marine structures or structures buried in soil.

The main fact is that the concrete structure cannot, in any way, contain 100% water inside its pores. This problem is clear if the concrete around the anode is completely dry, which requires increasing the potential volts by using 10 to 15 V instead of 1 to 5 V. On the other hand, the dry conditions around the steel bars prevent the causes of corrosion. This is only seen in summer, without high relative humidity.

9.8.2.1 Source of Impressed Current

Most design methods use a current of about 10 to 20 mm amps on the surface meter for steel reinforcement and take into consideration the lower layer of steel reinforcement. It is always best to use potential voltage of about 12 to 24 V to ensure that the effect of a shock to a human or animal is very little.

The source of the electric current is chosen in the design stage, and it must be enough to stop the corrosion process. It is often estimated high, with 50% added as a safety factor. This increase in current will produce little heat and it is assumed that some part of electric current will be lost.

9.8.2.2 Anode System

The most important element in the cathodic protection system is the anode. The anode is also the most expensive element of the system, and experience must be a factor in choosing it and its installation. The choice of the anode system depends on the type of the structure and its shapes and other requirements.

There are two main types of anodes. The first type, which is used on bridge decks and in fixing the deck from the top surface, requires special properties to accommodate the moving vehicles and being covered by an asphalt layer. The second type is used in vertical buildings. These anodes do not need high resistance to abrasion, as is the case in the first type.

9.8.2.2.1 Anodes for Bridge Decks

This type of anode is installed by making holes in the concrete surface and inserting the anode inside it. By this method, you can overcome the increase of dead load on the structure. On the other hand, you will face the challenge of cutting part of the concrete, as the anodes should be spaced at 300-mm intervals to maintain the distribution of electricity in a suitable manner.

Figure 9.9 illustrates making holes on the upper concrete surface by a diameter of 300 mm. The main target is to reduce the dead load on the bridge, which consequently will reduce the overall construction cost. It will also reduce the need for bridge surface leveling.

9.8.2.2.2 Vertical Surface Anodes

For this type of concrete structure member, it is always best to use titanium mesh and shotcrete is often used on the surface to cover the anode. A large degree of caution is needed when using shotcrete. This will require very competent contractors, workers, and supervision. Special curing is required to guarantee a good quality concrete and good adhesion between it and the old concrete surface. Precautions must be taken to ensure that the layer is 100% bonded with the old concrete and the electric conductivity will be through the pieces of titanium sheeting that will be welded on the mesh as shown in Figure 9.10.

We may use a rod of titanium or platinum, placed inside the coke of a concrete member of a subject, to reduce the impact of acid generated on the anode. One of the most important precautions during implementation is to make sure that there is no possibility of a short circuit between the anode and steel reinforcement. The anodes should be distributed by the design sys-

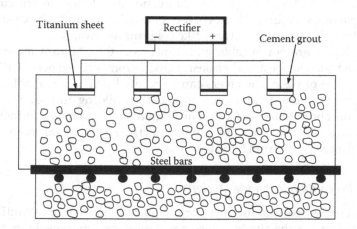

FIGURE 9.9
Sketch presenting anode holes at the surface.

FIGURE 9.10
Fixing titanium on a concrete column.

tem to ensure the protection of the steel reinforcement in the structure. This method is also used in horizontal surfaces.

9.8.2.3 Conductive Layer

In the most commercially successful applications, the anode is a main pole of inert metal and the secondary poles consist of layers of mortar, asphalt, and a coating and it is this coating that will be conductive to the electric current by carbon particles. Figure 9.11 shows the conductive layers in the case of vertical reinforced concrete structures like columns and walls.

These layers are not durable with time, as is the titanium mesh, but are considered the cheapest and easiest method from a construction and maintenance point of view. The layers can be painted with a reasonable match to the architectural design and decoration of the building and used for ceiling, vertical, and horizontal surfaces. The main disadvantage is that they cannot be used on surfaces exposed to abrasion, as their capacity for that is low and their lifetime is only from 5 to 10 years.

9.8.2.4 Precaution in Anode Design

The most expensive and most important part of a system of cathodic protection techniques is the anode. Therefore, it requires some special precautions and specifications, including the selection of the appropriate type for the structure, as every type has advantages and disadvantages and needs to be

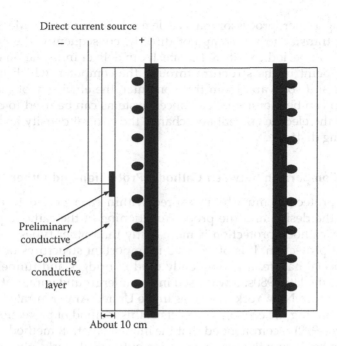

FIGURE 9.11
Anode conductive layer.

studied based on the application. It should be noted that, as a result of chemical reactions, the alkalinity of the concrete on the surface decreases and this means an increase in the acidity. The materials used in the anode might have a high resistance to acid, but the concrete around it might not. The amount of accumulated acid, proportional to the density of the electric current, must be considered.

For example, a current density of 1.1 amp/m^2 with a surface area equal to that of steel and concrete results in reactions that will cause corrosion of about 0.3–0.4 mm from concrete in general. If using shotcrete with a cement to aggregate ratio of 1:5, increase the current density or the deterioration rate will increase in addition to that of the existing chlorides and increasing it has an effect on the increasing acid rate. Therefore, increasing the area of anode to concrete surface area will be better for anode consumption and acid formation.

9.8.2.5 Follow-Up Precaution

Note that the cathode protection needs to be followed up periodically to ensure that the cathode protection system runs continuously without any obstacles. This is done by putting a cell pole near the steel reinforcement and connecting a microprocessor via a cable modem connection. Measurements and data transfer to the computer directly, consequently, the system can work without periodic visits to the site. It can deliver information from more than one point in the structure through the computer, which has special hardware and software. From this computer, the efficiency of the cathodic protection can be monitored. Advanced systems can be used to control the source of the electric current and change the current density and potential from a long distance.

9.8.3 A Comparison between Cathodic Protection and Other Methods

Cathode protection must be measured within each period to ensure the safety of the design and the proper functioning of the cathode protection. Therefore, cathode protection is more costly than other types of steel reinforcement protection. It is often used in important structures or structures with a special nature. It is frequently used in bridges and tunnels and has been since the late 1980s. It was used in an underground garage at the World Trade Center in New York as well as in the United Arab Emirates.

After studying the economic aspects of this method of protection, Kendell and Daily (1999) recommended that the implementation method of cathodic protection be during the implementation of the civil works, therefore eliminating the need for excess labor, as the anodes are installed with the reinforcing steel, as illustrated in Figure 9.12, along with the development of titanium mesh with the reinforcing steel and how to stabilize the titanium mesh by a strong plastic stuff.

In Figure 9.13 shows the bars of a titanium with steel, and as shown in the photo, the method does not need workers, especially because the workers are that placing the steel reinforcement will place it also.

FIGURE 9.12
Fixing titanium mesh by plastic binder.

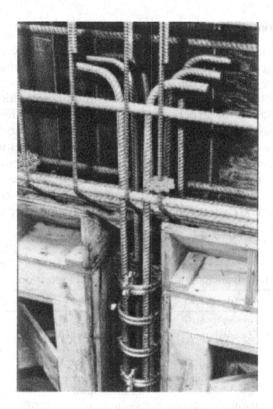

FIGURE 9.13
Placing of titanium bar.

9.8.4 Cathodic Protection for the Prestressed Concrete

Currently, there are many LNG and NGL projects worldwide, and the main component of these projects is the prestress tank with the steel shell inside. Most specifications state that you must not use cathode protection when using prestressed concrete because the movement of electrons results in the formation of hydrogen ions on the surface of the steel reinforcement and the presence of hydrogen affects the composition of iron atoms for the high strength steel that is usually used in prestressed concrete.

However, cathodic protection can in fact be used, taking into account that hydrogen will not affect the nature of iron and the quantity of hydrogen resulting from the cathode protection process does not pose a threat to the prestressed steel.

There are some prestressed underground structures for which cathode protection should not be used. Also Kessler et al. (1995) mentioned that there are some cases where using sacrificial anodes in the bridge piles, as in Florida, is done by using depleted anode.

9.8.5 Bond Strength in Case of Cathodic Protection

The impact of electric current on the bond strength between the steel reinforcement and concrete is considered to be a very important subject and researchers have studied it for several years. From a practical point of view, there are no notes for any impact of the cathodic protection on the bond strength. There are some structures that have been studied for about 20 years, and there has been little impact on the bond strength, as most bond strength is carried out by the ribs that exist in the deformed steel reinforcement.

When using cathode protection, no other protection method, such as membranes, additives, or painting, is needed. This reduces the total cost of the project and eliminates the cost of a deteriorating concrete structure or some of its members. The cost of repairing the concrete member may actually be higher than the cost of constructing it.

TABLE 9.3

Advantages and Disadvantages of Different Types of Protection

Method	Advantages	Disadvantages	Compatibility
Increase concrete cover	No cost increase	Limited	Yes
	Obtained from good design	Increased cover will increase the shrinkage crack	
Nonpermeable concrete	Dependent on good design	Requires additives for good curing	Yes, but silica fume must be <10% cement weight
Penetrating sealer	Relatively low cost	Complicated	Yes
Waterproof membrane	Well-known technology	Can be damaged	Yes, but only cathodic protection under the membrane
		Difficult to fix without fault	
Epoxy-coated steel bars	Known technology	Problems in quality control under certain weather conditions	Yes, without continuous electric current
	Little maintenance		
Galvanized steel	Easy use on site	Easily damaged by contact with nongalvanized steels	Yes, with higher level of protection
Stainless steel	Excellent for corrosion prevention	High cost	Yes
Corrosion inhibitor	Calcium nitrate effect for small chloride concentration	Must know chloride amount	Yes
		Long-term effect not known	
Cathodic protection	Known technology	Regular inspection of the rectifier	
	Long life of titanium anode		

Source: K. Kendell, and F. S. Daily. 1999. *Concrete International Magazine* 21(6).

In some structures, because of the special nature of their construction, you will find it difficult to make necessary repairs, so cathode protection is the best and only solution. In Table 9.3, you will find a summary of the methods of protection prepared by Keven in 1999, which gives the advantages and disadvantages of each method and the appropriateness of use with the cathode protection at the same time.

References

ACI Committee 318. 1988. Revisions to building code requirements for reinforced concrete. *American Concrete Institute Structures Journal*, 85(6):645–674.

Andrade, C., J. D. Holst, U. Nurnberger, J. J. Whiteley, and N. Woodman. 1994. Protection system for reinforcement. Prepared by Task Group 7/8 of Permanent Commission VII CEB.

Arnon, B., S. Diamond, and N. S. Berke. 1997. Steel Corrosion in Concrete: Fundamental and Civil Engineering Practice. London: E & FN Spon.

Broomfield, J. P. 1995. *Corrosion of Steel in Concrete: Understanding, Investigation and Repair*. London: E & FN Spon.

Building Research Establishment. 1969. Zinc-coated reinforcement for concrete. *BRE Digest* 109.

Cairns, J. 1992. Design of concrete structures using fusion-bonded epoxy-coated reinforcement. *Proceedings of the Institute of Civil Engineering Structures and Buildings* 4(2):93–102.

El-Reedy, M. A., M. A. Sirag, and F. El-Hakim. 1995. Predicting bond strength of coated and uncoated steel bars using analytical model. M.Sc. thesis: Cairo University.

Gustafson, D. P., and T. L. Neff. 1994. Epoxy-coated rebar, handled with care. *Concrete Construction* 39(4):356–369.

Kendell, K., and F. S. Daily. 1999. Cathodic protection for new concrete. *Concrete International Magazine* 21(6).

Kessler, R. J., R. G. Powers, and I. R. Lasa. 1995. Update on sacrificial anode cathodic protection of steel reinforced concrete structures in seawater. *Corrosion 95*, Paper 516, NACE International, Houston, TX.

Miller, J. B. 1994. Structural aspects of high-powered electrochemical treatment of reinforced concrete. In *Corrosion Protection of Steel in Concrete*, ed. R. N. Swamy. Sheffield, England: Sheffield Academic Press, pp. 1400–1514.

Pike, R. G., et al. 1972. Nonmetallic coatings for concrete reinforcing bars. *Public Roads* 37(5):185–197.

RILEM Committee 124-SRC. 1994. Draft recommendation for repair strategies for concrete structures damaged by steel corrosion. *Materials and Structures* 27(7): 415–436.

Satake, J., M. Kamakura, K. Shirakawa, N. Mikami, and R. N. Swamy. 1983. Long-term resistance of epoxy-coated reinforcing bars. In *Corrosion of Reinforcement in Concrete Construction*, ed. A. P. Crane. London: The Society of Chemical Industry/Ellis Horwood Ltd, pp. 357–377.

Yeomans, S. R. 1994. Performance of black, galvanized and epoxy-coated reinforcing steels in chloride-contaminated concrete. *Corrosion* 50(1):72–81.1

10

Repair of Industrial Structures

10.1 Introduction

In most cases, repair and retrofit of a concrete structure are much more complicated than executing a new concrete structure, as new constructions do not require engineers with large amounts of experience and can be handled by junior engineers with reasonable amounts of experience, two to three years, for example. However, the repair of concrete structures is more challenging and requires competent engineers and consultants. If the building already exists, you need to define new solutions for the problem that must match the nature of the building and owner requirements while also satisfying the safety and economic requirements. The repair process for reinforced concrete structures is both very important and dangerous. Extreme care must be taken in choosing suitable repair methods and tools.

The reasons for structural deterioration can be summarized as follows:

- Codes, specifications, and precautions were not considered in the design phase.
- Codes, specifications, and precautions were not considered in the execution phase.
- A suitable protection method was not selected in the design phase.
- A suitable protection method was selected, but the execution was bad.
- No maintenance monitoring was performed.

In all previous cases, the results will be the same: The structure will deteriorate and this will become obvious by cracks and falling of concrete cover of the different reinforced concrete elements of the structure. When cracks are present and the concrete cover is falling, the situation is very bad. Shortages occur in cross-sectional areas of the concrete member, and this situation requires repair as soon as possible to avoid the structural condition worsening and likely collapsing with time.

The first and most important step for any repair is to determine accurately what should be repaired and this depends entirely on the assessment of the structure and the answer to these questions:

- What is the reason for the corrosion?
- Are the cracks and the deterioration of the structure increasing?
- What is the expected extent of the deterioration of concrete and how far does the corrosion extend into the steel?
- What is the impact of the current and expected future deterioration to the safety of the structure?

Answering these questions will require the use of structural assessment methods and methods of measurement to decide the cause of corrosion and the present and future deterioration of the structure. Chapter 4 discusses these methods of assessment and measurement in detail.

The process of repair and restoration of concrete structures, from the assessment phase to the execution phase, requires high levels of experience. Even the simplest stage, such as the evaluation stage, when it is conducted by inexperienced engineers, may lead to the wrong choice of repair method, and this is a big problem.

- When assessing structures, there are several strategies for repair mentioned in a special report by the RILEM committee (1994) that clarify the various repair strategies used in most cases.
- Reestablishment of the deteriorated member.
- Comprehensive repair of the concrete member to restore its ability to withstand full loads.
- Repair of a particular portion of the concrete member and follow-up over continuous periods.
- Strengthening of the structure by an alternative system to bear part of the loads.

In most cases, the strategy of repair is either a comprehensive repair of the concrete member or a partial repair of a concrete member. These strategies are common in the rehabilitation of concrete and depend on the structural system, external environmental factors, and the degree of structural degradation.

10.2 Main Steps to Execute Repair

There are several standard steps in the repair of all structures exposed to corrosion. The first, and very critical, step is to strengthen the structure by performing a structural analysis and deciding on the appropriate location of the temporary support. The second step is to remove the cracked and delaminated concrete. It is important to clean the concrete surface and also to clean the steel bars by removing rust. After removing the rust by brush or sand-blasting, paint the steel bars with an epoxy coating or replace the steel bars with new ones and then pour new concrete. The final step is to paint the concrete member to provide external protection. This is, briefly, the repair process. These steps will be explained in detail in the following sections.

10.2.1 Strengthening the Structure

Care should be taken as this is one of the most dangerous and important steps and is necessary for the repair. The temporary support selection depends on the following:

- Evaluation of the state of the entire structure.
- The method of transferring loads from equipment such as vessels, pipe supports, and others.
- The volume of repair that will be done.
- The type of the concrete member that will be repaired.

Therefore, as mentioned earlier, the repair process must be carried out by a structural engineer with a high degree of experience with the repair process, the capability to perform the structural analysis, and an extensive knowledge of the load distribution in the structure, according to the type of repair that will be performed, because it often follows the repair process of breaking the defective concrete. Therefore, that person has a responsibility to choose the right way to optimize the process of determining the ability of supports to carry the loads of the member or members that will be transferred. Therefore, the responsible engineer should design the temporary supports based on the collected data and the previous analysis and use caution in the phase of executing the temporary supports.

Choosing the ways of removing the defective parts will be based on the nature of the concrete member in the building as a whole. Any member with breaking concrete has a detrimental impact on neighboring members because the process of breaking will produce a high level of vibration. The temporary members must be strong and designed in an appropriate manner to withstand the loads transferred from the defective members easily and safely. You can imagine that the entire pool of the structure depends on the design and execution of the temporary supports and their ability to bear the loads with adequate safety. Figure 10.1 presents the temporary supports by section.

FIGURE 10.1
Temporary repair using steel sections.

10.2.2 Removal of Concrete Cracks

There are several ways to remove the part of the concrete member that has a cracked surface or steel corrosion on the surface. These methods of removing the delaminated concrete depend on the ability of the contractor, specifications, cost of the breakers, and the state of the whole structure.

Select concrete removal techniques that are effective, safe, economical, and that minimize damage to the concrete left in place. The removal technique chosen may have a significant effect on the length of time that a structure must be out of service. The selection of the breaker methods is based on the causes of corrosion. If it is due to carbonation or chlorides, it must also be decided whether to consider cathodic protection in the future. In this situation, the breaking work will be on the falling concrete cover and on cleaning and removing all of the delaminated concrete and the concrete parts and then pouring high-strength and nonshrinkage mortar.

If the corrosion in the steel reinforcement is a result of chloride propagation into the concrete, most specifications recommend removing about 25 mm behind the steel and making sure that the concrete on the steel has no traces of chlorides after the repair process.

The difference in the procedure of breaking the delaminated concrete is due to the difference in the causes of corrosion. Therefore, a study to assess the state of the structure and the causes of corrosion is very important to achieve high quality after the repair process. The evaluation process is the same as diagnosis of an illness. If there are any mistakes in the diagnosis, the repair process will be useless and a waste of time and money. Figure 10.2 shows the removal of the cracked cover from the tank ring beam and Figure 10.3 illustrates the removal of the concrete cover from the pump foundation.

FIGURE 10.2
Removal of delaminated concrete from ring beam.

It is necessary to define the work procedure and quantity of concrete that will be removed. This step is considered to be one of the fundamental factors in the design and installation of the wooden pillars of the building that will be used during execution. Therefore, the work plan must be clear and accurate for all the engineers, foremen, and workers who participate in the repair process.

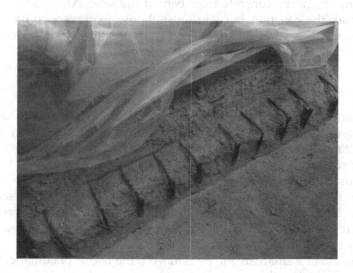

FIGURE 10.3
Removal of concrete cover from pump foundation.

After completion of the building assessment and design of the supports and ties, as well as ensuring the presence of competent staff, there may be some information that is still not available, for example, the construction procedures, workshop drawings, or specifications that were followed when the building was constructed. Therefore, the risks are still high. The only factor that can help us reduce this risk despite the steel corrosion is an increase in concrete strength with time. However, this compensation has limits as the steel is carrying most of the stresses and the risk will be very high in cases of spalling of the concrete cover due to a reduction in the concrete cross-sectional dimensions.

It is necessary to remove concrete for a distance greater than the volume required for removal of the defective concrete so that we can properly reach the steel, because it will be important later in the repair process. There are several methods commonly used for breaking and removing defective concrete. These will be explained in the following sections.

10.2.2.1 Manual Method

One of the simplest, easiest, and cheapest ways is to use a hammer and chisel. With the use of drawings, many specialists prefer to remove defective concrete this way. However, it is slow compared with mechanical methods. It should be noted that mechanical methods produce high noise and vibration and require special precautions and trained labor. Using this manual method, it is difficult to spare the concrete behind the steel. This method is used in the case of small spaces and is the preferred method in the event of corrosion due to carbonation and attacking chlorides from outside, which do not require breaking concrete from behind the steel. Any worker can manually break the concrete, but you must choose workers who have worked before in repair, as they must be sensitive in breaking the concrete to avoid causing cracks to the adjacent concrete members.

10.2.2.2 Pneumatic Hammer Methods

These hammers work by compressed air and weigh between 10 and 45 kg. If used on roofs or walls, the weight should be about 20 kg. They require a small, attached power unit, but in large areas may require a separate, larger air compressor.

This machine requires a properly trained operator. The compressed air hammer has a few initial costs, of which a few were discussed by the Strategic Highway Research Program (SHRP), and based on Vorster et al. (1992), a research program of highways that the terms of the contract are governed by the contractor, which will use the breaking machine.

In the case of a small area to be removed, the use of pneumatic hammers is more economical than in the case of large areas because, for large areas, it is preferable to use a water gun, which will be illustrated in the following section.

These hammers are usually used in removing the covers of retaining walls or reinforced concrete tanks. In the case of a client who does not specify the area (in square meters) that needs to be removed in the tender and cost calculation phase, the risk will be low because the machine's initial cost is low. In this situation, it is preferable to use pneumatic hammers. The performance rates indicate that some of the breakers are able to remove from 0.025 to 0.25 m^3/h using hammers weighing 10 to 45 kg, respectively. In summary, pneumatic hammers are useful in removing small areas of concrete but are not preferable in large areas.

10.2.2.3 Water Jet

This method has been commonly used since the 1970s, when it was introduced into the market. It relies on the existence of water at the work site. It is preferable for the removal of a significant depth of concrete in a large area and also for the removal of fragmented concrete and cleaning steel bars. It also removes part of the concrete behind steel bars.

The water jet is used manually by a technically experienced worker who has dealt with a pushing water using a high-pressure hose or, perhaps, with a mechanical arm. It is important to observe very high safety precautions for the worker who uses it and the site around it. The water used must not have any materials that can affect the concrete, such as high chloride ions. In general, it must be potable water.

The water gun consists of diesel engines and a pressure pump connected by a hose that can bear high-pressure water in the gun nozzle of 30 to 70 N/mm^2, as it requires at least 40 N/mm^2 to cut concrete. The rate of water consumption is about 50 l/min. The performance rate of a water jet is between 0.25 m^3/h (small pump) and ~1 m^3/h (two pumps or one big pump).

It is a reasonable option in industrial projects because it is a low-risk issue, as it removes a possible source of ignition that is present in the previous options. Moreover, in terms of safety, water itself reduces hazards during construction.

10.3 Cleaning the Concrete Surface and Steel Reinforcement

It is during this phase that any remaining broken concrete is removed with a cleaning process, and at the same time, assessing the steel and cleaning up and removing corrosion from the roof are carried out.

10.3.1 Concrete

After finishing the surface preparation, we start to pour the new concrete. The success of the repair depends on the good bond between new and old concrete. As the concrete element is carrying the stresses as one element.

Before applying the primer coating, which provides the bond between the existing old concrete and the new concrete for repair, the concrete surface must be well prepared and this depends on the materials used. In all cases, the concrete surface must be clean and must not contain any oils, broken concrete, soil, or lubricants. Any of these should be cleaned completely through sand blasting, water, or by manual brushing.

This stage is very important, regardless of the type of materials used to bond the new concrete with the old concrete, and lack of attention to the preparation of the surface might affect the repair process as a whole, taking into account that this is the less expensive stage in the repair process.

We do not need to prepare a surface when using a water gun to remove the delaminated concrete, as the surface will be wet enough and will be clean after the crushing process. It is beneficial to use a water gun because it uses air pressure to clean the surface of the concrete and remove any growth and fragmented concrete used as supplement for the surface material used for bonding with new concrete.

When using cement mortar or concrete for the repair, the concrete surface should be sprayed by water until saturation is reached. The stage of saturation can be reached by spraying water on the surface for 24 h or through wetting burlap. Water spraying should be stopped and the burlap should be removed for about 1 to 2 weeks, on the weather condition from low to high temperature, wind speed, and from high to low relative humidity, respectively. Then surface is coated with a mixture of water and cement, which is called *slurry*. This material should be applied by brush and also can use epoxy coating as an adhesive between new and old concrete and must follow the specifications and precautions stated by the manufacturer.

To achieve a better bond between the old and new concrete, the preparation of the concrete surface should be in accordance with the American specification ACI 503.2-79, which states that the surfaces that will receive epoxy compound applications must be carefully prepared, as the bonding capability of a properly selected epoxy for a given application is primarily dependent on proper surface preparation. Concrete surfaces to which epoxies are to be applied must be newly exposed and free from loose and unsound materials. All surfaces must be meticulously cleaned, as dry as possible, and at a proper surface temperature at the time of epoxy application.

The epoxy materials used for bonding should be compliant with the ASTM C881-78 code, which must be well-defined by the supplier specifications and compatible with the different circumstances surrounding the project, particularly temperature change that might occur in a resin as well as the nature of loads carried by the required repair member.

It is worth noting that during the execution of the repair process, epoxies will be used significantly. Therefore, we must take into account its safety for workers who use such materials. Therefore, the workers must wear their personal protective equipment (PPE) as gloves and special glasses for safety issues, as these materials are very harmful to the skin and cause many dangerous diseases if any one is exposed to it for long time.

There are also some epoxies that are flammable at high temperatures, and that must be taken into account during storage and operation.

10.3.2 Cleaning the Steel Reinforcement Bars

After removing the concrete covers and cleaning the surface, the next step is to measure the steel bar diameter. If it was found that the cross-sectional areas of the steel bars have a reduction $\geq 20\%$, additional reinforcing steel bars are required and before pouring new concrete, one must ensure that the development length between the new bars and the old steel bars is enough.

The preferred method in connecting the steel to the concrete is by drilling new holes on the concrete and putting the steel bars in the drilled hole filled with epoxy. However, in most cases, the steel bars are completely corroded and need to be replaced.

Figure 10.4 shows the pedestal under the pressure vessel during repair, which is done to increase the thickness of the two sides of the pedestal (Figure 10.5). The increase in thickness should be no less than 150 mm for each side and the new two sides and the existing concrete should be connected with a steel bar to ensure that the pedestal is working as one unit and

FIGURE 10.4
Repair of concrete pedestal.

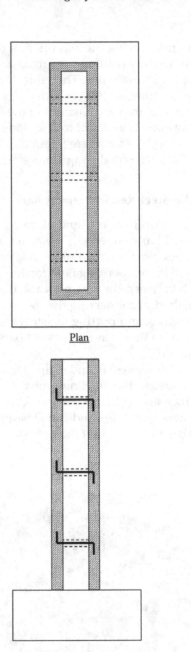

Plan

FIGURE 10.5
Pedestal repair.

that there is no likelihood for separation between the new concrete and old concrete, as the pedestal will be under low dynamic loads and probable thermal movements of the vessel. After connecting the two new sides with the existing concrete, as a normal procedure, apply epoxy coating as an adhesive between old and new concrete and then pour the new concrete manually or by shotcreting.

As for the columns, the repair is performed as shown in Figure 10.6, which shows how the casting is done at the site, but we must know that the minimum distance allowed to easily cast concrete is about 100–120 mm from each side.

Fixing the dowel for the first floor (Figure 10.6) does not require drilling holes on the base but making the legs of steel bars rest on the foundation. The steel bars will be distributed around the column circumference, as in the figure, to cope with the reduction of the steel reinforcement's cross-sectional area due to corrosion, and the cross-sectional area of the steel connected to the concrete column should have the same percentage as the old column member and not less than, as stated in the code.

Note that using a water jet in the breakers cleans the steel and removes corrosion, but sand blasting can also be used to remove corrosion. After cleaning steel bars completely (to remove the effects of chlorides), we then paint the old steel using epoxies. Slurry, which is a mix of cement and water, is used. When using mortar cement for repairs, the steel reinforcement is not painted with epoxy to obtain alkaline protection from the cement mortar.

It noted that the improved cement slurry can dry quickly, making it ineffective in repairs that require the installation of forms after painting the bars, but it works well in cases where the time between painting the steel and pouring cement mortar is short (i.e., this should not exceed 15 minutes, as stated in a U.S. Army manual, 1995).

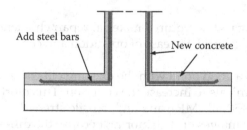

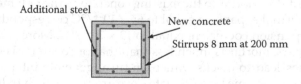

FIGURE 10.6
Repair of concrete column.

10.4 New Patches of Concrete

There are some mixtures available on the market, which, when mixed, can be used easily in repairs of small areas, as this mixture is very expensive.

In the case of repairs involving large surfaces, mixture preparation and mixing should be on-site to minimize cost, but this must be carried out by experts.

In addition to the concrete characteristics mentioned, using a pump for pouring or shotcreting needs a specially designed mix.

In the USA and Canada, most of the contractors who work on bridge repair use their own concrete mix, which has secret mixing proportions and are made from locally available materials.

Factories producing ready-mix concrete provide a guarantee in case of corrosion due to carbonation, but they do not provide a guarantee in case of corrosion resulting from the presence of chloride after repair processes.

It also noted that most manufacturers of materials used in mixing are field execution contractors, but even when they supply materials only, they should provide all the information and technical recommendations for the execution, performance rates to calculate the required amount of the materials, and appropriate method of operation.

Worldwide, all construction chemical companies have competent technical staff who can assist on site or supervise to avoid any error and define the responsibility in case of any defects.

There are two types of materials used as a new mortar for repair: polymer mortar and polymer-enhanced cement mortar. Both will be described in the following sections.

10.4.1 Polymer Mortar

This mortar is providing us the required material repair characteristic properties from the ability to control, the ease of operation and control the setting time.

This mortar provides a good bond with the existing old concrete and does not need any other chemicals to increase the cohesion. This mortar has high compression strength (~50–100 MPa) and high tensile strength.

Despite the many advantages of such mortar, because the difference in the properties of this mortar to the existing concrete as the polymers have a coefficient of thermal expansion equal to $65 \times 10^{-6}/°C$ corresponding to concrete thermal expansion coefficient equal to $12 \times 10^{-6}/°C$. Moreover, the modulus of elasticity for mortar is much less than that of concrete. These differences in properties lead to cracks, which, in turn, generate internal stresses, but this could be overcome by aggregate sieve grading, which reduces the difference in the natural properties of polymers.

There are many tests to define this mortar and non-compliance with the concrete (e.g., ASTM C 884-92 Standard Test Method for PCBs Comparability Between Concrete and Epoxy-Resin Overlay).

10.4.2 Cement Mortar

Polymer mortar gives a physical protection to the steel reinforcement; however, the cement mortar gives passive protection, as they increase the alkalinity around the steel reinforcement. The trend now is to use cement mortar for repair of damages that result from corrosion, as it has the same properties as the existing concrete and it gives steel passive protection from corrosion.

Use of polymer additives to the cement mortar, in liquid or powder form, improves some of the properties of cement mortar (Ohama and ACI Committee 548), increases flexural resistance, increases elongation, reduces water permeability, increases the bond between old and new concrete, and increases the effectiveness of its operation.

The polymer used is identical with the cement mortar mixture, and the properties that control the polymers in the cement mortar are stated in ASTM C1059-91 (Specification for Latex Agents for Bonding Fresh to Hardened Concrete).

Silka fume can be used with the mixture, as well as super plasticizers, to improve the properties of the used mortar and reduce shrinkage.

10.5 Execution Methods

There are several ways to implement the repair process, and these are dependent on the structure of the member to be repaired and the materials used in the repair.

10.5.1 Manual Method

The manual method is most usually used, especially in the repair of small delaminated parts.

10.5.2 Casting Way at the Site

This is done by making a wooden form and then pouring concrete in the damaged part (Figure 10.7) of concrete columns or vertical walls. One must take into account the appropriate distance to easily cast the concrete in the wooden form. This method is commonly used because of its efficiency and low cost and it does not require expensive equipment, but extensive

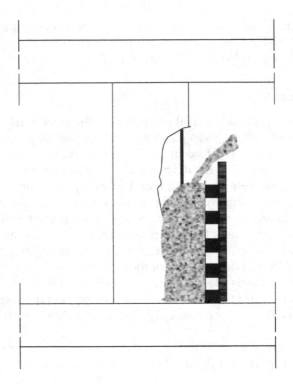

FIGURE 10.7
Casting concrete on site.

experience is needed for its fabrication, installation of the wooden form, and casting procedure.

10.5.2.1 Grouted Preplaced Aggregate

As shown in Figure 10.8, the aggregate with gap grading is placed in the area to be repaired. The next step is to fluid a grout by injecting it inside the aggregate using a pipe with pump to fill the gap. This method is used in repairing bridge supports and other special applications and uses special equipment such as pipe injection, pump, and other special miscellaneous equipment. Therefore, one can conclude that this method is used by private companies with high potential.

10.5.2.2 Shotcrete

This is used in repairs of large surfaces, such as reinforced concrete tanks, but the mix and components of concrete suitable for the use in shotcrete should be considered, as they need special additives and specifications.

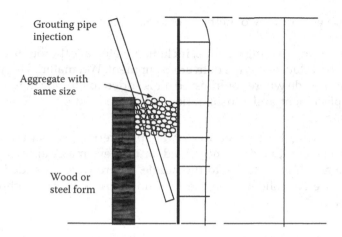

FIGURE 10.8
Injected preplaced aggregates.

Based on ACI 546 (1996), *shotcrete* is defined as concrete that is pneumatically conveyed at high velocity through a hose onto a surface. The high velocity of the material striking the surface provides the compactive effort necessary to consolidate the material and develop a bond to the substrate surface. The shotcrete process is capable of placing repair materials in vertical and overhead applications without the use of forms, and it can routinely place materials several hundred feet from the point of delivery.

Health and safety precautions must be carefully followed when using this method because the materials contain polymer resins and special additives that might harm the skin after prolonged exposure.

In the concrete mix design, the nominal maximum coarse aggregate size must be suitable to the shotcrete equipment's nozzle and pump to avoid any problem during casting the concrete.

10.5.3 Complete Member Casting

This is used in the total reconstruction of a concrete member whose steel reinforcement bars are depleted due to corrosion, and concrete is needed to be poured to ensure full depth casting.

It is frequently used in the repair of bathrooms because in some cases, the concrete slab will be in a very severe condition that ordinary repair will not be efficient. In this case, all the concrete slabs would be demolished, a new steel reinforcement installed, and then pour a concrete slab. Before deciding to use this method, the whole building condition must be considered. This method is usually easy to apply in the bathroom slab, as in residential buildings, the bathroom slab is usually designed as simply supported as it drops around 100 mm for plumbing pipes so when demolishing the slab will not affect the surrounding slabs.

There are generally two methods of application:

- Wet mixing: All ingredients, including water, are thoroughly mixed and introduced into the delivery equipment. Wet material is pumped to the nozzle where compressed air is added to provide high velocity for placement and consolidation of the material onto the receiving surface.
- Dry mixing: Preblended dry or damp materials are placed into the delivery equipment. Compressed air conveys material through the hose at a high velocity to the nozzle, where water is added. Materials are consolidated on the receiving surface by the high-impact velocity.

10.6 Repair Steps

Figure 10.9 presents a case study (as in the case study in chapter 8) for repairing reinforced concrete ring beams that support steel tanks containing oil.

The ring beam structurally carries the tensile strength due to earth load affecting the beam, which makes the ring beam under tension. Corrosion of the steel reinforcement bars reduces the capacity of the ring beam. So the best solution is to increase the number of steel bars to overcome the reduction on the steel cross-sectional area.

As in the following sketches, the first step in repair is to drill holes, put steel stirrups in the holes, and fix by epoxy.

In step 2, install the main steel bars and coat the concrete surface by epoxy to bond the existing old concrete with the new concrete.

In step 3, pour the new concrete with proper mixing design, which is improvement by adding polymer as described above.

10.7 New Methods for Strengthening Concrete Structures

Aside from the traditional method, another way to strengthen reinforced concrete structures is to use steel sections. This method has many advantages, one of which is that it does not add significant thickness to the concrete sections, and this is important in maintaining architectural design. As it is a quick solution to strengthen the concrete member, it is usually used in industrial structures and buildings.

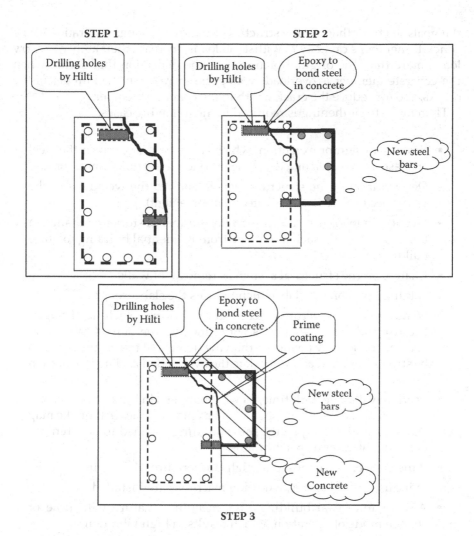

FIGURE 10.9
Steps of repair.

When there is corrosion in the steel reinforcement, using steel support by itself is not considered appropriate, but it can be used together with epoxy paints, which should be reapplied periodically.

Recently, there were many studies and researches that use fiber reinforcement polymer, which has many advantages over steel and the most important of which is it does not corrode and can therefore be used in any environment exposed to corrosion.

There are some modern ways to protect the steel reinforcement from corrosion, such as replacement of steel reinforcement bars with fiber polymer and using some epoxies that coat the concrete surface by painting. Generally, the goals of strengthening the structure are the following: restoration of the concrete member's capacity to withstand loads, increase their ability to carry loads more than the original plan, reduce the deflection that occurs when the concrete member is overloaded, stop the increase in the number of the cracks, and/or reduce the crack width in the concrete member.

Therefore, strengthening is required in the following cases:

- Error in design or execution, which causes a reduction in the steel cross-sectional area or a decrease in concrete section dimensions.
- Deterioration of the structure, which leads to the weakness of the resistance (e.g., corrosion in steel reinforcement).
- Increase in loads (e.g., increase in the number of stories or change in building function such as change from residential building to office building).
- Change on the structural system as removal of walls.
- Adding holes on the slab, which reduces the slab strength.
- Some special structures need strengthening, such as bridges, because traffic flow increases with time and the flow of traffic and heavier trucks may be higher than the design, so the bridge needs to be strengthened to accommodate the overloaded and new operation mode.
- Load may change with time, as what happens in buildings in earthquake zones; the same happened in Egypt when the earthquake map changed, and consequently, some buildings needed to be strengthened to match with specifications.
- If the structure is exposed to high temperatures due to fire.
- In industrial structures, when big machines are installed.
- Also in industrial buildings, the machines mature with time or change mode of operation, which results in high vibration.

10.8 Using Steel Sections

Traditionally, steel structure sections are used to strengthen the reinforced concrete structure and are preferred for industrial buildings because they are usually available in the field. In addition, repairs can be performed easily, and in some cases, the steel sections can be used as a temporary repair. As for example, it was used to overcome of moving the saddle for a horizontal vessel with respect to the pier.

In industrial structures, no need for aesthetics as the architectural design is not the main interest, on the other hand, the fast repair is very critical so the repair and strengthening is the best solution in most cases.

Strengthening the concrete member is varied and depends on the member that needs strengthening. The strengthening process is calculated based on the reduction in strength in the concrete section, and the goal is to compensate this reduction using steel sections.

To strengthen a concrete beam that lacks steel reinforcement, a C-shaped beam is used (as in Figure 10.10a). This offers an architectural advantage because it does not change the aesthetics and there is minor reduction in member capacity.

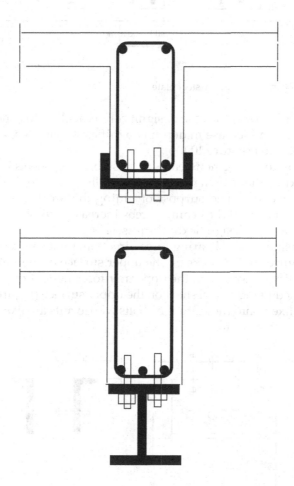

FIGURE 10.10
(a) Strengthening RC beam by C section and (b) strengthening RC beam by I-beam.

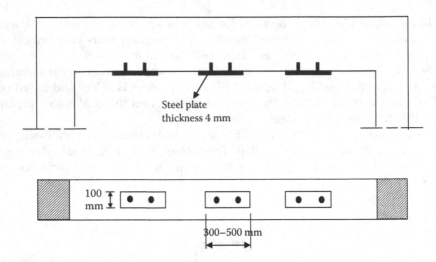

FIGURE 10.11
Strengthening of concrete beam by steel plate.

On the other hand, in the case of significant reduction in concrete member capacity and when increase in depth is significantly needed, an I-beam section should be used (Figure 10.10b).

Another way to strengthen beams is by stabilizing sheets of steel (Figure 10.11) and fixing them using mechanical bolts or through chemical substances such as epoxy. The purpose of adding this steel sheet is to increase the moment of inertia of the beam, thereby increasing its ability to withstand stress more than the design flexural stresses.

In case of beams with shearing force direction, the beam can be strengthened as in Figure 10.12. Cracks on the upper surface of the slab in the beam direction signify a shortage in the upper reinforcement of the slab and it can be strengthened using a steel sheet on the upper surface (Figure 10.13). These plates can be fixed onto the surface by bolts whose nuts are fixed on the lower part of the slab (Figure 10.13).

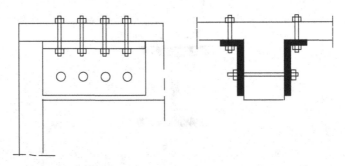

FIGURE 10.12
Beam strengthening in the direction of shear.

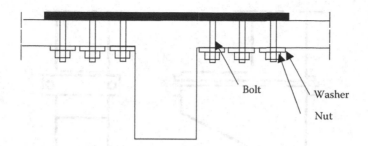

FIGURE 10.13
Slab strengthening in direction of upper steel.

To strengthen the flat slab, the steel angles should be fixed between slab and columns (Figure 10.14). Short cantilevers are used in the frames of factories or bridges, where the main stresses are due to shear and the others are due to flexural stress. To strengthen the cantilever, a steel sheet is added at its two sides, applying compression force with bolts, thereby reducing the likelihood of collapse due to shear (Figure 10.15).

In the case of a residential building, such as an administration building or a labor house, the steel with the smallest dimension is preferred so as to maintain the architectural design and can be covered with wire steel mesh, plaster, wood, or decor to provide an acceptable view as well.

10.9 Fiber-Reinforced Polymer

Modern science does not stop in the construction field, as we know that the development in computer technology is continuous and rapid, and this

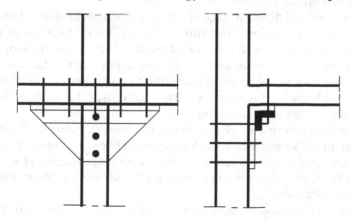

FIGURE 10.14
Strengthening of flat slab.

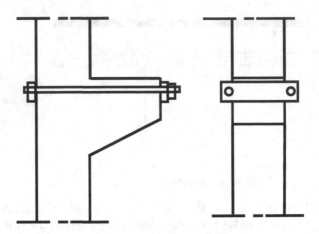

FIGURE 10.15
Strengthening of short cantilevers.

development is happening in all areas, including engineering constructions. At the same time, there has been an evolution in some materials such as plastic, which is based on epoxies, as explained in the previous chapters.

The use of such epoxies increases the quality of concrete, and the concrete compressive strength reaches 1500 kg/cm². The competition between international companies led to the development of different methods for strengthening reinforced concrete structures.

The traditional way of strengthening concrete structures is done increasing the concrete sections or through steel sections, and concrete members of various constructions and special structures such as bridges use steel sheets that are fixed on the concrete through bolts.

The fundamental disadvantage of using steel sheets to strengthen the concrete member is corrosion. Therefore, steel sheets need to be protected from corrosion, particularly in the region of contact between surface of concrete and steel; thus, use of steel sections in severe environmental conditions is not ideal and should not be used. In addition, steel is heavy, and this leads to a higher cost because it is more expensive to transport and store than fiber-reinforced polymer (FRP) solution.

The growing interest in FRP systems for strengthening and retrofitting can be attributed to many factors. Although the fibers and resins used in FRP systems are relatively expensive compared with traditional strengthening materials like concrete and steel, labor and equipment costs to install FRP systems are often lower.

FRP systems can also be used in areas with limited access and where traditional techniques would be very impractical, for example, a slab shielded by pipe and conduit. FRP systems also have lower life-cycle costs than conventional strengthening techniques because the FRP system is less prone to corrosion.

The presence of such defects in steel led to the development of plastic fiber bar (FRP) as an alternative to steel sheets.

FRP is now being used in a wide range of their advantages and makes alternative to the steel used in the strengthening of their distinctive characteristics, such as its resistance to corrosion, which operate under any circumstances environmental caution as the concrete member is usually exposed to heat, salt, and different chemicals, which makes it a solution of suitable of the industrial structures.

There are different types of FRP, the most famous of which is the carbon fiber–reinforced polymer (CFRP), which is usually used because of its practical applications and its unique properties in terms of resistance to stress and resistance with time. As stated (Roaster 1990), the performance is good at increased temperature.

10.9.1 CFRP Types

There are different commercial dimensions of CFRP in the market, and the most common widths used are 1–1.5 and 50–150 mm. Both widths consist of 60% to 70% carbon fiber in one direction, with diameter of about 10 µm embedded in epoxy resin.

Its mechanical properties are different from one type to another, but generally, the modulus of elasticity is between 165 and 300 N/mm^2 and the tensile strength ranges from 2800 to 1300 N/mm^2, and we find that the lower value of the modulus of elasticity corresponds with the maximum tensile strength.

The carbon fiber sheet, as shown in Figure 10.16, is embedded in resin of epoxy.

Another important property is density. The density of CFRP is about 1.6 t/m^3, compared with steel, which is 7.8 t/m^3, or about 10 times heavier than CFRP; this big difference in weight makes the transfer of CFRP easier to manage.

The adhesive materials used to bind the sheets of carbon fiber resin are a mixture of high resistance and the rule of filling quartz and its tensile strength is about 30 MPa, up from the concrete tensile strength to about 10 times as well as the rate of shrinkage and creep is small and bear the high temperatures and bear exposure to chemicals.

The mechanical properties of carbon fiber sheets in the longitudinal direction are often controlled by the fiber and its behavior is linear elastic until it collapses.

10.9.2 Application on Site

Before strengthening concrete structures, it is necessary to fully examine all the circumstances surrounding the structure in order to determine the type

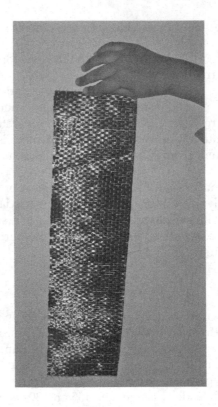

FIGURE 10.16
Carbon fiber sheet.

of strengthening needed. It is preferred that the company supplying the plastic fiber perform the installation because they are more experienced with the design and can easily identify the required thickness.

If the loss of strength is the result of steel corrosion or chemical attack, it is important to first remove the rust from the steel bars and the chemical substances that attack the concrete before strengthening the structure.

The concrete must be settled through sand blasting to level the surface and eliminate any gaps or delaminated concrete.

The next step is to collect all the damaged concrete and sand by air suction, air pressure, or water pressure to clean the surface.

Before fixing the CFRP, it is strongly recommended that the surface must be leveled and cleaned, and in case of cracks or delaminated concrete, the surface must be repaired. The mixture that provides the bond between the segments of CFRP and the concrete surface is very important because it is responsible for the transfer of the stresses from the concrete to the CFRP. Therefore, the preparation of the epoxy mix and the percentage of mixture

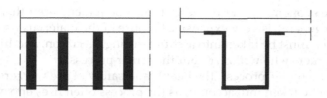

FIGURE 10.17
Strengthening the beam for shear failure.

component must be precise and under the supervision of the manufacturer, as it needs an extensive experience in the application, transportation, and storage.

This epoxy material is a blend of resin and hardener mixed by an electric hand mixer for about 3 minutes, until mixture homogeneity find clear when either the bottom and sides of the container containing a homogenous mixture and generally greater speed allowed is about 500 rpm.

The epoxy should be applied accurately on the surface, with a thickness of 0.5–2 mm, and all the gaps must be filled, making sure there are no air voids.

The repair of a beam damaged by shear stress or with reduced shear stress capacity is also shown in Figure 10.17.

10.10 General Precaution

Note that the safety of the structure relies primarily on the quality of the repair process in terms of execution, design, and planning, the repair process, the identification of the necessary temporary support of the building and the location, which are dependent on the nature of the structure and its structural system.

Also, determining the method in breaking off and removing the defective concrete and choosing the materials for the repair must be in full conformity with the state and nature of structure and location of the retrofit members within the overall structural system, and all these elements require special expertise. Without these experiences in the repair process, the possibility of risk in terms of safety of structure is very high. An ineffective repair process also has serious economic consequences (e.g., performing a complete repair of a member due to chloride that is suspected to come from the outside but is actually inside the concrete).

From the economic point of view, it must be remembered that the repair process, in general, is very expensive because of the materials used, the precautions that must be taken into account during execution, and highly experienced workers who will carry out the repair process.

During the repair process, the health and safety of the workers doing the repair must be taken into account, as the process often uses epoxy materials or polymers that cause breathing and skin problems over time; in addition, some other tools require careful attention. Unfortunately, although health and safety precautions are being followed in developed countries, they do not have the same degree of attention in developing countries, despite the fact that the interest in this matter provides a lot of time and money for the whole project.

After the repair process, taking into consideration the repair methods, the different ways of execution, and the workers' health and safety precautions, the equipment should be cleaned very well; otherwise, the epoxy polymers and other materials left will damage the equipment and lead to subsequent operation problems.

References

Abdul-Wahab, H. M. 1989. Strength of reinforced concrete corbels with fibers. *ACI Struct J.* 86(1):60–66.

American Concrete Institute (ACI) Committee 440. September 2000. *Guide for the Design and Construction of Externally Bonded FRP Systems for Strengthening Concrete Structures.* Farmington Hills, MI: ACI.

American Concrete Institute (ACI) Committee 506. 1994. Proposed revision of specification for materials, proportioning and application of shotcrete. *Am Concrete Inst Mater J.* 91(1):108–115.

American Concrete Institute (ACI) Committee 548. 1994. *State of the Art Report on Polymer Modified Concrete. ACI Manual of Concrete Practice.* Farmington Hills, MI: ACI.

American Concrete Institute (ACI) Committee 546. 2001. *Concrete Repair Guide.* Farmington Hills, MI: ACI.

U.S. Army Corps of Engineers. 1995. *Engineering and Repair of Concrete Structures,* Manual 1110-2-2002. Washington, DC: U.S. Army Corps of Engineers.

Mufi, A., M. On Ofei, B. Benmok, N. Banthica, M. Boulfiza, J. Newhook, B. Bakht, G. Tedros, and P. Brett. 2005. *Durability of GFRP Reinforced Concrete in Field Structure.* 7th International Symposium on FRP for Reinforced Concrete Structure (FRPRC5-7), USA, November 7, 2005.

Neubauer, U., and F. S. Rostasy. 1997. Design aspects of concrete structures strengthened with externally bonded FRP plates. *Proc 7th Int Conf Structural Faults and Repair.* 2:109–118. Edinburgh: Engineering Technics.

Ohama, Y. 1995. *Handbook of Polymer-Modified Concrete and Mortars.* Park Ridge, NJ: Noyes Publication.

RILEM Committee 124-SRC. 1994. Draft recommendation for repair strategies for concrete structures damaged by steel corrosion. *Mater Struct.* 27(171):415–436.

Vorster, M., J. P. Merrigan, R. W. Lewis, and R. E. Weyers. 1992. *Techniques for Concrete Removal and Bar Cleaning on Bridge Rehabilitation Projects.* SHRP-S-336. Washington, DC: National Research Council.

11

Economic Study for Maintenance Plan

11.1 Introduction

In general, the economic factor is one of the most influential factors in an engineering project. In most industrial companies, part of the annual budget is allocated for the maintenance of concrete and steel structures to ensure their reliability during plant operations. The main challenge is to define the maintenance budget and how to spend this budget in the right manner.

Maintenance philosophies are usually initiated by the mechanical engineering section as part of its efforts to increase the reliability and availability of machines and rotating equipment. Recently, these methods have been applied to structures.

These philosophies stem from corrective failures that fail to perform any maintenance until partial failure occurs such as spall of concrete cover. This method is very costly and cannot be applied to critical and large structures. As an alternative, we apply preventive maintenance, which consists of painting the steel structure regularly.

Nowadays, maintaining a reliable structure entails adopting predictive maintenance by monitoring the foundation vibration or hair cracks or settlement, which we will focus on in our discussion regarding periodic inspection plans. This type of maintenance philosophy features a proactive maintenance stance, which focuses on monitoring and addressing of root causes of failures (e.g., concrete contamination with chloride).

To apply the predictive and proactive maintenance philosophies, a group must be formed that will oversee this program from the planning phase up to its implementation phase through the integrity management organization.

Recently, based on the total integrity management system, large corporations have started adopting this strategy to maintain the structural integrity of their facilities.

Thus, there are different techniques that can be used to define inspection and maintenance plans for onshore and offshore structures. One prime example is the risk-based inspection (RBI) methodology, which provides a complete set of inspection and maintenance plans. Using this approach, a company can set aside an annual budget to perform inspection and maintenance and at the same time draw up resources management plans to achieve its maintenance targets.

By knowing the whole structure lifetime, it can be easy to calculate the expected number of times that routine maintenance and inspection must be performed throughout the structure's lifetime.

The economic study to choose alternatives in the case of a new structure is very important in determining the appropriate method for protection by taking into account the life span of the protection method. In addition, it is necessary to take into account the initial costs of periodic maintenance.

Generally, cost calculation is based on a summation of the initial costs of protection with the cost of maintenance that will be performed in different periods. The number of maintenance times during the whole life of a structure varies depending on the method of protection and the method used in the repair procedure during maintenance.

To maintain an existing old building and restore the structure to its original strength, the repair method of rehabilitation is also governed by the initial costs of repair and the number of times the maintenance will be performed and its costs over the remaining life of the structure.

The previous chapters focused on how to choose among the various alternatives for the protection methods that can be used as well as the appropriate repair methods and the materials that are usually used in repairs. The previous discussion is presented from a technical point of view only. In this chapter, however, the way of comparing between different alternatives will be discussed from an economic point of view to assist us in the decision-making procedure.

The lifetime of maintenance or the protection of the structure must be taken into consideration, so the required time to perform the maintenance must be discussed thoroughly, and the first step is to calculate this time.

The process of determining the time of maintenance depends on the maintenance cost estimate versus the structure probability of failure; therefore, it is important to decide on the selection of the appropriate time to perform maintenance.

The process of decision-making will be clarified and how to use it from the standpoint of determining the right time for maintenance in a less-expensive verification process method called "optimization procedure."

11.2 Basic Rules of Cost Calculation

There are several basic rules for calculating the economic costs of any engineering project and for applying those rules to select the type of protection required for a structure as well as the selection method of repair.

The most popular methods of economic analysis tools are the present value, future value, and interest rate of return. Here, the method of calculating the present value is described briefly, as it is the easiest way to select the appropriate method of economic repair and an appropriate system to protect the structure from effects of corrosion.

11.2.1 Present Value Method

The cost of protecting reinforced concrete from corrosion consists of the preliminary costs of this method of protection and the money paid with the beginning of construction. On the other hand, the cost of maintenance and repair will extend over the lifetime of the structure.

In many cases, the cumulative cost of maintenance and repair is higher than initial costs. In many projects, cost calculation is often based on the initial costs only, but the result is that the total cost is very high compared to those of the structure's cost estimate, which covers only the initial cost.

A method of present value is used to calculate the present value of future repair, including the cost of equivalent current value with the assumption that the repair will take place after a number of years, n.

$$\text{Present value} = \text{Repair cost } (1 + m) - n \qquad (11.1)$$

where m is the discount rate, which is the interest rate before the inflation rate (e.g., when one assumes that the interest rate is 10% and the inflation rate is 6%, the discount rate (m) is equal to 4% or 0.04), and n is the number of years.

The total cost of the structure consists of the initial cost, which is called capital cost (CAPEX), in addition to the sum of the present values of future costs due to maintenance, which is called operating cost (OPEX). When the rate of inflation increases, the cost of future repair will not be affected. But when it decreases, present value of future repair will increase. Note that the inflation rate varies depending on a country's general economic conditions and this information is usually available.

11.3 Repair Time

The time required to repair the structure as a result of steel corrosion is the time at which corrosion begins in the steel reinforcement bars and the time needed to spall the concrete cover with signs of concrete deterioration. This essentially requires work with repair.

Tuutti (1982) offered a simple explanation of a process of corrosion with time; the steps were for all types of corrosion and there was no difference if corrosion happened as a result of chloride attack or carbonation propagation. However, the invasion of chlorides or the carbonation of concrete takes a long time to break the passive protection layer on the steel bars and start corrosion.

After that, the span between the beginning of corrosion to a significant deterioration in the concrete, when repair will be necessary, will take more time. Raupach (1996) pointed out that in the case of concrete bridges, degradation occurs in about 2–5 years. Therefore, the time for repair is the total time required for the protection of depassivation in addition to 3 years.

The process by which corrosion affects the concrete structure is illustrated in Figure 11.1 (El-Reedy, 2006). Note that after construction, it will take time before chloride concentration or carbonation starts to accumulate on the surface and eventually spread into the concrete.

The next step is the propagation of chlorides or carbonation until the steel bars are reached. The third step is the start of corrosion on the steel bar, which at this time will have an impact on concrete strength via reduction of steel diameter and appearance of cracks on the concrete surface. The last step is an increase in crack width until spalling of the concrete cover.

From the preceding analysis, we find that the time required for repair depends on the time needed to increase the percentage of chloride concentration to the limit that will initiate corrosion, in addition to the rate of corrosion, which will happen after that.

The previous chapter discussed several types of methods for the protection of the structure from corrosion. These methods will delay the start of

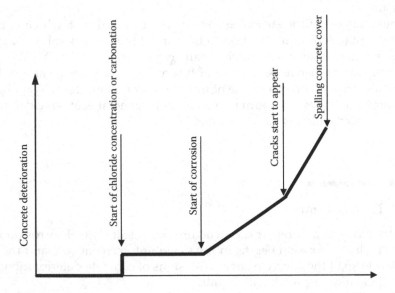

FIGURE 11.1
Deterioration of a concrete structure.

corrosion for a longer period, as they reduce the rate of the chloride or carbonation propagation in concrete. They will also reduce the rate of corrosion
after that. Note that the preceding analysis relies on the noninterference of
hair cracks on the concrete in the rates of spread of chlorides or carbonation
within the concrete. It is assumed that the design was based on the absence
of an increase in cracks more than that permissible in codes.

The time required to start the structure repair depends on the nature of
the surrounding weather and environmental factors affecting the beginning
and the rate of corrosion. This time is determined by knowing the rate of corrosion and the required time to spall the concrete cover. The deterioration of
concrete increases the probability of the collapse of the structure with time;
several studies have identified the probability of failure, which should not be
beyond the structural reliability classified in various specifications.

A study on residential buildings by El-Reedy et al. (2000) focused on how
to determine the appropriate time to perform repair due to corrosion on the
concrete columns. This study considered environmental conditions around
the structures—they were affected by humidity and temperature, which
have a large impact on the increasing rates of corrosion. A corrosion rate
of 0.064 mm/yr reflects the dry air, whereas a corrosion rate of 0.114 mm/
yr is based on a very high rate of moisture. This study took into account the
increasing resistance of concrete over time, as well as the method of determination in the case of higher steel or low steel columns with different times
required for the repair process.

11.3.1 Capacity Loss in Reinforced Concrete Sections

Based on the fundamentals of design of any reinforced concrete member,
the member's capacity is dependent on the cross-sectional dimensions (concrete and steel area) and material strength (concrete strength and steel yield
strength).

In the case of uniform corrosion as shown in Figure 11.2, the total longitudinal reinforcement area can be expressed as a function of time, t, as
follows:

$$As(t) = \begin{cases} n\pi D^2/4 & \text{for } t \leq T_i \\ n\pi \left[D - 2C_r \left(t - T_i \right) \right]^2 / 4 & \text{for } t > T_i \end{cases} \tag{11.2}$$

where D is the diameter of the bar, n is the number of bars, T_i is the time of
corrosion initiation, and C_r is the rate of corrosion. Equation (11.2) takes into
account the uniform corrosion propagation process from all sides.

The curves in Figures 11.3 and 11.4 show that, over time, the collapse of a
structure is more likely with the increase in the rate of corrosion.

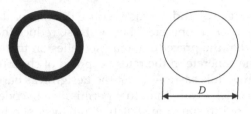

Uniform corrosion on steel bars   Steel bars without corrosion

FIGURE 11.2
Reduction in steel diameter due to uniform corrosion.

Khalil et al. (2000) study revealed that concentrically loaded reinforced concrete columns will be due for maintenance in about 4 to 5 years from the initial time of corrosion with the lowest steel ratio and higher corrosion rate. In cases of higher steel ratios and lower corrosion rates, this period may increase to about 15 to 20 years. Moreover, this study expects that in the case of columns with eccentricity, the moment on the column will increase so the design should increase the percentage of steel. In such a case, it is expected that the deterioration of a structure toward criticality will ensue after a couple of years in the case of the lowest reinforcing steel ratio and very high corrosion rate. On the other hand, in the case of a higher reinforcing steel ratio and lower corrosion rate, the maintenance will be due after about 6 years.

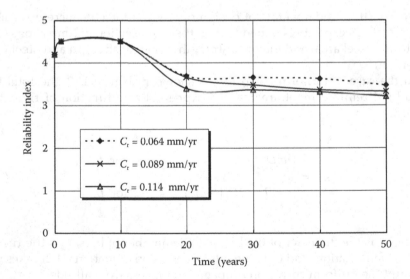

FIGURE 11.3
Effects of corrosion rate on reliability index at a reinforcement ratio = 1%.

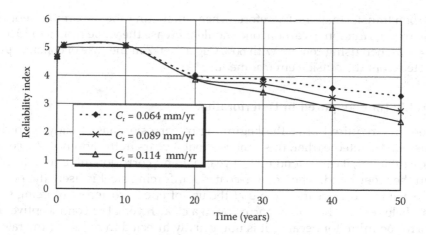

FIGURE 11.4
Effects of corrosion rate on reliability index at a reinforcement ratio = 4%.

As we expected, the time required for repair is closely related to the nature of the structure and the method of design, as well as to the importance of the building. For example, vital structures such as nuclear facilities require protective measures with a time much different from that of other structures such as residential buildings.

11.3.2 Required Time to Corrosion

In the case of structures exposed to chlorides, the time required for the spread of chloride in the concrete until it reaches the steel is the time required for corrosion.

In the case of structures exposed to carbonation spread inside the concrete cover, several equations can be used to calculate the time required for this propagation. To calculate the depth of the carbon transformation, the following equation can be used:

$$d = A(t)^{-0.5}, \tag{11.3}$$

where A is a fixed amount depending on the permeability of concrete as well as the quantity of carbon dioxide in the atmosphere and several other relevant factors.

$$A = (17.04(w/c) - 6.52)SW, \tag{11.4}$$

where w/c is the water/cement ratio (less than 0.6), S, is the effect of cement type, and W is the weather effect. This equation gives the average depth of carbon transformation. Therefore, when calculating the maximum depth of the transformation of carbon, one should increase the value from 5 to 10 mm. $S = 1.2$ when using cement with 60% slag. $W = 0.7$ in the case of concrete protected from the outside environment.

11.3.3 Time Required to Deterioration

The time required after the beginning of corrosion has already been discussed. As noted earlier, this time is about 3 years in the absence of a corrosion inhibitor; it can extend up to 4 years when a corrosion inhibitor is added. On the other hand, when epoxy-coated reinforcing steel is used, the period is extended to 15 years. Note that the use of epoxy-coated reinforcing steel can help reduce the rate of corrosion in a clear reversal of contraceptive use corrosion inhibitor because it is not actually affected in reducing the rate of corrosion.

In general, the time of collapse of concrete cover from the beginning of corrosion depends on the rate of corrosion occurring in steel reinforcement. Several studies have calculated the rate of corrosion. It was found that corrosion is closely related to relative humidity. As stated by Tuutti (1982), in the case of steel corrosion due to carbonation, the corrosion rate decreases gradually at a relative humidity of 75% or lower. This rate increases quickly when relative humidity reaches 95%. It is noted that when temperatures drop below 10°C, the rate of corrosion will be reduced by about 5–10%.

Broomfield (1997) stated that the rate of corrosion is affected by the relative moisture and a dry situation where concrete is poured and the proportion of chlorides. Morinaga (1988) stated that the rate of corrosion was totally prevented when the relative humidity was less than 45%, regardless of the chloride content and the temperature or oxygen concentration.

Broomfield also stated that cracks in concrete occur when a lack of steel is about 0.1 mm or lower. That depends on the oxygen concentration and distribution and the concrete's ability to withstand excessive stresses. The rebar distribution can increase the cracks by concentrating stresses, if the spacing of the steel bar is not enough or at a corner where there is a less confinement to the steel bar to resist cracks. It has been observed that reduction in a section of about 10–30 μm can result in a fragile layer of corrosion and is enough to cause concrete cracks.

The following equation of Kamal et al. (1992) is used to calculate the time necessary from the beginning of corrosion to the emergence of the effects of corrosion on the concrete:

$$t_s = 0.08(C - 5)/(DC_r), \tag{11.5}$$

where t_s is the time from the beginning of corrosion until the concrete cover falls, C is the concrete cover thickness, D is the steel bar diameter, and C_r is the corrosion rate (mm/year). The total time expected to perform the first repair depends on the time of the beginning of corrosion in addition to the time needed to increase the rate of corrosion to cause concrete deterioration, and to perform the repair.

11.4 Repair and Inspection Strategy and Optimization

The decision-making methods of engineering projects have been the focus of numerous studies—in terms of their economic importance to the cost of the project as a whole—because a wrong decision could result in spending huge amounts of money to reverse it. Therefore, some studies have used decision trees to determine the appropriate time for repair, particularly in some bridges that require periodic inspections.

When performing the maintenance plan strategy based on cost, we must consider all the factors affecting corrosion in general, and other factors affecting structure durability.

Any method upon which a decision is based should define the time for regular maintenance and inspection (e.g., 10, 15, or 20 years). This decision is made first and foremost through the identification of the least expensive option.

Figure 11.5 summarizes the reliability of any structure during its lifetime. As shown, the structure in the beginning will have a higher capacity; the probability of failure value is based on the code of standard on which the structure design is based (in most codes it is 1×10^{-5}). After a certain period, the structure will deteriorate, thereby increasing its probability of failure. After time Δt, when inspection and repair have been conducted, the structure will recover its original strength, as shown in the figure. After another period, the probability of failure will increase until it reaches a certain limit and then further maintenance is performed, and so on.

Inspection alone does not improve reliability unless it is accompanied by a corrective action in the event that a defect is discovered. Several policies

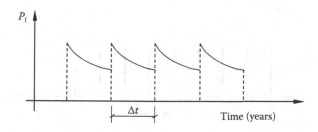

FIGURE 11.5
Concrete structure performance.

and strategies used in a wide range for concrete structures, which have been programmed, include:

- Monitoring until the crack depth reaches a certain proportion of the material thickness and then repairing
- Immediately repairing upon detection of indications of damage
- Repairing at a fixed time (e.g., 1 year) after detection of indications of damage
- Repairing as new (i.e., repair welding)

In general, it is assumed that inspections are performed at constant time intervals as shown in Figure 11.6, since the inspection authorities often prefer a constant inspection interval to facilitate planning. From this point, the inspection intervals are chosen so that the expected cost of inspection, repair, and failure is minimized.

Although most concrete structure inspection techniques are visual or involve various nondestructive testing (NDT) inspection methods, the ability to detect damage depends on the quality of inspection performed. A higher-quality inspection method will provide a more dependable damage assessment. No repair will be made unless the damage is detected. In other structures such as steel structures, general inspection usually includes visual inspection and performance of NDT such as ultrasonic tests (UT) if section reductions are detected or UT and magnetic particle inspection to check the welds if the structure is under dynamic loads.

For offshore structures, inspection will be performed on the topside similar to steel structures onshore, but the jacket of the platform must undergo underwater inspection. Underwater inspection, which incurs high costs, can be performed by divers or remote operating vehicles (ROVs). ROVs or divers perform visual inspection to detect the flooded member, define the location of damaged or bent members, define the thickness of marine growth and scour, and measure the cathodic protection reading.

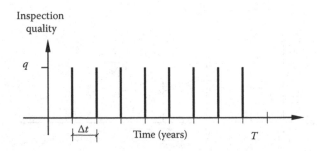

FIGURE 11.6
Inspection strategy.

11.4.1 Repair

Inspection may not affect the probability of failure of the structure. After an inspection, a decision must be made regarding repair if damage is found. The repair decision will depend on the quality of inspection. With advanced inspection methods, even a small defect can be detected and repaired. Higher quality inspection may lead to higher quality repair, which brings the reliability of the structure closer to its original condition. Aging has an effect on the structure, so the reliability of the structure is decreased. In this chapter, it is proposed that after inspection and repair, the structure capacity will be the same as the design conditions, which are shown in Figure 11.5.

11.4.2 Expected Total Cost

As noted by El-Reedy and Ahmed (1998), the first step is to determine the service life of the structure. Assume that it is 75 years and routine maintenance is scheduled once every 2 years (starting at $t = 2$ years and continuing until $t = 74$ years). Consequently, preventive maintenance work will be performed 37 times during the life of the structure. Therefore, the lifetime routine maintenance cost becomes:

$$C_{FM} = C_{m2} + C_{m4} + C_{m6} + \cdots + C_{m74}, \tag{11.6}$$

where C_{FM} indicates the total maintenance cost, and the total expected cost in its lifetime (T) is based on the present value. The expected lifetime preventive maintenance cost becomes

$$C_{IR} = C_{IR2}\frac{1}{(1+r)^2} + C_{IR4}\frac{1}{(1+r)^4} + C_{IR6}\frac{1}{(1+r)^6} + \cdots + C_{IR74}\frac{1}{(1+r)^{74}}, \tag{11.7}$$

where C_{IR} is the periodic inspection and minor repair and r is the net discount rate of money.

In general, for a strategy involving m lifetime inspections, the total expected inspection cost is:

$$C_{ins} = \sum_{i=1}^{m} C_{ins} + C_R\frac{1}{(1+r)^{T_1}}, \tag{11.8}$$

where
C_{ins} = Inspection cost based on inspection method
C_R = Repair cost
R = Net discount rate

Finally, the expected total cost C_{ET} is the sum of its components including the initial cost of the structure, the expected cost of routine maintenance, the expected cost of preventive maintenance, which includes the cost of the inspection and the repair maintenance, and the expected cost of failure. Accordingly, C_{ET} can be expressed as:

$$C_{ET} = C_T + (C_{ins} + C_R)(1 - P_f) + C_f P_f \qquad (11.9)$$

The objective remains to develop a strategy that minimizes C_{ET} while keeping the lifetime reliability of the structure above a minimum allowable value.

11.4.3 Optimization Strategy

To implement an optimum lifetime strategy, the following problem must be solved:

$$\text{Minimize } C_{ET} \text{ subjected to } P_{f\,life} \leq P_{max},$$

where P_{max} is the maximum acceptable lifetime failure probability. Alternatively, considering the reliability index,

$$\beta = \phi^{-1}(1 - P) \qquad (11.10)$$

where ϕ is the standard normal distribution function. The optimum lifetime strategy is defined as the solution of the following mathematical problem:

$$\text{Minimize } C_{ET} \text{ subjected to } \beta_{life} \geq \beta_{min}$$

The optimal inspection strategy with regard to costs is determined by formulating an optimization problem. The objective function (C_{ET}) in this formulation is defined as including the periodic inspection and minor joint repair cost, and the failure cost, which includes the cost of major joint repair. The inspection periodic time (Δt) is the optimization variable constrained by the minimum index β specified by the code and the maximum periodic time.

The optimization problem may be mathematically written as:

Find Δt, which minimizes the objective function:

$$C_{ET}(\Delta t) = (C_{IR})(1 - P_f(\Delta t)) \left(\frac{(1+r)^T - 1}{\left((1+r)^{\Delta T} - 1\right)(1+r)^T} \right) + C_f P_f(\Delta t) \left(\frac{(1+r)^T - 1}{\left((1+r)^{\Delta T} - 1\right)^T (1+r)^T} \right)$$

$$(11.11)$$

subject to $\beta(t) \geq \beta^{min}$, $\Delta t \leq T$, where C_{IR} is the periodic inspection and minor repair cost per inspection; C_f is the major repair cost; i is the real interest rate; and β^{min} is the minimum acceptable reliability index. C_{IR} and C_f are assumed constant with time.

Even though the failure cost is minimized (as part of the total cost), it is often necessary to put a constraint on the reliability index to fulfill code requirements. Since the period for the proposed repair is T, and n is the total lifetime of the building and C_{IR} is the cost of inspection and repairs, and they are costs resulting from the collapse.

The value of the expected cost at each period is a curve that follows Figure 11.7, which defines the period required for the process of periodic maintenance, which is achieved at less expense, as in the curve that follows.

Δt indicates the period required for the process of the periodic maintenance, which achieved at the highest cost. The previous equation can generally be used for comparison between different types of repair with different time lengths and different costs. It is calculated by adding the initial cost of the repair and design curve. The decision can be determined in terms of the expected cost and risk of collapse and is calculated by using different methods to calculate the possibility of the collapse of the latest member or the use of the approximate method. For the possibilities imposed at the beginning of the age of the structure, the probability of collapse at the age of 0 and the structure lifetime is assumed to be 60 years. The possibility of collapse is 100% at this age, and calculating the probability of collapse is shown at every period in Table 11.1.

A number of special software for the management of bridge systems (BMS), as identified by Abd El-Kader et al. (1998), defines the time required to

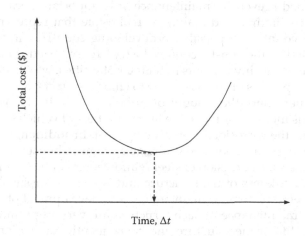

FIGURE 11.7
Optimization curve to obtain optimum maintenance period.

TABLE 11.1

Relation between Structure Lifetime and Probability of Failure

Life (years)	0	10	20	30	40	50	60
Probability of failure	0%	9%	25%	50%	75%	91%	100%

conduct inspection and maintenance for each bridge. These programs take into account the rate of deterioration in bridges with the lifetime of the structure. One such program includes a previous section that has been studied, a limitation on the time required for periodic inspection and maintenance work, and how to determine the cost, which depends on the budget planning for the repair of bridges. The program was developed by the Federal Highway America.

The maintenance plan and its implementation depend on the criticality and the importunacy of the building. To proceed with maintenance, one should consider the sequences of failure. For example, there is a difference between concrete columns and concrete slabs. The cost of repairing these two members may be similar, or in some situations the repair of the concrete slab on the grade will cost more than repairing the columns. Which should be repaired first? It is obvious that repair will start on the concrete columns because failure will consequently cause bad consequences from an economic point of view. Therefore, as noted earlier, the previous equation is a general equation, so all factors affecting decision-making must be taken into consideration.

In general, conducting maintenance on a regular basis within a defined period is called preventive maintenance. It is important because it extends the structure's lifetime and maintains real estate that represents economic wealth for a country—especially in developing countries, in which all the materials used in the construction industry are imported. Moreover, some countries have huge investments in their coastal cities, which are highly vulnerable to the process of corrosion due to chloride effects.

On the other hand, the amount of carbon dioxide in the atmosphere in urban areas is increasing due to the large numbers of vehicles moving daily; this increases the corrosion of reinforcing steel. In addition, there are other problems such as sanitation and drainage systems in houses, considered one of the key factors in corrosion of steel reinforcement of bathroom slabs. This exposes the weakness of the structure and is a serious issue for the whole building. Therefore, regular maintenance and development of an integrated plan for the maintenance of each structure are very important in order to preserve the life of the structure and consequently, the national wealth of the country.

11.5 Maintenance Plan

In any organization, the maintenance team is responsible for maintaining the structural reliability of the buildings. The responsibility of this team is clearly in the ministry of transportation that is responsible for the maintenance of bridges. For instance, in the United States, the famous Golden Gate bridge is maintained by a team that regularly starts to repaint the first section of the bridge as soon as it has finished painting the last section.

When we discussed the time spent performing maintenance earlier, we took into account the probability of structural failure. From a practical viewpoint, it is complicated to calculate this accurately as it is needed for research rather than for practical reasons. The calculation of the probability of failure and its consequences to obtain the structural risk is called the *qualitative risk assessment*. Another popular method is the *quantitative risk assessment*, which is easy and can be handled through the maintenance team without the need for outside resources. The quantitative RBI will be the main tool used to maintain structural integrity.

11.5.1 Assessment Process

The assessment of the concrete structure depends on its structure type, location, and existing load and operation requirements. For studying the risk assessment, the different critical items that have an economic effect must be considered. The general definition of risk is summarized in the following equation:

$$\text{Risk} = \text{probability of failure} \times \text{consequences}$$

To define the risk assessment of any concrete structure, the factors that may affect the business's economics are worth considering. The structural risk assessment is represented by the probability of failure and its consequences; failure may involve the whole building, a part of it, or a concrete member only.

The first step in defining and calculating the quantitative risk assessment is to hold a meeting with the maintenance team, and perhaps with a consulting engineering company when numerous structures are involved. Within this meeting, which could take several hours, different discussions define the factors that affect the probability of structural failure, such as corrosion on steel reinforcement bars, overload compared to the design, new data revealing lower concrete strength than what the design and project specifications called for, and the redundancy of the structure itself.

The required data about the corrosion of steel bars, concrete compressive strength, and other factors have been discussed before and can be obtained

FIGURE 11.8
Comparison between structural redundancies.

by collecting the data and performing visual inspection, and performing detailed inspection using measurement equipment such as ultrasonic pulse velocity technique, rebound hammer, or other available techniques. The factor that has the largest influence on estimating the probability of structural failure is the structure's redundancy, which can be explained by Figures 11.8 and 11.9.

A beam needs to be designed and the following two solutions are offered: (1) a structural system to be fixed at two ends or (2) a structural system hinged at two ends. Which system should be chosen? Take 5 minutes to think about this.

Many factors influence your decision in selecting a suitable system. The following list gives the advantages and disadvantages for these two systems.

(1) Structural system 1
  - The beam cross section will be smaller.
  - The connections will be big and complicated as they have a shearing force and moment.
  - It is reasonable from an architectural point of view.
  - The construction is complicated in connection.
(2) Structural system 2
  - The beam cross section will be large.
  - The connection will be small as it is designed for shear force only.
  - It is easy to construct because the connection is simple.

After discussing these two systems, perhaps the decision will rest on in-house engineers and workers to do the construction. To avoid any problems, the simple beam option will be selected. This type of thinking is always

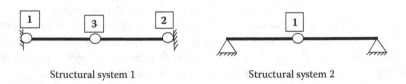

FIGURE 11.9
Comparison between structural redundancies.

happening as the maintenance point of view is often forgotten. The steps of collapse failure are shown in Figure 11.9.

Structural system 2, which is the simple beam, assumes that the load is gradually increased and the beam can accommodate the load until a plastic hinge is formed in the middle of the beam at the point of maximum bending moment; then, collapse will occur. On the other hand, for structural system 1, which is fixed at two ends, when the load is increased gradually, the weaker of the left or right connection will fail first. As shown in the figure, the plastic hinge will form on the left connection and increase the load; the other connection will be a plastic hinge, so it is now working as simple beam. By increasing the load, plastic hinge 3 will be formulated and then collapse will occur.

From the preceding discussion, one can find that structural system 1 will take more time to collapse because it failed after three stages. Structural system 2 failed on the first stage, so system 1 is more redundant than system 2. Moreover, when a comparison is made between different reinforced concrete members, such as slabs, beams, columns, and cantilevers, one can observe that some members are more critical than others. A cantilever is the most critical because any defect in it will have a high deflection and then failure. In addition, when a column fails, the load distributed to other columns until the whole building fails. Thus, the column has low redundancy and is very critical as any because failure on it will result in the whole building failing. However, cantilever failure will be a member failure only, so we can go through the consequences from this approach.

In the case of beams and slabs, when a slab is designed, a simple beam is assumed and the maximum moment is calculated in the middle. Next, the concrete slab is designed by choosing the slab thickness and the steel reinforcement. The selected steel reinforcement will be distributed along the whole span. Theoretically, by increasing the load, the failure will be at a point, but actually the surrounding area will carry part of this load, so the redundancy of the slab is very high. Some studies have noted that the reinforced concrete slab can accommodate loads that are twice the design load.

For the steel structure as a whole, it is easy now to perform pushover analysis to obtain the redundancy of the structure. This analysis is nonlinear and is now available for any structure analysis software in the market. From this analysis, one can obtain how much more load the structure can carry in excess of the design load until failure. In addition, one can know the location of the first plastic hinge to formulate and, from this, one can know the critical member in the structure.

Therefore, it is necessary to put a maintenance plan in place for a concrete structure. The first step is to discuss all the structural parts that need to be inspected and repaired and arrange this information in a table. The team will supply all the factors affecting the probability of structural failure into this table. Different factors will be assigned different values. For example, in the case of a redundancy factor, values will range from 1 to 10, and a column will have a value of 10 and a slab will bear a value of 2. All team members must agree on these numbers, based on their experience. The same values or higher values will be used for other factors: age, the engineering office that produced the design, the contractor who handled the construction, and the code used in the design. As discussed earlier, in some countries, those in charge of construction used seawater in mixing water, and some codes in the past have allowed the use of 6-mm-diameter steel in stirrups. Therefore, if corrosion does occur, these stirrups could not be seen because complete corrosion would have taken place.

Table 11.2 shows the probability of structural failure factors. (If you decide to use this table, make sure that it matches your structural requirements.) Note that the table is not limited to these factors only.

Table 11.2 is a simple example of calculating the quantity of the probability of failure. However, when a meeting is held, it is necessary to draw up a set of criteria about the number. For example, consider the engineering office that produced the design. If it is professional and well known in the market, and one has had no problems in dealing with this firm, then lower values should be assigned. For an incompetent firm with which one has had problems involving the same type of structure, higher values are given. This procedure will also apply to the contractor.

The code values are based from the experience of the team considering problems not only for its own buildings, but also for other buildings all over the country. Therefore, for complex structures or huge facilities, it is worth contacting a competent engineering office for an optimum maintenance plan.

If no inspection has been conducted, the building is a "black box" and should be given the highest value. The things that one knows are bad are less critical than the things one knows nothing about. Thus, it is necessary to go through the consequences using the same approach, considering the economic impact. Economic impact is the answer to the question of what the effect will be from an economic viewpoint if the whole structure or part of it fails.

TABLE 11.2

Factors Effects on Structure Probability of Failure

Structure	Redundancy 1–10	Age 1–10	Designer 1–10	Contractor 1–10	Code 1–10	Last Inspection 1–10	Total Score
S1	8	10	6	6	9	10	49
S2	5	2	7	3	3	10	30
S3	3	5	8	10	5	2	33
S4	7	4	2	5	5	1	24
S5	1	3	4	2	5	10	25
S6	3	1	4	5	6	7	26

TABLE 11.3

Consequences Weight

Structure	Impact on Person 1–10	Impact on Cost 1–10	Impact on Environment 1–10	Impact on Repetition 1–10	Total Score
S1	8	10	6	6	30
S2	5	2	7	3	17
S3	3	5	8	10	26
S4	7	4	2	5	18
S5	1	3	4	2	10
S6	4	5	3	5	17

There are many hazard factors that will guide structure risk assessment. The first factor is the location itself, as well as its expenses and if it carries a processing facility for industry that must be taken into consideration because failure may cause hazards or stop production. For example, consider the foundation under machines; any failure in it will influence the performance of the machine and will be the main cause of the processing plant's shutting down and stopping production.

The weight of the consequences will be calculated by the same procedure used to calculate the probability of failure (Table 11.3).

Now that we have the weight of the probability of failure and the consequences, we can calculate the quantity risk assessment using Table 11.4.

11.5.2 RBI Maintenance Plan

After we calculate the quantity risk assessment as shown in Table 11.4, we can classify the maintenance plan. The top third of the top risk structures will be red, the second third will be yellow, and the remaining third will be green; the priority list is shown in Table 11.5.

From Table 11.5, the structures S1 and S3 are considered the critical structures upon which inspection would start. Structures S2 and S6, being the

TABLE 11.4

Risk Weight Lists for All Structures

Structure	Probability of Failure	Consequences	Risk
S1	49	30	1470
S2	30	17	510
S3	33	26	858
S4	24	18	432
S5	25	10	250
S6	26	17	442

TABLE 11.5

Structures Priority List

Structure Priority	Color Code	Risk Values
S1	Red	1470
S3	Red	858
S2	Yellow	510
S6	Yellow	442
S4	Green	432
S5	Green	250

second priority, will be inspected next, and the last structures, S4 and S5, will be at less risk so inspection is not needed at this time.

As shown in the table, this simplified method can be used to plan your maintenance for inspection and repair, taking into account that the budget is a very important factor in the maintenance plan. From the previous example, the budget may be enough this year to inspect structure S1 only, so you will have to plan for the following years—perhaps for the next 5 years. After the inspection has been performed, the data will be analyzed through the system shown in Figure 11.10.

Nowadays, for international organizations with a number of different buildings and structural elements, the integrity management system is very important to maintaining structures during their lifetimes and accumulating all historical data.

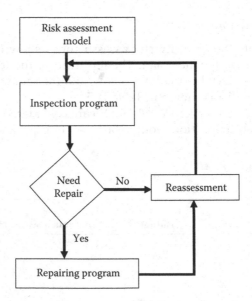

FIGURE 11.10
Flowchart for risk assessment system.

The closed loop of a structure risk assessment, an inspection program, a repair program, and reassessment is very important to maintaining the structure in a good condition and as a reliable match with operational requirements.

The total quality control management system is very important when reviewing design, construction, and maintenance to be matched with operational requirements in cases of proposed developments or adding extensions to existing structures or increasing the loads. Therefore, it must be considered by a system of change management procedures to upgrade the previous risk ranking table.

11.5.3 RBI Plan for Offshore Structures

Offshore structures are a special type of structures that use the same methodology as discussed in the previous sections, but the factors affecting the probability of structural failure are different.

The concept of ranking the platforms for topside and underwater inspection using a risk-based process is based on a similar approach developed by the American Petroleum Institute (API, 1996) for refineries and chemical plants. For refineries, the issue is the rationalization of inspection to determine the long-term degradation or damage of equipment, piping, and other elements of the facility's operations.

For offshore platforms, the issue addressed through RBI is the rationalization of inspections to determine the strength degradation/damage (if any) of the underwater portion and topside of the platform.

The RBI likelihood of failure corresponds to the probability that the platform will fail at some point in time through environmental overload. RBI, however, does not explicitly establish or use a quantitative probability of failure. Failure, in RBI, is defined as a collapse of the platform as a result of deterioration, extreme loading (storm or earthquake), or a combination of both.

Determination of the likelihood of failure requires information on a platform's structural configuration to determine its susceptibility to failure. For example, the fact that a platform has three, four, or eight legs, as well as its current state, based on inspection, may influence the extent of damage to its member(s).

As an example, a 1960s-era facility with an eight-leg, K-braced platform has a higher likelihood of failure than a 1980s-era eight-leg, X-braced platform. The newer platform is designed based on better standards, such as joint cans, and has an inherently more redundant structural configuration since it has eight legs and is X-braced.

However, if it is indicated that the newer platform has a track record of damage due to corrosion or fatigue cracking, then this may move the platform up the priority list, to a point where it is higher ranked in terms of risk compared to the older platform. The contribution of appurtenances such as risers and conductors to the likelihood of failure is also considered. Appurtenance failures may not necessarily lead to collapse of the platform, but may cause a higher impact in terms of environmental pollution and financial loss.

The consequence of failure corresponds to the safety, environmental, and financial issues that would arise should the platform fail at a future date. These are the standard consequential issues typically addressed in risk assessments for any type of facility, either onshore or offshore. As an example, a manned drilling and production platform would have a higher consequence of failure than an unmanned wellhead platform. Each of these consequences are converted into an economic value and then summed to result in the overall consequence.

11.5.3.1 Risk Matrix

Risk is a combined measure of both likelihood and consequence. By combining likelihood and consequence into a single quantity, it is possible to make rough comparisons of the effects of frequent inexpensive failures and infrequent catastrophic failures on the same scale. In a pure quantitative risk system, the likelihood is equivalent to the probability of failure and the consequence is equivalent to the dollar loss due to a failure.

Their product is the expected annual failure loss. The likelihood and consequence scores determined by RBI represent relative trends based on a comparison of a platform's operation, design, operation, and experience with other platforms. The categorization system results cannot be directly converted into a probability or a cost, making such a quantitative calculation impossible. Instead, the risk is expressed as a category that is based on the likelihood and consequence categorization results. These categories are usually summarized in the form of a risk matrix.

Figure 11.11 shows the risk matrix proposed by the API for the RBI assessment of refinery and petrochemical plants. In this figure, R refers to red, Y

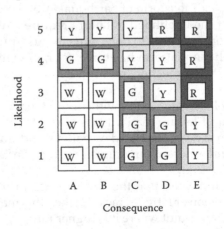

FIGURE 11.11
Risk matrix.

to yellow, G to green, and W to white; these colors indicate the degree of total risk. This same 5 × 5 matrix has been adopted for the RBI system, and although categorizing refinery equipment using this matrix differs from categorizing platform structures, the philosophy of the ranking is the same.

Other risk matrices have been used for offshore risk assessment. Many of these systems are based on 2 × 2 or 3 × 3 classification schemes. The 5 × 5 system shown was recommended by DeFranco et al. (1999) because it offers enough range in likelihood and consequence that it can distinguish the differences between platforms.

The combination of likelihood and consequence into the risk matrix results in four levels of risk ranging from insignificant to high as shown in Figure 11.11. The resulting risk value is used to rank the platforms in the fleet, with the platforms with the highest risk ranked at the top of the list and receiving priority for inspection.

11.5.3.2 Development of Likelihood

Likelihood is determined using a rule-based system that determines the likelihood score based on key platform information. In very simplified terms, the likelihood of structural failure is a function of two primary factors: the platform strength and the extreme load. The likelihood categorization system identifies the platform characteristics affecting the platform strength and loads, such as the year designed, the number of legs, the bracing system, and other factors, as shown in Table 11.6.

Factors indicating that the strength of the platform has deteriorated or is not up to current standards increase the likelihood. Factors indicating that extreme platform loads may increase in frequency or severity also increase the likelihood.

The RBI likelihood rules are based on the assumption that platforms designed according to modern structural detailing practice to resist present-day design environmental loads have the lowest likelihood of failure. The factors affecting the original strength, the maximum design loads, and the degradation of strength are used to measure any individual platform's failure likelihood against the ideal platform.

Table 11.6 lists most of the factors that were identified as key indicators of the likelihood of failure; these factors should be defined well by the team members with the appropriate expertise.

There are two types of data on the list. The first type is associated with the platform's original design configuration, such as the year designed, the number of legs, and the bracing scheme. This information is used in a set of rules to determine the baseline likelihood of failure, that is, this establishes the lowest possible likelihood of failure score for the platform.

The second type is previous inspection data such as the number of flooded members, excessive marine growth, scour, and others.

TABLE 11.6

Factors Affecting Offshore Platform Probability
of Failure

Rule	Weight
Design practice	5
Number of legs and bracing configuration	10
Foundations	10
Risers and conductors	5
Boat landing	5
Grouted piles	3
Damaged, missing, and removed members	20
Splash zone corrosion and damage	10
Flooded members	6
Cathodic protection (CP) surveys and anode depletion	6
Inspection history	5
Design loading	5
Marine growth	6
Scour	5
Topside weight change	10
Additional risers, caissons, and conductors	5
Wave-in-deck	20

There are no readily available simple rules that can be used in combination with this information to determine the likelihood of platform failure. In a fully quantitative risk analysis, a specific platform collapse pushover analysis for the platform as-designed and damage states or other types of complex study could be used to determine specific failure likelihood, but this work is time-consuming and may be cost-prohibitive to implement for a large fleet of platforms. Therefore, for RBI, a set of qualitative rules should be developed based on experts' recommendations and input from the study participants as well as results from past studies of their platforms.

It is important to note that these populations represent all three generations, a range of extent of inspection, and a large range of production rates, and hence provide a comprehensive testing and calibration data set. The combination of expert input plus calibration with actual platform performance was felt to provide an adequate level of accuracy for an approach such as this.

The equation for determining a risk ranking score S_{total} is

$$S_{Total} = \sum_i W_i S_i \tag{11.12}$$

where W_i is the ith individual weighting factor and S_i is the ith individual rule score. The general weights (W) for determining likelihood are shown in Table 11.1 (El-Reedy, 2006), whereas an example score (S) for the number of legs and bracing type are given in Table 11.6. As an example, the partial total score for a six-leg, K-braced platform, Bracing and Legs (i.e., the bracing configuration and number of legs interaction), would be 60 ($W_i = 10$, $S_i = 6$). Accordingly, the higher the score, the higher the relative risk. Should the risk be inspectable, then the higher-risk platforms may need a higher level or amount of inspection. The individual scores are then added to determine a total likelihood for a platform.

Table 11.7 shows an example score pertaining to the number of legs and bracing scheme. This score also shows that there may be interactions between the various parameters that must be accounted for in developing each rule (in this case, the number of legs and the bracing scheme) (Table 11.8). As another example, the effect of finding flooded members is dependent on the bracing scheme for the platform—an X-braced platform may be more tolerant to having flooded members than a K-braced platform because of the additional load paths in an X-braced structure. Finally, each rule is modified to account for the overall effect on the platform likelihood of failure.

This is as expected because, overall, an unmodified older platform typically has a higher probability of failure than a newer platform because of changes in the design procedure and hydrodynamic loading. However, in some cases, the effect of inspection is outweighed by the original design criteria. To make inspections more relevant, inspection-related data, such as the number of flooded members and damaged members, were assigned a higher weight on their scores so that these values are more relevant. This reweighting causes a damaged platform to move toward the top of the list quickly, depending on the amount of damage and its effect on structural integrity.

11.5.3.3 Development of Consequence

The consequence category is based on a simple analysis of the business, safety, and environmental losses, as measured in terms of an abstract dollar. A financial measure of loss for all types of failure consequences allows the combination of the three loss types into a single total loss. This total loss is then used to define a consequence category.

A default set of consequence rules was developed based on the work of Stahl (1986). When specific estimates of consequence based on detailed analysis or expert opinion are available, then the more accurate estimates may be used instead of the default values.

Although each type of consequence is measured in a common unit, abstract dollars, different factors determine the value of each type of loss. When possible, these consequence calculations are quantitative and are related to the actual economic values such as the price of oil and gas, the cost of offshore construction, and typical parameters associated with an operator's fleet. For

TABLE 11.7

Example Score for Determining Likelihood of Failure

Rule	Input/Purpose	Weight	Score
Design practice	Design year—accounts for how detailing practice has varied over the years.	5	0–50
Number of legs and bracing configuration	Number of legs, bracing system—accounts for how redundancy varies for basic structural bracing systems.	10	1–100
Foundations	To account for the effect that the pile system has on the likelihood of failure. Uncertainty results from lack of records and potential damage to the piles.	10	0–120
Risers and conductors	Current number of risers and conductors—accounts for the increased failure likelihood associated with the rupture or leak of a hydrocarbon carrying riser or conductor escalating to a platform failure.	5	0–78
Boat landing	Boat landings attract hydrodynamic loading increasing the fatigue loading. A failed boat landing may also damage the jacket.	5	0–50
Grouted piles	Location, whether piles are grouted—accounts for the strengthening of joints due to grouting the annulus between the pile and the leg.	3	0–30
Damaged, missing, and removed members	Number of damaged members, bracing legs score (BL)—penalizes platforms that have observed damaged members. The weight is multiplied by the bracing leg score to account for the damage tolerance of different structural systems.	20	0–200
Splash zone corrosion and damage	To account for the effect of corroded and damaged splash zone members on the likelihood of failure.	10	0–100
Flooded members	Number of flooded members, legs bracing system score—penalizes platforms where inspections have detected flooded members.	6	0–60
CP surveys and anode depletion	To identify the status of the CP system as an ineffective system will result in corrosion, increasing the likelihood of failure.	6	0–60
Inspection history	The years of the last level II, III, and IV inspections penalizes platforms that have not been inspected recently. This penalty is more severe for less detailed inspections.	5	0–50
Design loading	To account for the effect that design loading exerts on the likelihood of failure. This differs from the design practice rule, which accounts for design details of the structure such as joints and framing schemes.	5	0–50

(continued)

TABLE 11.7 Continued

Example Score for Determining Likelihood of Failure

Rule	Input/Purpose	Weight	Score
Marine growth	Measure marine growth, design marine growth—penalizes platforms where the observed marine growth has exceeded that used for design.	6	0–60
Scour	Measured scour, observed scour—penalizes platforms where the observed amount of scour exceeds the amount of scour assumed for the design.	5	0–50
Topside weight change	To account for the likelihood of failure of the platform due to changes in deck weight attributable to deck extensions, additional equipment, or change of use.	10	0–100
Additional risers, caissons, and conductors	To account for the effect that additional risers, caissons, and conductors, over and above those that the structure was originally designed for, have on the likelihood of failure of the platform.	5	0–50
Wave-in-deck	To account for the likelihood of failure of the platform due to storm wave crests impinging on and inundating the platform's lowest deck.	20	0–200

other consequences, the values are subjective and may not be actual economic values.

Similar to likelihood, the "scores" for the consequence rules, measured in terms of abstract dollars, are combined to determine the total consequence score for the platform. The platform is then assigned one of the five consequence categories as shown in Figure 11.11.

11.5.3.4 Inspection Planning for Offshore Structure

The effect of inspection and inspection results can be investigated in several different ways. Inspection affects only the likelihood of failure. Likelihood

TABLE 11.8

Example Rule for the Number of Legs and Bracing System

Bracing System	Number of Legs			
	3	4	6	8
K & Diamond	10	8	6	5
Diagonal	7	6	4	3
X	5	4	2	1
Unknown	9	7	5	4

rules associated with inspection, such as time to last inspection, level of inspection, the number of flooded and damaged members, and other anomalies, can be changed to model a trial inspection plan by entering future inspections and the expected amount of future deterioration.

RBI is then run to assess the fleet at different future dates. These inputs can be changed to reflect various future inspection plans and the results can be compared by assessing the risk at various future dates, with the intent of selecting an inspection program that provides the lowest risk.

After this has been performed, the risk matrix for the fleet can define the priorities of the platform—red for high risk and white for very low risk. Moreover, those components marked by red can be further prioritized on the basis of their risk weight value (as described in the onshore maintenance plan). The result is optimization of inspection costs to be delivered at the lowest risk.

This procedure can be used to adjust long-term future inspection plans (for 5 and 10 years) to account for the deterioration that is expected to occur based on past experience.

References

Abd El-kader, O., and A. Al-Kulaib. 1998. Kuwait Bridge Management System, Eighth International Colloquium on Structural and Geotechnical Engineering, 1998.

Broomfield, J. P. 1997. *Corrosion of Steel in Concrete*. London: E&FN SPON.

El-Reedy, M. A., and M. A. Ahmed. 1998. Reliability-Based Tubular Joint of Offshore Structure Based Inspection Strategy, Offshore Mediterranean Conference, OMC.

Khalil, A. B., M. M. Ahmed, and M. A. El-Reedy. 2000. Reliability analysis of reinforced concrete column. Ph.D. thesis, Faculty of Engineering, Cairo University.

Kamal, S., O. Salama, and S. Elabiary. 1992. *Deteriorated Concrete Structure and Methods of Repair*. Egypt: University Publishing House.

Morinaga, S. 1988. Prediction of Service Lives of Reinforced Concrete Buildings Based on Rate of Corrosion of Reinforcing Steel. Special report of Institute of Technology, Shimizu Corporation, No. 23, June 1988, Tokyo, Japan.

Raupach, M. 1996. Corrosion of steel in the area of cracks in concrete—laboratory tests and calculations using a transmission line model. In *Corrosion of Reinforcement in Concrete Construction*, Special Publication No. 183, eds. C. L. Page, P. B. Bamforth, and J. W. Figg. Cambridge, UK: The Royal Society of Chemistry.

Tuutti, K. 1982. *Corrosion of Steel in Concrete*. Stockholm: Swedish Cement and Concrete Research Institute.

API. 1996. Base Resource Document—Risk-Based Inspection—Preliminary Draft. API 581, June 1996.

Stahl, B. 1986. Reliability engineering and risk analysis (Chapter 5). In *Planning and Designing of Fixed Offshore Platforms*, ed. by B. McClelland and M. D. Reifel. New York, NY: Van-Nostrand Reinhold Company.

DeFranco, S., P. O'Connor, A. Tallin, R. Roy, and F. Puskar. 1999. Development of a Risk Based Underwater Inspection (RBUI) Process for Prioritizing Inspections of Large Numbers of Platforms, OTC, 10846.

El-Reedy, M. A. 2006. Risk based inspection for prioritizing repair and inspections for large numbers of platforms in Gulf of Suez, OMAE, 92109.

12

Overview of Fixed Offshore Structures

12.1 Introduction

Over the past 20 years, two major types of fixed platforms have been developed: the steel template type, which was pioneered in the Gulf of Mexico (GoM), and the concrete gravity type, which was first developed in the North Sea. A third type—the tension leg platform (TLP)—is currently in progress in response to the need for drilling wells in deep water and developments in gas projects in deep water.

As early as 1909 or 1910, wells have been drilled in Louisiana, where wooden derricks were erected on hastily built wooden platforms constructed on top of wood piles. In 1976, Exxon installed a platform in the Santa Barbara Channel at a water depth of 850 ft (259 m).

There are three basic requirements in the design of fixed offshore platforms:

1. They must withstand all loads expected during fabrication, transportation, and installation.
2. They must withstand loads resulting from severe storms and earthquakes.
3. They must function safely as a combined drilling, production, and housing facility.

Around 1950, while developments were taking place in the GoM and the Santa Barbara Channel, British Petroleum was engaged in similar explorations at the coast of Abu Dhabi in the Persian Gulf. Water depth was less than 100 ft (30 m) and the operation grew steadily over the years. In the 1960s, a series of hurricanes in the GoM caused serious damage to the platform, which prompted a critical reevaluation of the platform design criteria:

- Hurricane Hilda, classified as a 100-year storm, struck in 1964, bringing wave heights of 13 m and wind gusts of up to 89 m/s; as a result, 13 platforms had been destroyed.

- The next year, another 100-year storm event, Hurricane Betsy, destroyed 3 platforms and brought other forms of damage.
- As a result, designers abandoned the 25- and 50-year conditions and began designing for a storm recurrence interval of 100 years.

12.2 Types of Offshore Platforms

The function of these platforms is to produce or process oil and gas from offshore sources. The most traditional one is the fixed offshore fixed platform. As more investments were poured into deepwater operations, other platform types were also introduced: TLP and floating production, storage, and offloading (FPSO).

12.2.1 Fixed Offshore Platforms

There are different types of fixed offshore structure platforms worldwide and these are classified as follows:

- Drilling or well protector platforms
- Tender platforms
- Self-contained platforms (template and tower)
- Production platforms
- Flare jacket and flare tower
- Auxiliary platforms

Each type has its own set of characteristics from a functionality point of view and will be discussed in detail below.

12.2.1.1 Drilling or Well Protector Platforms

This drilling platforms are used to drill the oil and gas wells from it, so the drilling rigs will approach usually to these platforms to drill new wells or perform the work over along these platforms life time. The platforms that are built to protect the risers of the producing wells in shallow water are called well protectors or well jackets. Usually, a well jacket serves from one to four wells.

12.2.1.2 Tender Platforms

The tender platform is no longer as common as it was 20 years ago. This platform functions similar to the drilling platform, except that the drilling equipment is resting on the platform topside to perform the job; in this case, however, it is customary to use the jack-up rig. In this type of platform, the derrick and substructure, drilling mud, primary power supply, and mud pumps are placed on the platform.

As noted earlier, this type of platform is no longer popular, so it is unlikely that you will see it in new designs or new projects. However, information about this platform type is still useful in case you will be called upon to perform an assessment of the drilling platform.

12.2.1.3 Self-Contained Platforms

The self-contained platform is a large facility, usually with multiple decks, with adequate strength and space to support the entire drilling rig along with its auxiliary equipment, crew quarters, and enough supplies and materials to last through the longest anticipated period of bad weather when supplies cannot be brought in. There are two types: the template type and the tower type.

12.2.1.4 Production Platform

Production platforms support control rooms, compressors, storage tanks, treating equipment, and other facilities.

12.2.1.5 Quarters Platform

The living accommodations platform for offshore workmen is commonly called a quarters platform.

12.2.1.6 Flare Jacket and Flare Tower

A flare jacket is a tubular steel truss structure that extends from the mud line to approximately 10–13 ft above the mean water line (MWL). It is secured to the bottom by driving tubular piles through its three legs.

12.2.1.7 Auxiliary Platform

Sometimes small platforms are built adjacent to larger platforms to increase available space or to permit the carrying of heavier equipment loads on the principal platforms.

Such auxiliary platforms have been used for pumping or compressor stations, oil storage, quarters platforms, or production platforms. Sometimes they are free-standing, and at other times they are connected by bridges to the other structures.

12.2.1.8 Catwalk

A catwalk is a bridge 100–160 ft (30–49 m) in length that connects two neighboring offshore structures. The catwalk supports pipelines, pedestrian movement or handling of materials, and labor moving from one platform to another.

12.2.1.9 Heliport

The heliport is a landing area for helicopters so it must be large enough to handle loading and unloading operations. A square heliport must have a side length that is 1.5 to 2.0 times the largest helicopter length. The heliport landing surface should be designed for a concentrated load of 75% of the gross weight and the impact load is 2 times the gross weight of the largest helicopter. This load must be sustained over an area of 600 × 600 mm (24 × 24 in.) anywhere in the heliport surface.

Figures 12.1–12.3 show the different types of platforms. Figure 12.1 shows a sample of a production platform connected with a satellite platform.

12.2.2 Concrete Gravity Platforms

In 1973, the first concrete offshore oil platform was installed. This concrete gravity platform was constructed by CG Doris Company for Phillips Petroleum Co. (Figure 12.4). The Ekofisk storage tank remains one of the largest offshore gravity structures ever built. This platform is accompanied by a subsea tank that can store 160,000 m³ of oil. By the end of 1979, there were 14 concrete platforms installed in the North Sea alone.

FIGURE 12.1
Production platform with bridges.

(a)

(b)

FIGURE 12.2
(a, b) Complex platform.

Concrete gravity structures rest on the seabed by virtue of their own weight. Their advantages include: (1) no piles are required to be used and (2) use of traditional civil engineering labor and methods.

12.2.3 Floating Production, Storage, and Offloading

The first oil FPSO was designed for Shell Castellon and built in Spain in 1977. The first-ever conversion of a liquified natural gas (LNG) carrier (Golar LNG–owned Moss-type LNG carrier) into an LNG floating storage and regasification unit was carried out in 2007 by Keppel Shipyard in Singapore (Graff, 1981).

In the past few years, concepts for LNG FPSOs have also been launched. An LNG FPSO works under the same principles as an oil FPSO, but it only

FIGURE 12.3
Tripod platform.

FIGURE 12.4
Concrete gravity platform.

produces natural gas, condensate, and/or liquefied petroleum gas, which is stored and offloaded.

FPSO vessels are particularly effective in remote or deepwater locations, where seabed pipelines are not cost-effective (Figure 12.5). FPSOs eliminate the need to lay expensive long-distance pipelines from the oil well to an onshore terminal. They can also be used economically in smaller oil fields, which can be exhausted in a few years and do not justify the expense of installing a fixed oil platform. Once the field is depleted, the FPSO can be moved to a new location. In areas of the world subject to cyclones (NW Australia) or icebergs (Canada), some FPSOs are able to release their mooring/riser turret and steam away to safety in an emergency. The turret sinks beneath the waves and can be reconnected later.

The deepest water depth, operating FPSO is the Espirito Santo FPSO from Shell America and it is operated by SBM Offshore. The FPSO is moored in a water depth of 1800 m in Campos Basin, Brazil, and is rated for 100,000 bpd. The EPCI contract was awarded in November 2006 and was scheduled to draw its first oil in December 2008. The FPSO conversions and internal turrets were done at Keppel Shipyard Tuas in Singapore and the topsides were fabricated in modules at Dynamac and BTE in Singapore.

12.2.4 Tension Leg Platforms

A TLP or extended tension leg platform (ETLP) is a vertically moored floating structure normally used for offshore production of oil or gas, and is particularly suited for water depths greater than 300 m (about 1000 ft). The TLP is shown in Figure 12.6.

The first TLP was built for Conoco's Hutton field in the North Sea in the early 1980s. The hull was built in the drydock of Highland Fabricator's Nigg yard in the north of Scotland, with the deck section built nearby at McDermott's yard at Ardersier. The two parts were mated in the Moray Firth in 1984.

FIGURE 12.5
Floating production, storage, and offloading.

FIGURE 12.6
Tension leg platform.

The deepest (E)TLPs measured from the sea floor to the surface are:

- 4674 ft (1425 m) Magnolia ETLP, whose total height is about 5000 ft (1500 m)
- 4300 ft (1311 m) Marco Polo TLP
- 4250 ft (1295 m) Neptune TLP
- 3863 ft (1177 m) Kizomba B TLP
- 3863 ft (1177 m) Kizomba A TLP
- 3800 ft (1158 m) Ursa TLP, whose height above the surface is 485 ft (148 m), making a total height of 4285 ft (1306 m)
- 3500 ft (1021 m) Allegheny TLP
- 3300 ft (1006 m) W. Seno A TLP

12.3 Major Steps in Constructing an Offshore Structure

In terms of design and construction, similar steps and procedures are followed for all types of platforms. In this section, we shall discuss the main steps in constructing a new platform. These phases represent the life cycle of the process, and every phase has its own criteria and requirements in terms of special skills.

In general, there should be good communication between the engineering office, which is responsible for the project design, and the contractor, who will fabricate, erect, and construct the platform.

1. Preliminary phase front end egineering design (FEED)
   - Recognition of need and setting of operational criteria
   - Determination of environmental criteria
   - Feasibility studies and cost estimates
   - Financing arrangements and determining the budget
2. Design phase
   - Preliminary study and special investigations
   - Soils
   - Size selection for derrick and transportation barges
   - Condition relative to corrosion, ice, earthquakes, product transportation, and crew transportation
   - Design and engineering drawings
     - Foundation design
     - Structural design
     - Preparation of drawings
   - Document preparation
     - Specifications
     - Contracts
     - Bid reply sheet
     - Rental contracts for barges
     - Rental contract for boats
3. Bidding phase
   - Select bidders
   - Sending and receiving of bids
   - Evaluation of bids
   - Awarding of contracts
4. Construction phase
   - Fabrication onshore
   - Ordering and receiving materials
   - Fabrication of specialty items

- Welder qualifications
- Cutting, fitting, and joining of members into components
- Members coating and other corrosion protection
- Loading for transportation
- Erection offshore
- Placement of underwater component
- Installation of pile foundation
- Setting of above water components and equipment
- Miscellaneous associated construction
- Acceptance

12.4 Offshore Platform Design Overview

The main element to perform in any design project is to define the load and calculate it, and then perform the structure analysis and select the optimum sections. All the design steps depend on load calculation. In the following section, the different loads affecting the platforms are described.

12.4.1 Loads

In general, the loads acting on the platform are as follows:

- Gravity loads
- Wind loads
- Wave loads
- Current loads
- Earthquake loads
- Installation Loads
- Other loads

12.4.1.1 Gravity Load

The gravity load comprises the weight of the platform itself and the weight of equipment such as piping, pumps, compressors, separators, and other mechanical equipment used during the operation of the platform.

Dead load. The structural dead load is the weight in air of the overall platform structure including piling, superstructure, jacket, stiffeners, piping and conductors, corrosion anodes, decking, railing, grout, and other appurtenances (Table 12.1). Sealed tubular members are to be considered as either buoyant or flooded, whichever produces the maximum stress at the point under consideration. After a model of the structure is built (using a software), the structure dead weight is calculated automatically.

Live load. Live loads are the loads imposed on a platform during its use; these loads may change during a particular mode of operation or from one mode of operation to another. The live load usually includes the following items:

1. The weight of drilling and production equipment
2. The weight of crew living quarters, heliport, and other life support equipment
3. The weight of liquid in storage tanks
4. The forces due to deck crane usage

General deck area design loading. The topside deck structure shall be designed for the following specified imposed loads applied to open areas of the deck, where the equipment load intensity is less than the values shown (see Table 12.2).

Live load condition. All equipment loads but no future equipment, together with 2.5 kN/m² on the lay down areas, and a total additional live load of 50 tons. This live load shall be applied as a constant uniformly distributed load over the open areas of the deck. Deck loading on clear areas for extreme storm conditions for substructure design may reduced to zero as it will not be normally manned during storm conditions. A total live load of 200 kN at the topside center of gravity (CG) shall be assumed for the design of the jacket and foundations. Point load for the access platform beam is designed to be 10 and 5 kN for deck grating and stringers, respectively.

TABLE 12.1

Dead Load

	Uniform Load Beams and Decking (kN/m²)	Concentrated Line Load on Decking (kN/m)	Concentrated Load on Beams (kN)
Walkways and stairs	4.79	4.378	4.44
Areas over 37 m²	3.11		
Areas of unspecified light use	11.97	10.95	267
Areas where specified loads are supported directly by beams		7.3	

TABLE 12.2

Approximate Design Loads on the Platforms

Area	Loading (kN/m²) Member Category[a]				Point Load (kN)
	1	2	3	4	
Helideck areas	15	10	(Note 2)	(Note 3)	Bell 412
Laydown areas	12	10	(Note 2)	(Note 3)	30
Open deck areas and access hatches	12	10	(Note 2)	(Note 3)	15
Mechanical handling routes	10	5	(Note 2)	(Note 3)	30
Stairs and landing	2.5	2.5	1	–	1.5
Walkways and access platforms	5	2.5	(Note 2)	(Note 3)	5
					(Note 4)

Note:

1. Loading for deck plate, grating, and stringers shall be combined with structural dead loads and designed for the most onerous of the following:
   • Loading over the entire contributory deck area
   • A point load (applied over a 300 × 300 mm footprint)
   • Functional loads plus design load on clear areas
2. For the design of the main framing, two cases shall be considered:
   • Maximum operating condition
     All equipment, including future items and helicopter, together with 2.5 kN/m² on the laydown area.
3. Deck loading on clear areas for extreme storm conditions for substructure design may be reduced to zero in view of the not normally manned status of the platform during storm conditions. A total live load of 200 kN at the topside centre of gravity shall be assumed for the design of the jacket and foundations.
4. Point load for access platform beam design to be 10 kN and 5 kN for deck grating and stringers respectively.

[a] Member category: member 1, deck plate grating and stringers; member 2, deck beams; member 3, main framing; member 4, jacket and foundation.

12.4.1.2 Impact Load

For structural components carrying live loads that include impact, the live load must be increased to account for the impact effect (Table 12.3).

12.4.1.3 Wind Load

The wind data will be provided by the owner according to the metocean study, which defines the prevailing wind direction and the maximum wind speed for every 1, 50, and 100 years.

Figures 12.7–12.9 represent a sample of the wind data that will be in the metocean study report. The most important design considerations for an offshore platform are the storm wind and storm wave loadings, to which it will be subjected during its service life. This information is usually available

TABLE 12.3

Impact Load Factor

Structural Item	Load Direction	
	Vertical	Horizontal
Rated load of cranes	100%	100%
Supports of light machinery	20%	0%
Support of reciprocating machinery	50%	50%
Boat landings	200 kips (890 kN)	200 kips (890 kN)

to the owner and delivered to the engineering office; if this information is not available, it should be obtained from the appropriate national or international authorities.

The wind speed at any elevation above a water surface is presented as:

$$V_z = V_{10}(z/10)^{1.7}$$

From this equation, which is similar across different codes, we can calculate the wind force on the structure, where

V_{10} = Wind speed at a height of 10 m

z = Desired elevation (m)

The wind drag force on an object should be calculated as:

$$F = (\rho/2)V^2 C_s A$$

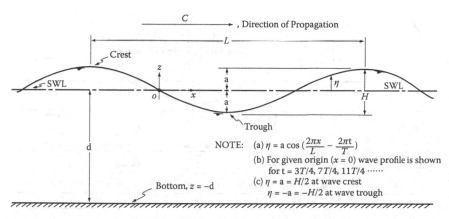

Definition of terms—elementary, sinusoidal, progressive wave

FIGURE 12.7
Definitions of sinusoidal progressive curve.

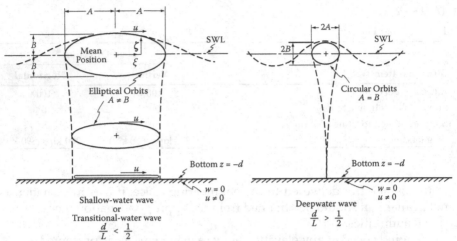

Water particle displacements from mean position for shallow-water and deepwater waves

FIGURE 12.8
Water particle displacement for shallow and deepwater waves.

where
F = Wind force
ρ = Mass density of air (0.0023668 slugs/ft³)
V = Wind speed (ft/s)
C_s = Shape coefficient
A = area of the object (ft²)

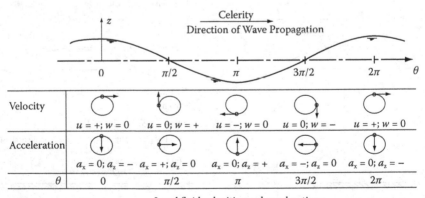

Local fluid velocities and accelerations

FIGURE 12.9
Local fluid velocities and acceleration.

TABLE 12.4

Design Wind Pressures at 125 mph for a 100-Year Period

Structure Member	Pressure (psf)
Flat surface such as wide flange beams, gusset plates, sides of building, etc.	60
Cylindrical structural members	48
Cylindrical deck equipment ($L = 4D$)	30
Tanks standing on end ($H \leq D$)	25

For the shape coefficient, C_s,

Beams	1.5
Sides of buildings	1.5
Cylindrical section	0.5
Overall projected area of platform	1.0

Table 12.4 lists some of the design wind pressures that have been used in conjunction with a 100-year sustained wind velocity of 125 mph.

Shielding coefficient. Use of these factors is based on the judgment of the designer and the following factors are usually based on API RP2A (Table 12.5).

12.4.1.4 Wave Load

The characteristics of waves in the ocean often appear as a confused and constantly changing sea of crest and troughs on the water surface because of the irregularity of wave shape and the variability in the direction of propagation. The direction of wave propagation can be assessed as an average of the direction of individual waves.

In general, actual water–wave phenomena are complex and difficult to describe mathematically because of nonlinearities, three-dimensional characteristics, and apparent random behavior. However, there are two classical theories—one developed by Airy in 1845 and the other by Stokes in

TABLE 12.5

Shield Factors on API

Component	Shielding Factor
Second in a series of trusses	0.75
Third or more in a series of trusses	0.50
Second in a series of beams	0.50
Third or more in a series of beams	0.00
Second in a series of tanks	1.00
Short objects behind tall objects	0.00

1880—that describe simple waves. The Airy and Stokes theories generally predict wave behavior better where water depth relative to wavelength is not too small. For shallow water regions, conoidal wave theory, originally developed by Korteweg and DeVries in 1895, provides a rather reliable prediction of the waveform and associates motions for some conditions. Recently, the work involved in using conoidal wave theory has been substantially reduced by the introduction of graphical and tubular forms of function by many researchers; however, application of the theory is still complex.

In 1880, Stokes developed a finite amplitude theory that is more satisfactory. Only the second-order Stokes equations are presented, but the use of higher-order approximation is sometimes justified for the solution of a practical problem.

Another widely used theory, known as the stream function theory, is a nonlinear solution similar to Stokes' fifth-order theory. Both use summations of sine and cosine wave forms to develop a solution to the original differential equation. The theory to be used for a particular offshore design is determined by the policy under which the designing engineers are working.

Metocean data will provide us the following data for the platform location as an example in Table 12.6.

W_s, 1 h mean wind speed at 10 m above sea level (m/s); H_s, significant wave height, estimated from the wave energy spectrum, $H_s = \sqrt{m_0}$, equivalent to the mean height of the highest one-third of the wave in a sea state (m); H_{max}, maximum wave height, highest individual zero crossing wave height in a storm of duration 24 h (m); T_z, mean zero crossing wave period, the average period of the zero-crossing wave heights in a sea state, and estimated from the wave energy spectrum, $T_z = \sqrt{(m_0/m_2)}$ (s); T_p, peak wave period, the period associated with the peak in the wave energy spectrum; estimated from the wave energy spectrum, $T_p = m - 2m_1/m_0$ (s); T_{ass}, wave period associated with maximum wave height (s); H_c, crest height, highest crest to mean-level height of an individual wave in a storm of duration 24 h (m); U, horizontal wave orbital velocity at 3 m above the seabed estimated from H_{max} and T_{ass} using stream function wave theory (m/s).

TABLE 12.6

Example of Metocean Data

Return Period (years)	W_s (m/s)	H_s (m)	T_z (s)	T_p (s)	H_{max} (m)	T_{ass} (s)	H_c (m)	U (cm/s)
1	16.1	2.2	4.7	6.2	3.2	6.3	1.9	1
10	17.7	2.8	5.4	7.1	5.3	7.2	3.2	5
50	18.7	3.2	5.9	7.7	6.6	7.7	4.0	11
100	19.2	3.3	6.0	7.9	7.2	8.0	4.3	15
10,000	22.0	4.5	7.1	9.4	10.9	9.4	6.6	51

The equation for wave velocity and acceleration are as follows:

$$F_1 = (2\pi(z + d)L) \tag{12.1}$$

$$F_2 = (2\pi d/L) \tag{12.2}$$

$$F_3 = (2\pi x/L) - (2\pi t/T) \tag{12.3}$$

- Velocity

$$U = [(H/2)(gT/L)\cosh F_1/\cosh F_2]\cos F_3 \tag{12.4}$$

$$W = [(H/2)(gT/L)\sinh F_1/\cosh F_2]\sin F_3 \tag{12.5}$$

- Acceleration

$$a_x = [+gpH/L \cosh F_1/\cosh F_2]\sin F_3 \tag{12.6}$$

$$a_z = [-gpH/L \sinh F_1/\cosh F_2]\cos F_3 \tag{12.7}$$

The theories concerning the modeling of ocean waves were developed in the nineteenth century (Figure 12.10). Practical wave force theories concerning

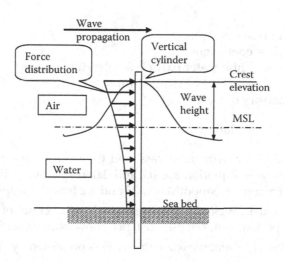

FIGURE 12.10
Wave forces distribution on a vertical pipe.

actual offshore platforms were not developed until 1950, when the Morison equation was presented

$$F = F_D + F_I \tag{12.8}$$

where F_D is the drag force and F_I is the inertial force.

- Drag force

  The drag force due to wave acting on an object can be obtained by:

$$F_D = 1/2\rho C_d V^2 A \tag{12.9}$$

  where
  F_D = Drag force (N)
  C_d = Drag coefficient
  V = velocity of object (m/s)
  A = projected area (m²)
  ρ = density of water (kg/m³)

- Inertial force

  The inertial force due to wave acting on an object can be found by:

$$F_1 = \rho C_m a D^2 / 4 \tag{12.10}$$

  where
  F_I = Inertial force (N)
  C_m = Mass coefficient
  a = Horizontal water particle acceleration (m²/s)
  D = Diameter of the cylinder (m)
  ρ = density of water (kg/m³)

Wave load calculation:

- C_d and C_m are dimensionless, and the values most often used in the Morison equation are 0.7 and 2.0, respectively. For API, they are 0.65 and 1.6 (smooth) or 1.05 and 1.2 (rough), respectively.
- Water particle velocity and acceleration are functions of wave height, wave period, water depth, distance above bottom, and time.
- The most elementary wave theory was presented by Airy in 1845.
- Another widely used theory, known as the stream function theory, is a nonlinear solution similar to Stokes' fifth-order theory.

12.4.1.5 Comparison between Wind and Wave Calculation

When we calculate the force that affects the structure due to wind, we take the drag force into consideration and neglect the inertial force; however, for waves we should consider drag force and inertial force, and the following example illustrates the reason why we neglect the inertial force in wind load calculations.

Example

Pipe diameter = 0.4 m

V_{air} = 25 m/s V_{water} = 1 m/s
a_{air} = 1 m²/s a_{water} = 1 m²/s
W_{air} = 1.3 kg/m³ W_{water} = 1000 kg/m³

$$F_d = (1/2)C_d wV^2 A$$

$$F_m = C_m w\pi(D^2/4)a$$

- Air

$$F_d = (1/2)(0.8)(1.3)(25)^2(0.4) = 130 \text{ N}$$

$$F_m = 2(1.3)(\pi)(0.4)^2/4(1) = 0.33 \text{ N}$$

- Water

$$F_d = (1/2)(0.8)(1000)(1)^2(0.4) = 160 \text{ N}$$

$$F_m = 2(1000)(\pi)(0.4)^2/4(1) = 251 \text{ N}$$

Conductor shielding factor. The conductor shielding factor is dependent on the configuration of the structure, and the number of well conductors can be a significant portion of the total wave forces. If the conductors are closely spaced, the forces on them may be reduced due to the hydrodynamic shield.

A wave force reduction factor will be taken from Figure 12.11, where S is the spacing in the wave direction and D is the conductor diameter.

12.4.1.6 Current Loads

The presence of current in the water produces minor effects; the most variable is the current velocity, which should be added vectorially to the horizontal water particle velocity. In design, the maximum wave height is sometimes increased by 5% to 10% to account for the current effect and the current is neglected.

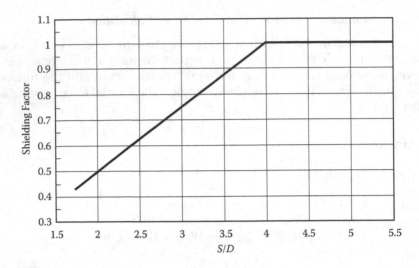

FIGURE 12.11
Shielding factor based on API RP2A.

12.4.1.7 Earthquake Load

The ductility requirements are intended to ensure that the platform has sufficient reserve capacity to prevent its collapse during rare intense earthquake motions, although structural damage may occur.

In areas of low seismic activity (less than 0.05 g), the design of environmental loading other than earthquake will provide sufficient resistance against earthquakes.

In case of seismic activity ranging from 0.05 to 0.1 g, the important factor is deck appearances (piping, facilities, etc.). The platform should be checked for protection against earthquake using dynamic analysis procedures such as spectrum analysis or time history analysis.

12.4.1.8 Other Loads

These loads are discussed in relation to the configuration of the platform and the following environmental conditions:

1. Marine growth
2. Scour

Marine growth. Marine growth increases the diameter of the jacket member, which in turn increases the drag force. Based on API, 1.5 in. from mean high high water level (MHHW) to –150 ft MHHW is 1 ft higher than mean low low water level (MLLW). Smaller or larger thickness values may be used from site-specific studies. A structural member is considered smooth if the value exceeds

MHHW or is lower than –150 ft where marine growth is light (and can therefore be ignored). The zone between MHHW and –150 ft is considered rough.

Scour. Seabed scour affects both the lateral and axial pile performance and capacity. Scour prediction remains an uncertain art. Sediment transport studies may assist in defining scour design criteria but local experience is the best guide.

Practical considerations in design assume that scour is 1.5D, where D is the pile diameter. The scour effect on pile is shown in Figure 12.12, which presents an underwater survey for the leg.

The scope of work for each discipline and the required knowledge and activity to start design will be as follow:

- Oceanography
  Wind, wave, current, tide, ice
- Foundation engineering
  Soil characteristics, vertical and lateral pile characteristics, scour
- Structural engineering
  Materials selection and corrosion, stress analysis, welding, structure analysis, design for fabrication and installation, welding
- Marine civil engineering
  Installation equipment, installation methods, navigation safety instrument
- Naval architecture
  Floating and buoyancy, towing, launching, controlled flooding

FIGURE 12.12
Photo for scour.

12.4.2 Platform Configuration

The three main components of a steel template platform are (1) topside facilities (decks), (2) jackets, and (3) piles. Topside facilities frequently comprise three decks: drilling deck, well head/production deck, and cellar deck. These decks are supported by a grid work of girders, trusses, and columns (Figure 12.13).

The topside structures are designed based on American Institute of Steel Construction in Allowable Stress Design (AISC-ASD) or Load and Resistance Factor Design specifications and the design procedure flow chart as shown in Figure 12.14. Usually, the main supporting element will be the plate girder or the tubular truss; in most cases, however, the tubular truss is preferred to carry a small load in case of wind loads. The floor is covered with steel plates, usually about 1-1/2 in. thick. The thickness of the deck framing is dependent on the spacing between floor beams and the anticipated load on the deck.

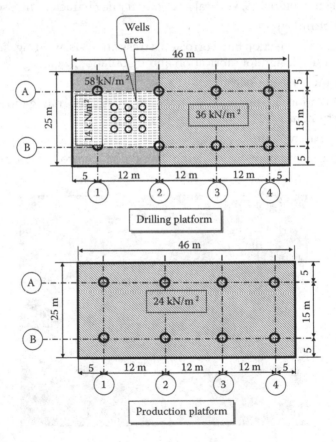

FIGURE 12.13
Load distribution on a drilling and production platform.

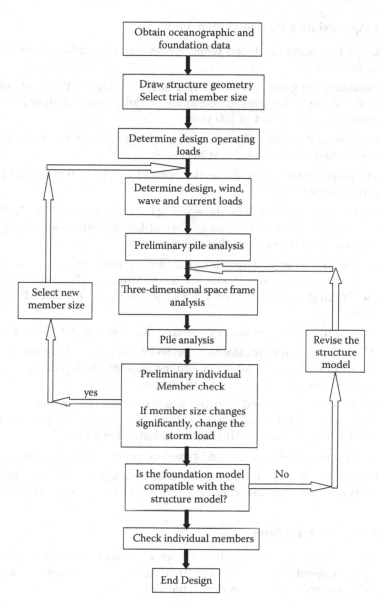

FIGURE 12.14
Platform design procedure.

12.4.3 Approximate Design Dimensions

- Large forces are produced when waves strike a platform's deck and equipment.
- Therefore, air gaps of at least 5 ft should be added to the crest height of the wave in the omnidirectional guideline wave heights with a normal return period of 100 years.
- For practical reasons, the sea deck level is usually stated at an elevation of 10–14 ft (3–4 m) above MWL.
- The jacket walkway is above the normal everyday waves that pass through the jacket.
- For the eight-leg platform, the spacing between legs is about 40–60 ft and is usually set by the availability of launch barges and the spacing of launch runners on these barges.
- In the transverse direction, the leg spacing is approximately 45 ft. This dimension is always constrained by the dimension of the drilling and/production packages that will be placed on the deck.
- The length of the cantilever overhang is usually about 12–15 ft.
- Allow a 1-in. annular clearance between the pile and the inside of a leg. In piles with 60 and 48 in. outer diameter, the legs will have an internal diameter of 62 and 50 in., respectively.
- Jacket legs are battered to provide a larger base for the jacket at the mud line and thus assist in resisting the environmentally induced overturning moments. The legs have batter for 1:8 or 1:7.
- Preliminary sizes are selected based on previous experience.
- Conductor and riser numbers and sizes will be provided. The conductor is always 18, 20, 24, or 30 in., and the risers are always 14–20 in.

12.4.4 Topside Structures

The main function of the topside is to carry the load from the facilities and drilling equipment. The percentage breakdown of the weights comprising the topside component is shown in Table 12.7.

12.4.5 Jacket Design

The function of the jacket is to surround the pile and hold the pile extensions in position all the way from the mud line to the deck substructure. Moreover, the jacket provides support for boat landings, mooring bits, barge bumpers, corrosion protection systems, and other platform components. Virtually all of the decisions about its design depend on the jacket leg. The soil conditions and foundation requirements often tend to control the leg size. The

TABLE 12.7

Weight and Weight Percentage of Various Elements for Eight-Leg Drilling/ Production Platform

1.	Deck	Weight (Ton)	Percentage (%)
	Drilling deck		
	Plate	72	11
	Production deck		
	Plate	52	7.8
	Grating	1.0	0.16
	Subtotal	125	18.8
2.	Deck beams		
	Drilling deck	174	26.3
	Production deck	56	8.5
	Subtotal	230	34.8
3.	Tubular trusses	146	22.1
4.	Legs	105	15.9
5.	Appurtenances		
	Vent stack	6	0.9
	Stairs	12	1.8
	Handrails	4	0.6
	Lifting eyes	2	0.3
	Drains	6	0.9
	Fire wall	4	1.7
	Stiffeners	14	2.2
Total		661	100

golden rule in design is to minimize the projected area of the member near the water surface (high wave zone), minimize the load on the structure, and reduce the foundation requirements (Table 12.8).

12.4.6 Bracing System

The bracing system consists of vertical, horizontal, and diagonal tubular members connected by jacket legs forming a stiff truss system. This system transfers the horizontal load, which acts on the platforms, to the piles. There are several variations in the platform bracing pattern. Every system has its advantages and disadvantages. For example, the K-brace pattern system results in fewer members intersecting at joints, thereby reducing welding and assembly costs. It lacks redundancy when compared to X bracing. This observation is based on the results of a joint industrial study.

It is recommended to choose a brace member diameter that has a slenderness ratio (KL/r) in the range of 70–90. Limiting the ratio to a range of 70–90

TABLE 12.8

Jacket Weight for Eight-Leg Platform Drilling/Production at 91 m Water Depth

ID	Component Description	Weight	% of Total Weight	Subtotal Weight	% of Total Weight
1.	Legs				
	Joint can	177	14.6		
	In between tubular and others	309	25.4		
2.	Braces				40
	Diagonal in vertical plan	232	19.1	19.1	
	Horizontal	163	13.4		
	Diagonal in horizontal plan	100	8.2		40.7
3.	Other framing				
	Conductor framing	35	2.9		
	Launch trusses and runners	82	6.7		
	Miscellaneous framing	2	0.2		9.8
4.	Appurtenances				
	Boat landing	28	2.3		
	Barge bumpers	29	2.4		
	Corrosion anodes	22	1.8		
	Walkways	16	1.3		
	Mud mats	5	0.4		
	Lifting eyes	2	0.2		
	Closure plates	2	0.2		
	Flooding system	7	0.6		
	Miscellaneous	4	0.3		9.5
					100

is an industry-accepted practice. As the slenderness of a brace increases, its allowable axial stress (F_a) starts to decrease. At $KL/r = 80$, the allowable axial stress (F_a) for A36 steel is 71% of that allowable for a nonslender member ($KL/r = 0$). For steel 50 KSi at $KL/r = 80$, the F_a will reduce to about 63% compared to the nonslender member.

High KL/r ratios would render the high-yield pipe less efficient compared to lower KL/r values. It should be noted that lower slenderness ratios also encourage higher D/t ratios for tubular members, which may compound local buckling problems.

For sizes up to 450 mm (18 in.), use the wall thickness for a standard pipe as a starter. For sizes up to 700 mm (27 in.), try 12 mm. For 750 to 900 mm, start with 16 mm. It is practical to keep the D/t ratio of the members between 19 and 60. Pipes with D/t values less than 19 are difficult to buy or make. A36 steel with D/t values exceeding 60 can present local buckling problems.

From a practical point of view, for water depths (h) measured in feet, begin to check for hydrostatic problems when D/t is higher than $250/(h)^{0.3333}$. In general, the legs of the jacket are interconnected and rigidly held by the following types of bracings: diagonals in vertical planes, horizontal, and diagonals in horizontal plans. Often, for plans of horizontal bracing spaced 12–16 m near the water surface, the span is approximately 12 m.

The benefits and the general function performed by the bracing system are as follows:

1. Transmits the horizontal load to the foundation
2. Provides structure integrity during fabrication and installation
3. Resists wrenching motion of the installed jacket-pile system
4. Supports the corrosion anodes and well conductors

For joint industrial projects (Report for Joint Industry Typical Frame Projects, 1999), Figures 12.15 and 12.16 illustrate buckling on the K-bracing member and failure in the joints, respectively (see also Figure 12.17).

FIGURE 12.15
Buckling in a member for K bracing.

FIGURE 12.16
Failure in K joint.

12.4.7 In-Place Structure Analysis

In-place structure analyses are performed through software packages such as SACS or SESAM, which are widely available in the market.

The steps in using software in design are as follows:

1. The designer must define the structure in terms of physical dimensions, member size, and material properties.
2. The designer must input the soil conditions as interpreted for him by soil specialists in some programs requiring the P–y curve.
3. Various loads must be entered into the program.
4. The design wave through the structure at several azimuth angles is required to determine the direction that produces higher reactions.
5. The program advances the wave through the structure at specified increments, calculating the total shear and overturning moment on the structure at mud line.

6. For each load condition, the analysis should provide the following:
   (a) Total base shear and overturning moment
   (b) Member end forces and moments
   (c) Joint rotation and deflection
   (d) External support reaction

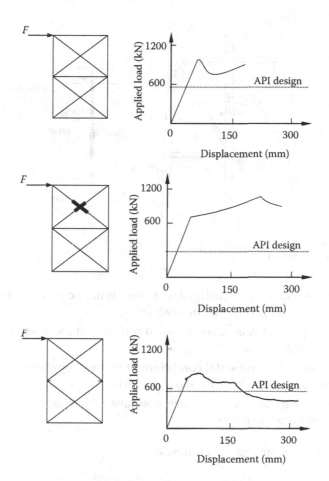

FIGURE 12.17
(a) X bracing with horizontal bracing without joint can. (b) X bracing with horizontal bracing with joint can. (c) X bracing with horizontal bracing without joint can.

7. After stresses have been calculated, they are compared with the allowable stresses defined by the AISC.

8. The pile is replaced by a lateral spring in two directions: an axial spring and a moment spring.

9. The piling interacts with the surrounding soil in an inelastic manner, and the program linearizes its response to generate the equivalent elastic spring (Figure 12.18).

12.4.8 Dynamic Structure Analysis

Dynamic analysis is becoming increasingly important for the following reasons:

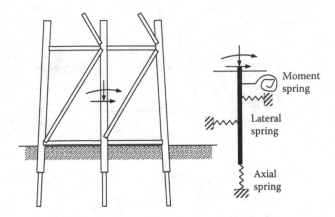

FIGURE 12.18
Foundation piling model.

1. For larger and more costly structures, dynamic analysis translates into substantial construction cost savings.
2. For complex offshore structures, an ordinary static analysis tends to incur a higher risk.
3. The harsh environmental conditions at many deepwater sites cannot be adequately modeled by static analysis.
4. Various software products are now available that make dynamic analysis easier to perform.

The equation of motion is as follows:

$$Mu + Cu + Ku = P(v, v)$$

where
  M = Diagonal matrix of virtual mass
  C = Matrix for structural and viscous damping
  K = Square linear structure stiffness matrix
  $P(v, v)$ = Load vector, where v and v are the water velocity and acceleration
  u = Structural acceleration
  u = Velocity
  u = Displacement

The output results for the dynamic analysis will be:

1. Time history of member-end forces
2. Time history of joint displacement

3. Maximum values of joint displacement
4. Time history of interstory shear
5. Time history of overturning moment
6. Time history of base shear
7. Time history of axial pile loads

12.4.9 Tubular Joint Design

The tubular joint is very critical and important in offshore structure. There are many researches working until in studying the capacity for this joint. The parameters that affect the joint design are as shown in Figure 12.19.

$$\tau = t/T$$

$$\beta = d/D$$

$$\gamma = D/2T$$

12.4.9.1 Tubular Joint Calculation

12.4.9.1.1 Punching Shear

$$V_p = \tau f \sin\theta$$

where f is the nominal axial, in-plane bending, or out-of-plane bending stressing the brace.

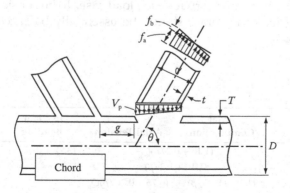

FIGURE 12.19
Geometric parameters for a tubular joint. t, brace thickness; g, gap between brace; T, chord thickness; d, brace diameter; D, chord diameter; θ, brace angle measured from the chord.

The allowable punching shear stress in the chord wall is lower than the AISC shear allowable value or

$$V_{pa} = Q_q Q_f \left(F_{yc}/0.6g \right)$$

$$Qf = 1.0 - \lambda \gamma A^2$$

where

$\lambda = 0.30$ for brace axial stress
$\quad = 0.045$ for brace in-plane bending stress
$\quad = 0.021$ for brace out-of-plane bending stress

$$A = \left[\left(f_{AX} \right)^2 + \left(f_{IPB} \right)^2 + \left(f_{OPB} \right)^2 \right]^{0.5} \Big/ 0.6 F_{ye}$$

$Q_f = 1.0$ when all extreme fiber stresses in the chord are tensile. The value of Q_q will be obtained from Table 12.9.

$$Q_\beta = 0.3 / \left(\beta (1 - 0.833\beta) \right) \qquad \text{For } \beta > 0.6$$

$$Q_\beta = 1.0 \qquad \text{For } \beta \le 0.6$$

$$Q_g = 1.8 - 0.1(g/T) \qquad \text{For } \gamma \le 20$$

$$Q_g = 1.8 - 4(g/D) \qquad \text{For } \gamma > 20$$

In any case, Q_g should be higher than or equal to 1.0.

Joint classification of K, T and Y, or cross should apply to individual braces according to their load pattern for each load case. To be considered a K joint, the punching load in a brace should be essentially balanced by loads on

TABLE 12.9

Values of Q_q

	Axial Compression	Axial Tension	In-Plane Bending	Out-of-Plane Bending
K (gap)	$(1.10 + 0.2/\beta)Q_g$		$(3.27 + 0.67/\beta)$	$(1.37 + 0.67/\beta)Q_\beta$
T and Y	$(1.10 + 0.2/\beta)$			
Cross without diaphragms	$(1.10 + 0.20/\beta)$	$(0.75 + 0.20/\beta)Q_\beta$		
Cross with diaphragm	$(1.10 + 0.20/\beta)$			

other braces in the same plane on the same side of the joint. In T and Y joints, the punching load is reached as a beam shear in the chord. In cross joints, the punching load is carried through the chord to the braces on the opposite side.

12.4.9.1.2 Allowable Joint Capacity

The allowable joint capacity will be calculated as follows:

$$P_a = Q_u Q_f F_{yc} T^2 / 1.7 \sin\theta$$

$$M_a = Q_u Q_f F_{yc} T^2 / 1.7 \sin\theta (0.8d)$$

P_a and M_a are the allowable capacity for the brace axial load and the bending moment, respectively. Values of Q_u are shown in Table 12.10.

12.4.9.2 Tubular Joint Punching Failure

The potential for punch-through in the compression of K joint members into the chord due to plastic deformations around the intersection under cyclic loads is shown in Figures 12.20, 12.21, and 12.22.

Limited deformation at $b = 1.0$ compression X joint as maximum load is attached—repeatable under cyclic loads.

Flattening of $b = 1.0$ tension X joint as chord yields—splitting of paint but no cracks in steel.

Joint can precautions and their details are shown in Figure 12.23 as adopted from the API RP2A WSD (2000).

12.4.10 Fatigue Analysis

Experience over the past 60 years and many laboratory tests have proven that a metal may fracture at a relatively low stress if that stress is applied a great number of times (Table 12.11). Failure is expressed as damage or life fatigue

TABLE 12.10

Values of Q_u

	Axial Compression	Axial Tension	In-Plane Bending	Out-of-Plane Bending
K (gap)	$(3.4 + 19\beta)Q_g$		$(3.4 + 19\beta)$	$(3.4 + 7\beta)Q_\beta$
T and Y	$(3.4 + 19\beta)$			
Cross without diaphragms	$(3.4 + 19\beta)$	$(3.4 + 13\beta)Q_\beta$		
Cross with diaphragm	$(3.4 + 19\beta)$			

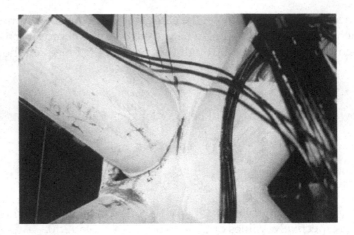

FIGURE 12.20
Punching shear.

damage—number of cycles of a particular stress range divided by the allowable number of cycles for that range from the S–N curve (see Figure 12.24).

Assume that a point is subject to five cyclic stress ranges (due to wave).

$$D_5 = 10 \times 10^6 / 50 \times 10^6 = 0.2 \quad D_{10} = 0.1$$

$$D_{20} = 0.2 \quad D_{50} = 0.05 \quad D_{90} = 0.025$$

$$\text{Total damage} = 0.575$$

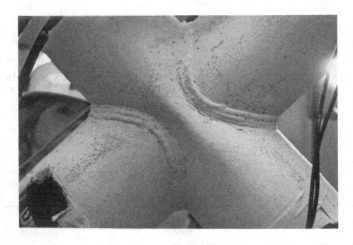

FIGURE 12.21
Buckling in tubular joint.

FIGURE 12.22
Tearing in tubular joint.

If these waves occurred over a period of 10 years, then

$$\text{Life} = 10/0.575 = 17.4 \text{ years}$$

$$\text{Tubular joint S–N curve}$$

12.4.11 Boat Landing

The design of a boat landing is mainly influenced by loads due to impact from the vessel or impact of the boat to the structure.

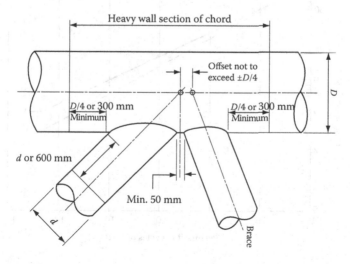

FIGURE 12.23
Details of a tubular joint.

TABLE 12.11

Fatigue Analysis

Stress Range	Occur	Allowable
5	10×10^6	50×10^6
10	4×10^5	4×10^6
20	6×10^4	3×10^5
50	5×10^2	1×10^4
90	25	1×10^3

To absorb the impact load, the usual practice is to fix a fender (usually a car tire) attached to the boat landing. The connection between the boat landing frame structure and the platform jacket will be through a shock absorber like a piston.

12.4.11.1 Calculation of Collision Force

The calculations are based on an 850-ton boat impacting at 0.5 m/s, using Regal Shock cell model SC1830:

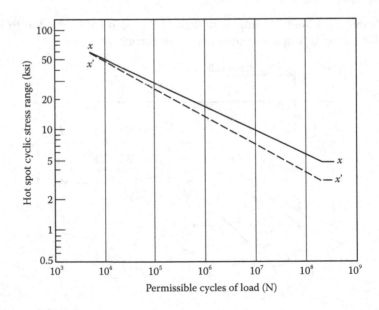

FIGURE 12.24
S–N curve adopted from API.

$$M = (850 \times 9810) \times 2200/2000(9.81 \times 1000) = 9.35 \text{ N s}^2/\text{mm}$$

$$\text{Approaching velocity, } v = 500 \text{ mm/s}$$

$$E = 1/2 \times 935 \times (500)^2 = 11.7 \times 10^7 \text{ N mm}$$

From the shock cell type, choose a suitable curve for the relation between energy versus the deflection curve as shown in Figure 12.25.

$$\delta = 208.3 \text{ mm for } E = 11.7 \times 10^7 \text{ N mm}$$

For the force-versus-deflection base on the model type curve as shown in Figure 12.26,

$$F = 1.16 \times 10^6 \text{ N.}$$

Some of the energy is absorbed by the vessel (assume that 30% is absorbed by the vessel and 70% by the structure):

$$E = 11.7 \times 10^7 \times 0.7 = 8.2 \times 10^7 \text{ N mm}$$

$$\delta = 177.8 \text{ mm}$$

$$F = 9.81 \times 10^5 \text{ N}$$

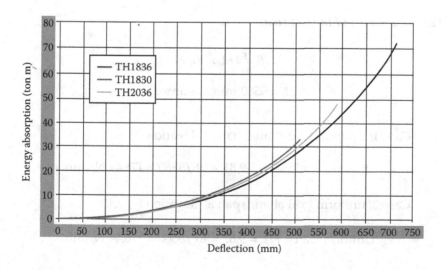

FIGURE 12.25
Energy absorption versus deflection.

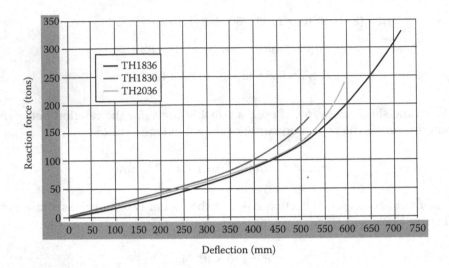

FIGURE 12.26
Reaction force versus deflection.

This force will not be concentrated at one point because the collision will affect a wider area depending on the dimensions of the vessel and its position during the impact, and also because the fender system distributes the load (Figure 12.27).

12.4.11.2 Cases of Impact Load

$$F = 9.81 \times 10^5 \, \text{N}$$

$$L = 5560 \, \text{mm (assumed)}$$

Case 1: uniform load at midspan at elevation (+) 300

Uniform load = $F/L = 9.81 \times 10^5/5560 = 176.44 \, \text{N/mm}'$

Case 2: uniform load at midspan at elevation (–) 900

Uniform load = $F/L = 9.81 \times 10^5/5560 = 176.44 \, \text{N/mm}'$

Case 3: uniform load at midspan at elevation (+) 300 and elevation (–) 900

(a)

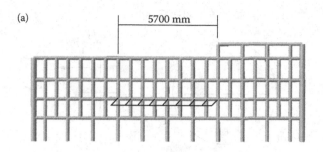

(b) L = Boat impact length

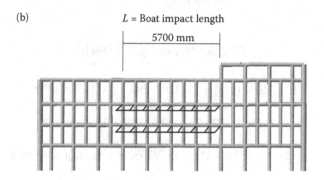

(c)

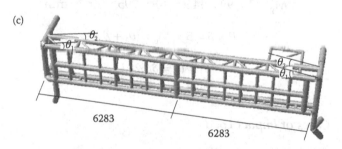

FIGURE 12.27
(a) Boat landing. (b) Boat impact length. (c) Proposed deformation of boat landing.

Uniform load = $0.5F/L$ = $4.91 \times 10^5/5560$ = 88.22 N/mm'

Case 4: assume impact load distributed as a concentrated load at the midspan at elevation (+) 300 and elevation (−) 900 (six concentrated loads)

Force at each joint = $F/6$ = $9.81 \times 10^5/6$ = 1.635×10^5 N

Riser Guard

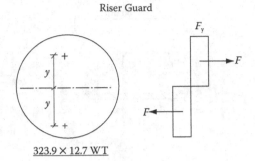

323.9 × 12.7 WT

$$F_y = 240 \text{ N/mm}^2 \text{ (mild steel)}$$

$$F = 240 \times (\pi/8) \times [323.9^2 - 298.5^2] = 14.9 \times 10^5 \text{ N}$$

$$R_{av} = \tfrac{1}{2} [323.9/2 + 298.5/2] = 155.6 \text{ mm}$$

$$y = (2/\pi) \times 155.6 \times [1 + (12.7/155.6)^2 \times (1/12)] = 99 \text{ mm}$$

$$M_p = 2 \times 99 \times 14.9 \times 10^5 = 295 \times 10^6 \text{ N mm}$$

$$P \times \delta = 5 \times M_p \times (\theta_1 + \theta_2)$$

$$\theta_1 = \theta_2 = \delta/6283$$

12.4.11.3 Cases of Impact Load

Case 1: uniform load at midspan at elevation 0.0 (MWL)

$$L = 7000 \text{ mm (assumed)}$$

Uniform load = 67.2 N/mm

L = Boat impact length

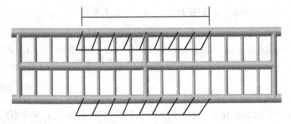

Case 2: uniform load at mid-span at elevation (+) 1200 and elevation (–) 1200

$$L = 6500 \text{ mm (assumed)}$$

Uniform load = 33.6 N/mm (for each elevation)

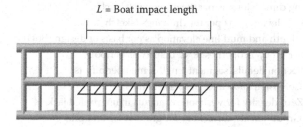

12.5 Design Quality Control

Designing an offshore structure is a complicated and difficult task, so it needs multiple ways to check the quality. One of these method checks is to use a checklist to ensure that all factors are taken into account for the design.

Table 12.12 is a checklist for jacket in-place analysis and Table 12.13 is a checklist for topside in-place analysis.

12.6 Construction Procedures

The design of a jacket is constrained by water depth and the lifting, launching, or self-floating, which depends primarily on the available offshore installation barge capability. In general, the preference is to lift the jacket in place. The size of offshore structure jackets has being increasing as offshore lifting capacity has grown. Nowadays, the lifting capacity can reach up to 14,000 tons.

In the case of jackets in shallow water, where the height of the jacket is approximately equal to the plan dimensions, erection is usually carried out vertically in the same attitude as the final installation. In such cases, the jacket may be lifted or skidded onto the barge.

Jackets designed for use in higher water depths are usually erected on their side. Such jackets are loaded by being pulled out into a barge. Historically,

TABLE 12.12

Checklist for Jacket in Place Analysis

Project:

Client:

Items	Check Points	Check (Yes/No)
	Computer Model	
1	Framing dimensions as per the drawings/sketch	
2	Framing elevations as per the drawings/sketch	
3	Water depth and mud line elevation as per basis of design (BoD)	
4	Member group properties: E, G, density as per BoD	
	(a) Section details, segment lengths, member offsets	
	(b) Corrosion allowance in splash zone	
	(c) Zero density for wishbone elements in ungrouted jackets	
	(d) Grout density corrections for grouted leg and pile	
5	Member properties: K_x, K_y, L_x, L_y	
6	Member end releases where applicable	
7	Flooded members (legs, risers, J tubes, caissons, etc.)	
8	Dummy members (relevant joints kept, the rest deleted)	
9	Plate/membranes modeled correctly	
10	Boundary conditions	
11	Drag and inertia coefficients (smooth and rough members)	
12	Marine growth data as per BoD	
13	Member and group overrides	
	(a) No wave load and marine growth on piles and wishbones in jacket legs	
	(b) Enhancement of C_d, C_m for anode supported members	
	(c) Enhancement of C_d, C_m for jacket walkway members	
	(d) Enhancement of C_d, C_m	
14	LDOPT card and OPTION card in the analysis input file	
15	Hydrostatic collapse check selected with redesign option	
16	Allowable stress modifiers (AMOD) for extreme storm load cases	
17	Unity check ranges (UCPART) in the analysis input file	
	Loads	
18	Load description, calculations, and distribution	
19	Wave theory, wave, current, and wind directions	
	(Nonlinear current stretch with apparent wave period calculation)	
20	Equivalent C_d, C_m calculations for items mentioned above	
21	Load contingencies as per BoD	
22	Load combinations for operating and extreme cases	
23	Load summations	
24	Load summation verification against weight control data	
	PSI Data and Input File	

(*continued*)

TABLE 12.12 Continued

Checklist for Jacket in Place Analysis

Project:

Client:

Items	Check Points	Check (Yes/No)
	Loads	
25	Units for T–Z, Q–Z, and P–Y data from the client supplied data	
26	Pile segmentation data, end-bearing area	
27	With reference to the input format	
	With reference to the input format, check T, Q, and P factors	
	Analysis Results	
28	Enclose sea-state summary (check against load summation)	
29	Enclose member check summary: review for overstressed members	
30	Enclose joint check summary: review overstressed joints (check $F_y = 2/3F_u$ for chords of high strength members)	
31	Enclose PSI analysis results: check maximum pile compression and tension	
32	Enclose model plots: joints/group/section names, K_x, K_y, L_x, L_y, F_y, and loading	
33	Enclose deflection plots, member unity check ratio plots	
34	Enclose hydrostatic collapse check reports (check for need of rings)	
35	Enclose pile factor of safety calculations	
36	Enclose permissible deflection calculations	
37	Enclose plot plan and latest structural drawings	
38	Enclose relevant sections of weight control report	
	Dynamic In-Place Analysis (Optional)	
39	Determine dynamic amplification factor (DAF) based on single degree of freedom concept and apply on wave load cards	
40	Determine DAF based on inertial load distribution and apply on total structure	
	General	
41	Joint name range identified for each framing level and sequential	
42	Member group name specific to each framing level and sequential	
43	Check for future loads, loads due to specific requirement such as rigless interventions	

most large jackets have been barge-launched. This method of construction usually involves additional flotation tanks and extensive pipework and valves to allow the legs to be flooded for ballasting of the jacket into the vertical position on site. Nowadays, this method of construction is applicable to jackets up to 25,000 tons.

TABLE 12.13

Checklist for Topside In-Place Analysis

Items	Check Points	Check (Yes/No)
	Computer Model	
1	Framing dimensions as per the drawings	
2	Two framing elevations as per the drawings	
3	Member properties: K_x, K_y, L_x, L_y	
4	Member end releases where applicable	
5	Plate/membranes modeled correctly	
6	Boundary conditions	
7	Load option card and option card in the analysis input file	
8	Allowable stress modifiers for extreme storm load cases	
9	Unity check ranges in the analysis input file	
	Loads	
10	Load description and calculations	
11	Secondary structural item dead load calculations	
12	Equipment, piping operating, and dry load, electrical and instrument (E&I) bulk load calculations	
13	Wind load calculations, wind area considered	
14	Earthquake load calculation	
15	Load contingencies	
16	Load combinations for operating and extreme cases	
17	Load combinations for local checks	
18	Load summations	
	Installation and preservice loads may be applied as a separate load case	
	Crane load cases may be magnified and applied for local checks	
19	Check for future loads, loads due to specific requirement such as rigless interventions	
	Analysis Results	
20	Enclose sea-state summary (check against load summation)	
21	Enclose member check summary: review for overstressed members	
22	Enclose joint check summary: review overstressed joints (check $F_y = 2/3F_u$ for CHORDS of high-strength members)	
23	Enclose model plots: Joint/Group/Section names, K_x, K_y, L_x, L_y, F_y	
24	Enclose deflection plots, member unity check ratio plots	

Very large jackets (exceeding 25,000 tons) are constructed as self-floaters in a graving dock and towed offshore subsequent to flooding of the dock. The following method is used to reduce the construction time and cost:

- Subdivision of jackets into large components and modules, which would make their fabrication and assembly possible.

- Fabrication of major components in optimum locations and under the best conditions.
- Professional procurement plans to deliver the critical materials on time.
- Try to simplify the configurations and standardization of details, grades, and sizes.
- Avoidance of excessively tight tolerances.
- Selection of structural systems that utilize skills and trades on a relatively continuous and uniform basis.
- Protective coating should be applied in a fabrication shop as their on-site application will be difficult to manage.

Quality control, inspection, and testing should be performed during all phases of construction to ensure that specified requirements are being met. The most effective quality scheme is one that prevents the introduction of defective materials and workmanship into a structure, rather than addressing the problems after they occur.

12.6.1 Engineering of Execution

The engineering office, which performed the design, should oversee the execution during each phase to address site inquiries in a fast and accurate manner and to ensure that the design requirements are fulfilled on site accurately.

The construction procedure is included during the jacket design stage. The shape, geometry, and properties of the jacket require specific methods of loadout, transportation, and installation. Noting that this phase is under contractor responsibility, there is considerable interfacing of engineering requirements in these phases.

In the earlier stages of procurement through assembly and erection, the contractor, while being limited by design specification requirements, has a freedom of choice with regard to the method of execution. In all phases, the contractor is required to choose the methods and ensure that they are compatible with the specification requirements and do not affect the structure integrity.

Each stage of execution has its own specific engineering requirements, which are determined by the construction execution plan. Accordingly, the engineering group that supports procurement and shop fabrication is a large operation, but the good news is that the work it performs is usually repetitive (e.g., material take-offs, shop drawings, cutting plans).

The volume of contractor construction engineering work on a large jacket is usually about 130,000/150,000 man-hours. When designing larger elements, consideration must be given to their subdivision into elements, which should not distort when fabricated and which can be relatively easily assembled without welding or dimensional problems. For instance, tubular joints are categorized as either complex or simple from the execution viewpoint, based on the number of separate fitting-welding and nondestructive testing cycles required during fabrication.

Node stubs can be classified as simple or overlapping. Overlapping stubs add at least one complete cycle to node fabrication and should therefore be avoided where possible. The minimum separation between the weld toes of adjacent simple stubs is typically specified as 50 mm. However, this distance is too small to allow simultaneous welding of adjacent stubs—150 mm is a more practical distance.

12.6.2 Fabrication

The engineering firm will deliver the specifications for fabrication of offshore jackets. In general, there are two main codes that are used extensively for establishing specification requirements: API RP2A and AISC *Design, Fabrication, and Erection of Structural Steel for Buildings*. The quality control for welds usually entails visual, magnetic particle inspection, and ultrasonic test inspection by the quality control engineer.

The location and orientation of circumferential and longitudinal welds during construction is based on minimizing interferences and ensuring the minimum distance between circumferential welds. Special attention is required on items such as pile sleeve shear plates, launch runners, and mud mats.

According to the American Welding Society (AWS), if welds are found to be defective, they should be rectified by grinding, machining, or welding as required. Welds of insufficient strength, ductility, or notch toughness should be completely removed before repair.

In general, subassemblies are executed so that at least one of the two edges that will mate during subsequent assembly/erection has a cutoff allowance. This procedure provides flexibility in that the subassemblies can be sent to the field with the cutoff allowance in place, and cut to fit on location. Alternatively, they can be cut to exact dimensions during subassembly where the as-built dimension has already been determined.

12.6.2.1 Joint Fabrication

The primary structure joints are frequently geometrically complex. Accordingly, their fabrication presents particular problems, especially in terms of dimension accuracy control and welding.

On a complex jacket, the designer may specify the joint cans or the whole joint including stubs and ring stiffeners, in materials with specified through-thickness properties. This requirement is introduced because of tearing or punching effects are likely to be sustained by these elements during their design life and, indeed, during fabrication.

The designer usually increases the thickness or reinforces the cans so they can withstand local stresses. Finally, in an effort to ensure that joint welds contain minimal levels of residual stress due to fabrication, thermal stress relieving or post-weld heat treatment of the heavier, more restrained welds may be prescribed.

12.6.3 Jacket Assembly

The jacket is assembled in the fabrication yard because it requires a heavy crane and lifting tools. For a large jacket, the assemblies are typically of four types:

- Jacket levels incorporating conductor guide frames
- Top frames
- Jacket rows, that is, bents or partial bents
- Pile sleeve clusters

The assembly and erection phases are based on the following objectives:

- Maximize on-the-ground assembly (as opposed to erection) and maximize access around the jacket during execution.
- Minimize erection joints in principal structural elements, such as jacket legs, launch runners, rows, and levels. Align critical areas such as conductor guides, pile sleeves, and launch runners.
- Subassemble principal structural elements of jacket such as jacket legs, rows, levels. Subassemble, and, where possible, pretest systems such as grouting and ballasting. Include maximum quantity of secondary items such as anodes, risers, J tubes, and caissons. Coat or paint required areas (top of jacket, risers) before erection.
- Minimize the use of temporary items that require subsequent removal, such as scaffolding, walkways, and lifting aids, and preinstall such aids where they are necessary.

The overall assembly sequence and plan requires that each assembly be completed before lifting. It is normal to determine their exact location, orientation, and attitude. Face-up or face-down of each assembly in the field is determined in anticipation of its lifting procedure.

Dimensional control of the assembly before and after welding can be achieved via a series of self-checking measurements on the structure itself. Provided cross checks are adequate, the time-consuming exercise of referring measurements to an external benchmark can be avoided.

An outlined sequence of events that applies to all types of assemblies is as follows:

- Preparation of assembly support and staging.
- Rough setting of the assembly main structure and position tacking. Dimensional control of assembly main structure.

- Dimensional control of the assembly and the secondary structure.
- Installation of appurtenances (e.g., anodes, supports, walkways, risers, J tubes, caissons, grouting, and ballasting) and scaffolding, lifting, aids, erection guides, temporary attachments.
- Test (e.g., hydrotest) if required. Overall nondesctructive test, dimensional control.
- Blasting and painting or touch-up. Removal of temporary assembly supports and staging.
- Preparation for transport, lift, and erection.

12.6.4 Jacket Erection

This phase includes the assembly and subassembly of the fabricated structure. As shown in Figure 12.28, Figure 12.29 also shows the erection of the topside.

For jacket structure, Figure 12.30 presents the fabrication of one side of the jacket, and Figure 12.31 shows the fabrication of the horizontal bracing and the conductor guide. The outlined sequence for the erection of all major components is as follows:

- Technical appraisal of lift methods. Calculations for crane configuration, rigging accessories, and other tools.
- Preparation of cranes for lift. Preparation for rigging. Transport assembly to lift location. Roll up into position with scaffolding and staging in position, if possible.
- Preparation of fixing system and wind bracing.

FIGURE 12.28
Topside erection.

FIGURE 12.29
Jacket fabrication on the yard.

12.6.5 Loads from Transportation, Launch, and Lifting Operations

The topside structure and jacket components are subjected to critical loadings during construction operations. Some jacket members and joints may be subjected to high bending and punching shear loads as the braces and bents are assembled into a jacket in the fabrication yard. Analysis of such assembly loading conditions would require sequential simulation of jacket geometry and loads, and knowledge of the jacket assembly as shown in Figures 12.32 and 12.33.

As the jackets are transported by barge to the site, the jacket and tie down braces, their connections, and the transportation barge are subjected to significant dynamic accelerations and inclined self-weight loads. These motions

FIGURE 12.30
Jacket assembly.

FIGURE 12.31
Fabrication of conductor guide.

and resulting dynamic loads must be simulated in incremental loading sequences to determine and dimension the highest stressed components. Note that bracing may be needed only for the jacket transportation phase; some of these braces may have to be removed before the jacket is installed on site, to reduce in-place wave loads.

During its launch to sea, the jacket will be subjected to significant inertial and drag loadings. In general, most critical loadings would occur as the jacket starts tilting around the launch beam and rapidly descend into the sea. At this position, the tilting beams would exert high concentrated loads on the stiff bracing levels. These require a launch bracing system specially designed to distribute and reduce the launch forces. As the jacket hits the water plane and rapidly descends into the sea, leading jacket braces may experience high drag and inertial forces.

FIGURE 12.32
Assembly conductor guide to the jacket.

FIGURE 12.33
Assembly for topside with facilities.

Crane lifting of the deck or jacket from the transportation barge is another critical loading condition that requires simulation, analysis, and careful design. In such lifting operations, deck and jacket members and connections are loaded in directions different than their in-place loading directions. Additionally, redundant or shorter/longer lifting sling than planned may result in substantially different loads than those calculated for idealized conditions, noting that in the case of a four-sling lift, if one sling is shorter than planned, three instead of four slings may carry the entire deck loads. Such an unplanned load distribution may also be caused by a CG that is located somewhere different than calculated. Lifting padeyes and lugs are components with high consequences of failure. A single padeye failure may result in the loss of the entire deck, jacket, and the crane. Such critical components and their connections to the structures lifted must be designed for higher safety factors. Safety factors for four or more against the ultimate capacity are commonly used for padeyes, their connections to the structure, and the associated lifting gear.

12.6.6 Lifting Forces

Lifting forces are calculated based on the weight of the structural component being lifted, the number and location of lifting eyes used for the lift, the angle between each sling and the vertical axis, and the conditions under which the lift is performed as shown in Figure 12.32. All members and connections of a lifted component must be designed for the forces resulting from the static equilibrium of the lifted weight and the sling tensions.

The API-RP2A recommends that in order to compensate for any side movements, lifting eyes and the connections to the supporting structural members should be designed for the combined action of the static sling load and a horizontal force equal to 5% this load, applied perpendicular to the padeye at the center of the pin hole. All these design forces are applied as static loads if the lifts are performed in the fabrication yard.

On the other hand, if the lifting derrick or the structure to be lifted is on a floating vessel, the dynamic load factors should be applied to the static lifting forces. In particular, the API-RP2A recommends for lifts made offshore two minimum values of dynamic load factors—2.0 and 1.35—for designing padeyes and all other members transmitting lifting forces, respectively.

According to the API-RP2A, in case of loadouts at sheltered locations, the corresponding minimum load factors for the above should be 1.5 and 1.15, respectively.

12.6.7 Loadout Forces

Loadout forces are the loads required to move the jacket from the fabrication yard onto the barge. If the loadout is carried out via direct lifting, there is no need to design for lifting. In cases where the structure is pulled out, however, structure analysis needs to be redone to consider these new forces, a move that could lead to a change in the design (if required), although this is expected to happen only in few minor cases.

The coefficients of friction for the calculation of skidding forces are:

- Steel on steel without lubrication   0.25
- Steel on steel with lubrication   0.15
- Steel on Teflon   0.10
- Teflon on Teflon   0.08

Figure 12.34a and b shows the loadout for the topside.

12.6.8 Transportation Forces

These forces affect the platform jacket and the topside during offshore transport on barges. They depend on the weight, geometry, and support conditions of the structure (by barge or by buoyancy), and also on environmental conditions that are predicted during the transportation period.

The types of motion affecting the floating structure (in our case, the barge) and the topside and jacket on the barge are as shown in Figure 12.35.

To minimize the associated risks and secure safe transport from the fabrication yard to the platform site, it is important to plan the operation carefully by following these recommendations from the API-RP2A:

1. Previous experience along the tow route
2. Exposure time and reliability of predicted "weather windows"

(a)

(b)

FIGURE 12.34
(a) Loadout for topside. (b) Loadout for topside.

3. Accessibility of safe havens

4. Seasonal weather system

5. Appropriate return period for determining design wind, wave, and current conditions, taking into account characteristics of the tow such as size, structure, sensitivity, and cost

Transportation forces are generated by the motion of the tow on the structure by the supporting barge. They are determined from the design winds, waves, and currents. If the structure is self-floating, the loads can be calculated directly. According to the API-RP2A, towing analyses must be based on the results of model basin tests or appropriate analytical methods, and

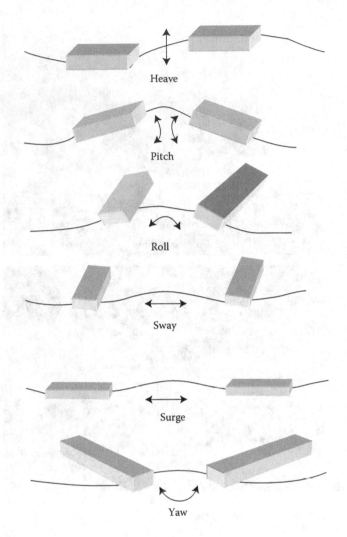

FIGURE 12.35
Types of motion of barge or vessel.

must consider wind and wave directions parallel, perpendicular, and at 45° to the tow axis.

Inertial loads may be computed from a rigid body analysis of the tow by combining roll and pitch with heave motions, when the size of the tow, magnitude of the sea state, and experience make such assumptions reasonable.

For open sea conditions, the following may be considered typical design values:

Single-amplitude roll: 20°

Single-amplitude pitch: 10°

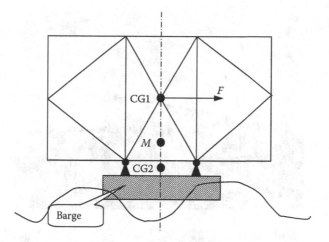

FIGURE 12.36
Launching of jacket on a barge. CG1, jacket center of gravity; CG2, tow center of gravity; *M*, metacenter of tow.

Period of roll or pitch: 10 s

Heave acceleration: 0.2 g

It is important to calculate the CG of the jacket or the topside before transport as the location of the structure CG is important to barge stability.

A special computer program or software module for offshore structure analysis is available to assist in computing the transportation loads in the

FIGURE 12.37
Jacket laid on a barge.

FIGURE 12.38
(a) Step 1—launching of jacket. (b) Step 2—launching of jacket. (c) Step 3—laying the jacket on the sea by floating.

FIGURE 12.39
Lifting and installing of jacket.

FIGURE 12.40
Pile installation.

structure–barge system and the resulting stresses for any specified environmental condition.

Figure 12.36 presents the CG and load effect on the jacket during launching and transporting of the jacket at sea.

12.6.9 Launching and Upending Forces

These forces are generated during the launch of a jacket from the barge into the sea and during the subsequent upending into its proper vertical position to rest on the seabed. The launching of the jacket is shown in Figures 12.37 and 12.38.

(a)

(b)

FIGURE 12.41
(a) Lifting topside offshore. (b) Connection between topside and piles.

The loads, static as well as dynamic, induced during the launching of the jacket and the force required to set the jacket into motion can be evaluated using appropriate analyses, which also consider the action of wind, waves, and currents expected during the operation.

Software products are available to perform the stress analyses required for launching and upending, and also to portray the whole operation graphically.

As shown in Figure 12.39, the jacket is uplifted towards its location. After this stage, the piles will be placed into its legs and will be hammered in.

12.6.10 Installation

After the jacket is installed into position, piles are placed into its legs and the hammering starts until the piles reach the required depth, as shown in Figure 12.40.

Pile refusal is defined as the minimum rate of penetration beyond which further advancement of the pile is no longer achievable because of the time required and the possible damage to the pile or to the hammer. A widely accepted rate for defining refusal is 300 blows/ft (980 blows/m).

After the pile is installed, the topside is lifted and then installed on the jacket as shown in Figure 12.41a and b. The final shape of the platform is shown in Figure 12.42.

FIGURE 12.42
Final shape of platform after construction is completed.

References

Graff, W. J. 1981. *Introduction to Offshore Structures*. Houston, TX: Gulf Publishing Company.

API RP2A WSD, Recommended Practice for Planning, Designing and Constructing Fixed Offshore Platforms-Working Stress Design, 2000.

Report for Joint Industry Typical Frame Projects. 1999. Bomel Engineering Consultants.

AISC Standard 316-89. 1989. *Manual of Steel Construction: Allowable Stress Design*. Chicago, IL: AISC. *Specification for the Design, Fabrication and Erection of Structural Steel for Buildings*, latest edition.

AWS Structural Welding Code AWS D1.1-88.

13

Soil Investigation and Pile Design

13.1 Introduction

Site investigation is the first essential operation to be performed in the analysis, design, and choice of a foundation. This process, which is a prerequisite for an economical construction of civil engineering projects, provides the necessary information on the geological, physical, and geotechnical properties of soil.

Insufficient information regarding the characteristics and bearing capacity of the underlying soil and the engineering structure may result in serious damage or collapse of the structure.

In general, site investigation is conducted:

1. To provide the civil engineer with a reliable and detailed information about soil and groundwater conditions for the selection of the most efficient and economical type of foundation.
2. To plan for a suitable method of construction and efficient equipment to be used.
3. To investigate whether any harmful subsoil water or organic impurities exists in the soil.
4. To study other conditions that may affect the structure (groundwater level, future excavation, etc.).

Because soil is a very important construction factor in the process by which civil engineers build structures, it is doubtful if any major civil engineering project will push through without any site investigation being undertaken. Soil investigation should provide data on the following items:

1. Location of the groundwater level
2. Bearing capacity of soil
3. Settlement predictions
4. Selection of alternative depths of foundation

5. Earth pressure data on retaining structures and excavation supports
6. Necessary data for construction purposes (permeability, stratification, etc.)

13.2 Soil Exploration Methods

Various methods of subsurface investigation are used to obtain information about the subsoil. These methods include test pits and open cuts, borings (wash boring, auger boring, rotary, and percussion drilling), soundings, load tests, and geophysical methods.

Some methods will obviously yield more information than others and the soil engineer should constantly keep this in mind when planning the exploration program, the purpose of the project, and the relative costs involved. Two main points, in particular, are to be considered in soil investigation:

1. For large projects, extensive geological studies will be necessary to determine the type and extent of other methods of investigation; for smaller projects, some borings and the study of the neighboring foundation may be sufficient.
2. The type of structure and its sensitivity to future settlement. In many cases, this study requires a complete change of the statical system of the superstructure.

The choice of the exploration method is also determined from the location of the site. For a compacted site in a town where local information is available, borings and soundings will provide the required data. For new sites, site survey and geological studies are necessary to determine the program for soil investigation (Figures 13.1 and 13.2).

13.2.1 Planning the Program

The actual planning of a subsurface exploration program includes some or all of the following steps:

1. Assembly of all available information (dimensions, columns spacing, type and use of structure, basement requirements, bridge span and pier loadings, height or retaining structure, etc.).
2. Reconnaissance of the area: This entails a field trip to the site to examine the topography, type, and behavior of adjacent structures such as cracks or noticeable settlement, erosion in existing cuts, and

FIGURE 13.1
Soil core.

soil stratification; the reconnaissance may also be in the form of a study of geological maps or aerial photographs.

3. Preliminary site investigation: Few borings or a test pit to establish the types of soil, stratification, and possibly the location of ground water table (GWT) (Figure 13.3).

4. Detailed site investigation: For complex projects and/or erratic soil conditions, samples are collected for shear strength determination and settlement analysis.

13.2.2 Organization of Fieldwork

After studying the geological data, maps, and photographs, the engineer can now consider the number, type, and depth of exploratory borings required. The engineers in charge of soil investigation have to utilize the results as soon as samples have been extracted and do not have to wait until all borings have been completed; otherwise, doing so would lead to a considerable waste of resources and efforts (because chances are high that duplicate information has been collected while some data are still lacking).

To obtain the most useful information in a cost-effective fashion, the soil engineer in charge of the investigation should adhere to the following rules:

1. He must know exactly what type of information the investigation is expected to provide. This means that he must be completely familiar with the type of construction being planned for the site. If the investigation concerns a proposed building, he must know the value and type of loads on the foundation and the evaluation of the lowest floor level in relation to the ground surface. If future extensions

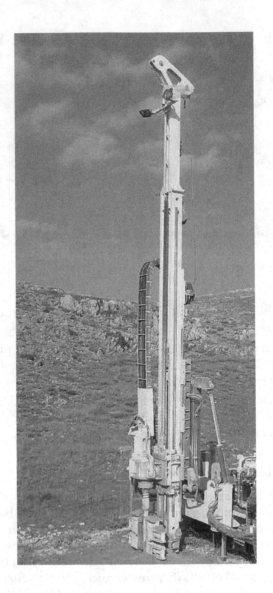

FIGURE 13.2
Soil rig.

or additional floors are planned, he should know their location and loads. He should also know if there are social requirements regarding the allowable total or differential settlements.

2. He must write down the minimum requirements for bore hole spacing and depth while taking into consideration the type of structure and estimated soil condition. The borings should provide sufficient data to allow reliable soil profiling along two principal axes. The minimum depth of borings should be related to the extent to which

FIGURE 13.3
Open pit.

the subsoil will be affected by the proposed construction. This can be evaluated by using the concept of stress distribution in soils. A useful approximation rule to determine the depth of the investigation is finding the depth at which the increase in vertical stress caused by the proposed structure is 10% less of the original vertical stress.

3. For cuttings, retaining walls, and other structures where lateral stability is the main concern, the investigation should extend below the depth of the deepest possible surface of failure. In soils, the frequency of sampling is usually governed by the maximum allowable spacing of 1–1.5 m plus the condition(s) under which the sample was taken. Changes in soil conditions and investigation for dams, reservoir, and landslides require continuous sampling using a special apparatus.

4. The engineer must select the appropriate drilling and sampling equipment, with attention to local preferences for certain types of equipment.

5. The engineer should select the field supervisor and must be able to lay out boring locations with reasonable accuracy.

13.2.3 Soil Boring Methods

The main purpose of drilling a test hole or boring is to obtain samples from various strata and to acquire an accurate picture of the subsoil profile at one location. Different methods of soil boring are available and will be discussed below.

- Test pits and open cuts

  This method is ideal for soils requiring little support for excavation. The test hole is made large enough to permit an observer to descend to log the soil profile visually, and the succession of strata can be examined in the wall of the pit. Block samples may be taken from the excavation sites in cohesive soils for laboratory testing.

Trial pits are generally used for exploration of shallow foundation (depth of up to 4 m). As depths of pits increase, the cost increases very rapidly, particularly in water-bearing strata. Trial pits are also used to explore depth, cables, and other underground public works and old structures.

- Hand and powered augers

  Auger drilling is developing as the most common method of soil exploration to depths of up to 60 m. The speed of operation in most types of soil is greater compared to other drilling methods. Distributed samples may be obtained from the soil brought to the surface and undistributed samples may be taken from the bottom of the hole at any required depth. Powered auger with large machines is capable of making borings up to 1.0 m in diameter. In appropriate soils where no casing is required, the size of the hole can be examined directly as in trial pits.

13.2.3.1 Wash Borings

In this method, the soil is loosened by a high-pressure water jet from a pipe passing down the borehole. The washings are brought to the surface in the water that travels upward via the outside of the jet pipe. This method is cheap and fast. However, the washings are usually disturbed and the samples obtained are used only for identification of soil layers.

13.2.3.2 Sampling Methods

The two methods of sampling are discussed below.

- Disturbed samples

  There are certain instances when sample disturbance is of little importance, because only the soil type and stratifications need to be determined. Disturbed samples are more satisfactory when classification of the soil is the prime objective since they are cheaper and faster to obtain (wash boring, auger boring). Disturbed samples are adequate for the investigation of borrowed materials to be used in fills, highway projects, and other grading works.

- Undisturbed samples

  Undisturbed samples are necessary for studying soil behavior and performing laboratory tests under conditions that are nearly similar to those in the original environment. Shear strength, consolidation, and permeability tests must be carried out in the laboratory using undisturbed samples. The major problems associated with undisturbed samples are:

1. Samples are unloaded of their in situ overburden pressure.
2. Friction of the sites of samplers that creates some disturbance.
3. Samples from levels below GWT may drain during the recovery process.
4. Changes in bore pressure will disturb the sample.

Undisturbed soil samples are obtained by forcing a thin-walled seamless, stainless steel sampling cylinder into the soil at the bottom of a bore, or at the bottom or in the wall of a test pit. The forcing is accomplished by jacking or a continuous push.

In its simplest and more frequently used form, a thin-walled sampler consists of a 60- to 90-cm-long pipe with a wall thickness of about 1.5 mm and a diameter of 5 to 8 cm. At one end, a cutting edge is provided, whereas at the other end the type is attached to an adapter filled with a ball valve and vents. After samples have been taken, the sampling tube can be detached from the adapter and the ends are sealed off with wax for transport to the laboratory. This may be stored or the wax may be removed and the sample extruded for immediate testing.

13.2.3.3 Spacing of Borings

The spacing of borings cannot be determined with absolute accuracy. They depend on many factors such as the nature and condition of soil, the shape and extent of the structure, etc. Furthermore, the spacing of borings should conform to the importance, size, and system of the structure. The spacing values shown in Table 13.1 are often used in planning boring work.

13.2.3.4 Boring Depth

The depth explored is generally the depth beyond which the effect of structure load is negligible. It should include all stressed zones of soil involved in the foundation system and should cover all layers of soil that affect the

TABLE 13.1

Soil Boring Spacing

Structure or Project	Spacing or Borings, m
Highway (subgrade survey)	300–600
Earth dams, dikes	30–60
Borrow pits	30–120
Multistory buildings	15–30
Single-story manufacturing plants	30–90

settlement of the structure. The depth of boring may be taken as 1.5 times the breadth of the raft or from 3 to 5 times the breadth of the footing. If the sand layer is reached, it must be penetrated sufficiently to ensure its continuity.

A better simple rule for structures such as hospitals and office buildings relates the estimated boring depth, d, to the number of floors.

$$\text{Light steel or narrow concrete} \quad d = 3(s)^{0.7} \text{ m}$$

$$\text{Heavy steel or wide concrete} \quad d = 6(s)^{0.7} \text{ m}$$

where s is the number of floors.

13.2.3.5 Boring Report

A boring report should contain the following:

1. Situation plan of the construction site drawn to scale and oriented with respect to the north direction.
2. Identification, data, place, location of boreholes and their coordinates from a reference axis, and drilling equipment used.
3. Description of soil profiles over the full depth of the borehole along with the classification of soils encountered in each boring. The boring log should be drawn to a stated scale and contain information on soil types and thickness encountered.
4. Description of the technique used for advancing and stabilizing the borehole.
5. Location of the groundwater table, the groundwater condition, and surface drainage observation.
6. Information on any difficulties or obstruction(s) encountered during boring operations (boulders, roots, drains, telephone and electric cable pipes, old foundation, etc.).
7. Listing of samples obtained and report on any testing of soil in place.

13.2.4 Standard Penetration Test

The standard penetration test (SPT) is a well-established and unsophisticated method that was developed in the United States about 1925. It has since undergone refinements with respect to equipment and testing procedure. The testing procedure varies in different parts of the world.

Therefore, standardization of SPT was essential to facilitate the comparison of results from different investigations. The equipment is simple, relatively inexpensive, and rugged. Another advantage is that representative but disturbed soil samples are obtained.

The reliability of the method and the accuracy of the result depend largely on the experience and care of the engineer on site (Figure 13.4).

A split-barrel sampler is driven from the bottom of a prebored hole into the soil via a 63.5-kg hammer, dropped freely from a height of 0.76 m. The diameter of the prebored hole varies normally between 60 and 200 mm. If the hole does not stay open by itself, casing or drilling mud should be used. The sampler is first driven to a depth of 15 cm below the bottom of the prebored hole, then the number of blows required to drive the sampler another 30 cm into the soil, the so-called N30 count, is recorded. The rods used for driving the sampler should have sufficient stiffness. Normally, when sampling is carried out to depths greater than about 15 m, 54-mm rods are used.

The quality of test results depends on several factors: actual energy delivered to the head of the drill rod, dynamic properties (impedance) of the drill

FIGURE 13.4
Rig for the standard penetration test.

rod, method of drilling, and borehole stabilization. The actually delivered energy can vary between 50% and 80% of the theoretical free-fall energy. Therefore, correction factors for rod energy (60%) are commonly used (Seed and De Alba, 1986). The SPT can be difficult to perform in loose sands and silts below the groundwater level (typical for land reclamation projects), as the borehole can collapse and disturb the soil to be tested. The following factors can affect the test results: nature of the drilling fluid in the borehole, diameter of the borehole, configuration of the sampling spoon, and frequency of delivery of the hammer blows.

Therefore, it should be noted that drilling and stabilization of the borehole must be carried out with care. The measured N value (blows/0.3 m) is the so-called standard penetration resistance of the soil. Penetration resistance is influenced by the stress conditions at the depth of the test. Peck et al. (1974) proposed, based on settlement observations of footings, the following relationship for the correction of confinement pressure. The measured N value is to be multiplied by a correction factor C_N to obtain a reference value, N_1, corresponding to an effective overburden stress of 1 ton/ft^2 (approximately 107 kPa)

$$N_1 = N \cdot C_N \tag{13.1}$$

$$C_N = 0.77 \cdot \log 10(20/p) \tag{13.2}$$

where C_N is a stress correction factor and p' is the effective vertical overburden pressure. SPT is performed based on ASTM D1586-08a.

13.2.5 Cone Penetration Tests

The cone penetration test (CPT) was invented and developed in Europe but has gained increasing importance in other parts of the world, especially in connection with soil compaction projects. Different types of mechanical and electric cone penetrometers exist but the electric cone is the most widely used. A steel rod with a conical tip (apex angle of 60° and a diameter of 35.7 mm) is pushed at a rate of 2 cm/s into the soil. The steel rod has the same diameter as the cone.

The penetration resistance at the tip and along a section of the shaft (friction sleeve) is measured. The friction sleeve is located immediately above the cone and has a surface area of 15,000 mm^2. The electric CPT is provided with transducers to record the cone resistance and the local friction sleeve.

A CPT probe equipped with a pore water pressure sensor is called CPTU. It is important to ensure a complete saturation of the filter ring of the pore water (piezo) element. Otherwise, the response of the piezo transducer, which registers the variation in pore water pressure during penetration, will be slow and may give erroneous results. The CPTU offers the possibility to determine hydraulic soil properties (such as hydraulic conductivity—permeability) but is most widely used for identification of soil types and soil stratification.

The CPT can also be equipped with other types of sensors, for example, vibration sensors (accelerometer or geophone) for determination of vibration acceleration or velocity. The "seismic cone" is not yet used on a routine basis but has, because of the relative simplicity of the test, potential for wider applications especially in soil compaction projects.

The CPT is standardized and follow ASTM D3441-05 and the measurements are less operator-dependent than those of the SPT, thus giving more reproducible results.

The CPT measures the cone resistance (q_c) and the sleeve friction (f_s) from which the friction ratio, F_R, can be determined. F_R is the ratio between the local sleeve friction and the cone resistance (f_s/q_c), expressed as a percentage. In spite of the limited accuracy of sleeve friction measurements, the valuable information, which can be obtained in connection with compaction projects, has not yet been fully appreciated. As will be discussed below, the sleeve friction measurement reflects the variation in lateral earth pressure in the ground and can be used to investigate the effect of soil compaction on the state of stress, as will be discussed later. Cone and sleeve friction measurements are also strongly affected by the effective overburden pressure. It is necessary to take this effect into account, similar to the SPT.

One important objective of the CPT investigations in connection with soil compaction is to obtain information concerning soil stratification and variation in soil properties both in horizontal and vertical directions. The friction ratio is often used as an indicator of soil type (grain size) and can provide valuable information when evaluating alternative compaction methods.

Measurement of the excess pore water pressure with the CPTU can detect layers and seams of fine-grained materials (silt and clay). It is also possible to obtain more detailed information concerning soil permeability and thus, soil stratification.

13.2.6 Vane Test

Vane shear testing is one of the most common in situ methods for the estimation of the undrained shear strength of the soil.

Figure 13.5 illustrates the vane shear tip used in the tests. The tip has a diameter of 75 mm and a height of 112 mm. Typically, the height-to-diameter ratio should be equal to 2.

The vane is introduced into the borehole at the depth where the measurement of the undrained shear strength is required. Then it is rotated and the torsion force required to cause shearing is calculated. In Figure 13.6, a manual vane shear is shown. The blade is rotated at a specified rate that should not exceed 0.1°/s (practically 1° every 10 s). The amount of rotation is specified in the green arrow, whereas the red arrow has a device that measures the required torque. The procedure and the equipment typically should follow the procedures suggested by the ASTM D2573-72.

FIGURE 13.5
Cone penetration test.

The shear strength of the material is calculated from the torque by dividing by a constant K, which depends on the dimensions and the shape of the vane. More can be found in ASTM D 2573-72.

The results of the corrected vane shear strength measured in situ are shown in the following plot. The plot is shear strength (corrected) versus time, and the test was performed in a rate of 1° per 10 s. The peak shear strength and the residual strength can be seen in the plot. After the test was completed, the vane was rotated twice and then the residual strength was measured.

13.2.7 Cross-Hole Test

For very critical projects such as new facilities with large storage tanks, the phenomenon of the soil through extensive depths is important in making an engineering decision.

FIGURE 13.6
Vane test.

This test is important in defining the wave propagation through soil depth, critical information especially in earthquake-prone regions.

The objective of this test is to estimate the profile of the seismic velocity of longitudinal and transversal waves along boreholes at the selected location using the cross-hole seismic technique. This test consists of two boreholes 95 m deep and drilled using rotary drilling. The borehole will be cased with 97.6-mm-diameter PVC pipes to accommodate the source and the receiver probes. The boreholes are spaced about 5 m from center to center at the ground surface.

The inclination of the two boreholes will be measured using an inclinometer setup. The coordinates obtained from the inclinometer readings are then used to accurately calculate the distance between the two boreholes at 1.0-m depth intervals starting at the ground surface and down to the end of the boreholes.

The raw data obtained from the cross-hole survey are the travel times of the compression, P, and shear, S, waves from the source hole, T, and the receiver hole, R, as shown in Figure 13.7. The travel times of the seismic waves are derived from the first arrivals identified on the seismic trace for each T and R

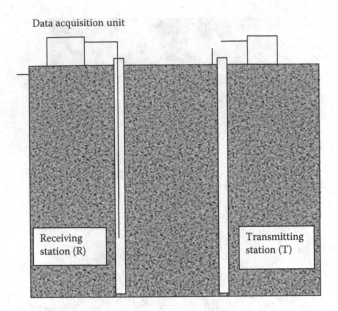

FIGURE 13.7
Schematic of cross-hole test setup.

position with the known distance to calculate the apparent velocities P and S for each depth interval.

The instrumentation and data acquisition consists of a probe, a control unit, and a cable with cable reel. The probe is characterized by a reliable clamping system, obtained through the progressive bending of a harmonic steel spring inside the probe. There is a continuous current motor, controlled via surface electronics, which moves the piston inside the probe, controlling the bending of the spring, and providing for the clamping and unclamping operations. Inside the probe there are three 10-Hz geophones, oriented along the x–y–z axis, which determine the arrival time of seismic waves of type S.

A specially designed down hole hammer that fits through the 10-mm casing will be used as the transmitting source. The hammer has gripper plates that expand out to make firm contact with the inside of the casing. The hammer operation is pneumatic and electronic.

The equipment for the cross-hole test has many design features that facilitate accurate shear wave velocity measurements in accordance with the procedures described in ASTM D4428/D4428M-91 for cross-hole seismic testing.

The velocity of propagation of stress waves through infinitive media is a function of the material properties of the media, and depends on the elastic modulus of the material, Poisson ratio, and the material density.

13.2.7.1 Body Waves

These waves can be primary (*P*) waves. A primary wave is a longitudinal wave in which the direction of the motion of particles is in the direction of propagation. This motion is nonrotational and the wave is one of dilation-propagated speed.

Body waves can also be secondary or shear (*S*) waves. This is a transverse wave in which the direction of motion of particles is perpendicular to the direction of propagation.

Compressive waves are the fastest, followed by the shear waves, and subsequently followed by various types of waves. A fluid such as pore water does not transmit shear waves. Accordingly, in soil, shear waves are conducted through the soil skeleton only.

13.2.7.2 Surfaces Waves

In a uniform, infinite medium, only *P* and *S* waves appear. If the medium is bounded or nonuniform, such as surface soils, other simple types of waves may appear. The most important are the surface waves that are propagated near the surface of a solid. Surface waves have depths of penetration depending on their wavelength. In nonuniform media, surface waves travel at a velocity that depends on their frequency.

13.3 Deep Foundation

Pile foundation is the most common type of deep foundations used to transmit the structural loads into the deeper layers of firm soil in such a way that these layers of soil or rock can sustain the loads. A pile foundation, in general, is more expensive than an ordinary shallow foundation, and is used where the soil at or near the surface is of poor bearing capacity or settlement problems are anticipated.

The main functions of piles are:

1. To carry more load from the superstructure to the lower, more resistant soil strata, thus increasing the load capacity of the site.

2. To reduce the settlements to the minimum value and consequently the differential settlements. They are most effective in sensitive structures, which by virtue of their sensitive structural statical system, cannot undergo appreciable differential settlements.

3. To avoid excavation under water for sites where GWT is high. This may represent an expensive item in the cost of the foundation and may also cause reduction in strength of certain soils.

4. To densify the soil by driving compaction piles in loose cohesionless soil deposits.

In general, piles are made of timber, steel, concrete, or a combination of these materials.

Technically, from the design point of view, there are two types of piles: end-bearing piles and friction piles. These two types have advantages and disadvantages.

Figure 13.8 presents an end-bearing pile, where most of the load is transferred to the rock layer by the end bearing of the pile. Thus, most of the load will be taken by its base and rest of the load is taken by the section friction. The skin friction along the system of the pile can be negligible and the bearing capacity of the pile is derived only from the point bearing resistance of the soil under the pile tip.

To obtain the full benefit of the ultimate strength of the firm layer under the pile tip, the pile should penetrate the bearing stratum to a depth at least three times the pile diameter.

Figure 13.9 shows friction piles that transfer their loads to the surrounding soil by friction developed along their sides.

If the pile penetrates a clay layer, the skin friction is equal to the cohesion, C, of that layer. For granular materials, the skin friction is proportional to the intensity of earth pressure and can be considered to vary linearly with enough accuracy.

There are different types of piles used and the differences between them depend on the material and manner of execution.

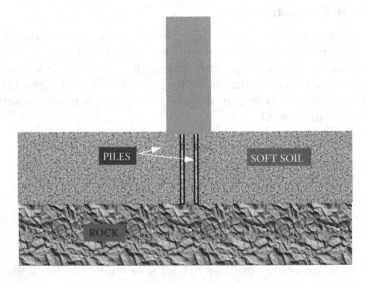

FIGURE 13.8
End-bearing pile.

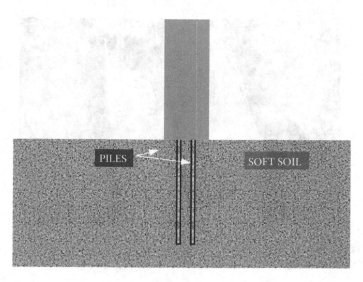

FIGURE 13.9
Friction pile.

13.3.1 Timber Piles

These piles, which are widely used in woody countries, are made of tree trunks of good quality and are of appreciable sizes (not less than 300 mm in diameter), as shown in Figures 13.10 and 13.11. After the timber pile is driven into the ground, the top end should be cut off in a square shape, so that the foundation is in contact with solid wood. If a timber pile is subjected to alternate wetting and drying, it should be treated with a wood preservative to increase its useful life. Timber piles can safely carry loads between 15 and

FIGURE 13.10
Preparation of wood piles.

FIGURE 13.11
Wood pile with steel cap.

25 tons under the usual conditions. They are relatively low cost and easy to handle.

13.3.2 Steel Piles

These are usually rolled sections of H-shaped steel pipes. Wide-flange and I beams may also be used.

Splices in steel piles are made in the same manner as in steel columns, that is, by welding, riveting, or bolting. Pipe piles are either welded pipes or seamless-steel pipes, which may be driven either open-ended or closed-ended. Pipe piles are usually used on offshore structure platforms.

13.3.3 Concrete Piles

These types of piles consist of two methods of construction: cast-in-place piles and precast concrete piles.

- Cast-in-place piles

  These piles are formed by making a hole in the ground and filling it with concrete. They may be drilled or formed by driving a shell. The steel piles shown in Figures 13.12 and 13.13 are driven into the ground to make a hole, which will later be filled with concrete.

- Precast concrete piles

  The steel shell is usually withdrawn during or after pouring of concrete and is sometimes left to protect the concrete from mixing with the mud or to prevent the cement from being washed away by the groundwater.

FIGURE 13.12
Steel pile with cone tip.

13.3.4 Precast and Prestressed Piles

Piles are formed to specified lengths, cured, and then shipped to the construction site. A primary consideration with precast piles is the handling stress. To cope with handling stresses, some of which are tensile, piles are reinforced and in some cases prestressed.

Precast reinforced concrete piles may have square or octagonal cross sections. They should be adequately reinforced to withstand driving and handling stresses. Long precast piles should be driven with care to prevent buckling during the operation. To overcome the driving stresses, lateral steel reinforcements should be closely spaced at the top and bottom of the pile to resist the stress wave concentration at the ends of the driven pile.

FIGURE 13.13
Raymond shell piles.

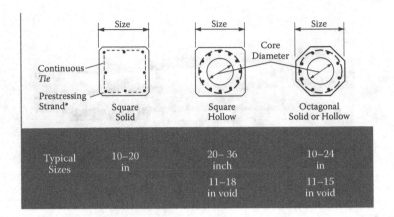

FIGURE 13.14
Prestressed pile sections.

To deal with bending moments due to handling, a relatively short pile (less than 12.0 m) is usually lifted from one end and is treated as a beam carrying its own weight. Longer piles should be lifted from two, three, or four points at the specified distances indicated in Figure 13.14 in order to reduce the bending stresses to a minimum. Lifting points should be marked by hooks or bolts, which will be removed later.

Special care should be given to ensure that concrete piles will remain intact and to prevent soil from attacking the pile (Figure 13.15). First of all, it is necessary to use siliceous aggregates and rich cement content (350 kg of cement per cubic meter of finished concrete) (Figure 13.16). In addition, concrete piles must be protected from dissolved sulfates or chemicals present in

FIGURE 13.15
Lifting of the concrete pile.

FIGURE 13.16
Composite section piles.

the underground water. If the ratio of sulfur trioxide (SO_3) in the soil water exceeds 0.03% (300 m g/L) in stagnant water, or 0.015% in running water, and if in addition the ratio of SO_3 in the soil itself is 0.2%, ordinary Portland cement should not be used. In such cases, special sulphate-resisting cements should be used. In all cements used, the presence of free lime or calcium traces should be minimized as much as possible.

Table 13.2 shows the summary of the comparison between different types of piles.

13.3.5 Pile Caps

Pile caps are a type of foundation that are affected by column loads from above and the piles' reaction at the point of contact between the piles to the caps (Figure 13.17). In this type of foundation, ignore the impact of the soil where soils are not contact with the caps in a rigid or flexible manner to allow them to carry any part of the column load and the stiffness of the piles is so high that the piles carry all the loads.

Often, the column load cannot be carried by one pile alone so the column needs more than one pile to carry the load, thereby requiring a pile cap to distribute the column load equally. This can be done by making the center of gravity of the column coincide with the center of gravity of the pile cap.

To ensure the transfer of load from the column to the pile, the pile steel reinforcement should extend inside the pile cap by at least 600 mm.

The pile caps are designed as a rigid foundation; for piles carrying part of the column load, the pile cap thickness should be able to resist the punching stresses and the tension at the top and bottom.

TABLE 13.2

Typical Use and Characteristics of Piles f_c'

Characteristic	Concrete-Filled Steel Pipe Piles	Composite Piles	Precast Concrete (Including Prestressed)	Cast in Place (Thin Shell Driven with Mandrel)
			Pile Type	
Maximum length	Practically unlimited	55 m	30 m for precast 60 m for prestressed	30 m for straight section, 12 m for tapered section
Optimum length	12–36 m	18–36 m	12–15 m precast 18–30 m prestressed	12–18 m for straight section 5–12 m for tapered section
Application material specifications	ASTM A36 for core ASTM A252 for pipe ACI 318 for concrete	ASTM A36 for core ASTM A252 for pipe ACI 318 for concrete ASTM D25 for timber	ASTM A15 for reinforcing steel ASTM A82 for cold drawn wire ACI 318 for concrete	ACI
Recommended maximum stress	$0.4f_y$ reinforcement, $0.5f_y$ or core $0.33f_c'$ for concrete	Same	$0.33f_c'$, $0.5f_y$ for reinforcement unless prestressed	$0.4f_y$ if steel gauge = 14, $0.35f_y$ if shell thickness = 3 mm
Maximum load for usual conditions	1800 kN with cores 18,000 kN for large sections with steel core	1800 kN	8500 kN for prestressed, 900 kN for precast	675 kN
Optimum load range	700–1100 kN with cores 4500–14,000 kN with cores	250–725 kN	350–3500 kN	250–550 kN

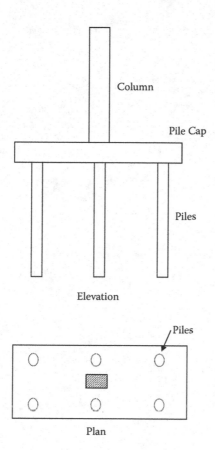

FIGURE 13.17
Sketch of pile cap.

References

ASTM D1586-08a Standard Test Method for Standard Penetration Test (SPT) and Split-Barrel Sampling of Soils.

ASTM D3441-05 Standard Test Method for Mechanical Cone Penetration Tests of Soil.

ASTM D2573-08 Standard Test Method for Field Vane Shear Test in Cohesive Soil.

ASTM D4428/D4428M-07 Standard Test Methods for Cross Hole Seismic Testing.

Seed, H. B., and P. De Alba. 1986. Use of SPT and CPT tests for evaluating the liquefaction resistance of sands. *Proceedings, In Situ '86, ASCE*, pp. 281–302.

Peck, R. B., W. E. Hanson, and T. H. Thornburn. 1996. *Foundation Engineering*, 2nd edition. New York, NY: John Wiley and Sons.

Index

Printed in the United States
by Baker & Taylor Publisher Services